Chemistry of Iron

Chemistry of Iron

Edited by

J. SILVER
Department of Chemistry and Biological Chemistry
University of Essex

BLACKIE ACADEMIC & PROFESSIONAL
An Imprint of Chapman & Hall
London · Glasgow · New York · Tokyo · Melbourne · Madras

Published by
Blackie Academic & Professional, an imprint of Chapman & Hall,
Wester Cleddens Road, Bishopbriggs, Glasgow G64 2NZ

Chapman & Hall, 2–6 Boundary Row, London SE1 8HN, UK

Blackie Academic & Professional, Wester Cleddens Road, Bishopbriggs, Glasgow G64 2NZ, UK

Chapman & Hall Inc., 29 West 35th Street, New York NY10001, USA

Chapman & Hall Japan, Thomson Publishing Japan, Hirakawacho Nemoto Building, 6F, 1-7-11 Hirakawa-cho, Chiyoda-ku, Tokyo 102, Japan

DA Book (Aust.) Pty Ltd., 648 Whitehorse Road, Mitcham 3132, Victoria, Australia

Chapman & Hall India, R. Seshadri, 32 Second Main Road, CIT East, Madras 600 035, India

---

First edition 1993

Typeset in 10/12 pt Times by Thomson Press (India) Limited, New Delhi

Printed in Great Britain by Hartnolls Ltd, Bodmin, Cornwall

ISBN 0 7514 0062 9

A catalogue record for this book is available from the British Library

DEDICATION

To my late father Michael Silver and my cousin Morrie Bergson for always encouraging me and trying to answer my questions.

Preface

This book is designed to be of use to the reader in two different ways. First, it is intended to provide a general introduction to all aspects of iron chemistry for readers from a variety of different scientific backgrounds. It has been written at a level suitable for use by graduates and advanced undergraduates in chemistry and biochemistry, and graduates in physics, geology, materials science, metallurgy and biology.

It is not designed to be a dictionary of iron compounds but rather to provide each user with the necessary tools and background to pursue their individual interests in the wide areas that are influenced by the chemistry of iron. To achieve this goal each chapter has been written by a contemporary expert active in the subject so that the reader will benefit from their individual insight.

Although it is generally assumed that the reader will have an understanding of bonding theories and general chemistry, the book is well referenced so that any deficiencies in the reader's background can be addressed.

The book was also designed as a general reference book for initial pointers into a scientific literature that is growing steadily as the understanding and uses of this astonishingly versatile element continue to develop. To meet this aim the book attempts some coverage of all aspects of the chemistry of iron, not only outlining what understanding has been achieved to date but also identifying targets to be aimed at in the future.

I would like to thank those who reviewed initial synopses of this book, and colleagues who discussed sections, for their suggestions and criticism. Any omissions or shortcomings are the responsibility of the editor. I also thank the numerous authors and editors who have kindly allowed the chapter authors to reproduce diagrams from their papers. I am indebted to Mrs M. Cattrell for secretarial assistance in the preparation of the manuscript.

J.S.

Contents

Contributors

Professor F.J. Berry	Department of Chemistry, The Open University, Walton Hall, Milton Keynes, MK7 6AA
Dr B.W. Fitzsimmons	Unit for the Chemistry and Biochemistry of Iron, Department of Chemistry and Biological Chemistry, University of Essex, Wivenhoe Park, Colchester CO4 3SQ
Professor R.C. Hider	Department of Pharmacy, King's College, University of London, Manresa Road, London SW3 6LX
Professor J.G. Leigh	Unit of Nitrogen Fixation, University of Sussex, Brighton BR1 9RQ
Dr G.R. Moore	University of East Anglia, School of Chemical Sciences, Norwich NR4 7TJ
Professor P.L. Pauson	Department of Pure and Applied Chemistry, University of Strathclyde, Thomas Graham Building, 295 Cathedral Street, Glasgow G1 1XL
Dr A.K. Powell	University of East Anglia, School of Chemical Sciences, Norwich NR4 7TJ
Dr J. Silver	Unit for the Chemistry and Biochemistry of Iron, Department of Chemistry and Biological Chemistry, University of Essex, Wivenhoe Park, Colchester CO4 3SQ
Dr S. Singh	Department of Pharmacy, King's College, University of London, Manresa Road, London SW3 6LX
Professor E. Sinn	School of Chemistry, University of Hull, Cottingham Road, Kingston upon Hull HU6 7RX
Professor M.T. Wilson	Unit for the Chemistry and Biochemistry of Iron, Department of Chemistry and Biological Chemistry, University of Essex, Wivenhoe Park, Colchester CO4 3SQ

1 Introduction to iron chemistry

J. SILVER

1.1 Introduction

Iron (element 26) derives its name from the Anglo Saxons (iron, in German is *Eisen*). The atomic symbol for iron is Fe, this and words such as 'ferrous' and 'ferric' are derived from the latin ferrum, iron.

Iron has long been known and used by man; it appears that it was first smelted by the Hittites in Asia Minor in the third millenium BC. Its widespread use and the true start of the 'Iron Age' only occurred with the fall of the Hittite empire around 1200 BC (Childe, 1942). In fact, we date the dawn of modern man from when we could smelt and work iron into weapons.

More of the historical and present widespread use of iron in industrial chemistry is the subject of chapter 2 of this book. All of the industrial uses of iron depend directly on its availability, abundance and chemistry. These properties are discussed in this chapter.

Iron is the cornerstone metal of human technology. We use it to build our civilisation and produce more of it than all other metals combined. The iron age continues today, and in fact the atomic and information ages are only modern names to better describe a modern iron-based culture. As we examine the chemistry of iron, these facts should be borne in mind.

Iron plays an important role in the chemistry of living organisms being found at the active centre of many biological molecules. Many of these are discussed in chapters 6 and 7, and the need to control iron levels in humans is discussed in chapter 8.

1.2 Importance of iron

To illustrate the relative importance of iron to man it is useful to compare it to other common elements (Nelson, 1991). In the data presented by Nelson (1991), iron is seen to make up 1.5% of, and is the eighth most abundant element in, the lithosphere. 1.1×10^{13} moles of atoms of iron were consumed by the world manufacturing industries in 1983 making it the fifth most used element; only hydrogen, carbon, oxygen and calcium were consumed in greater quantities. The value of the element used in that year was around

Table 1.1 Overall importance of the elements to man

Elements	Overall ranking[a]
H, C, N, O	10
Si	9
Al, S, Fe	8
Na, P, Cl, Ca, Mn, Cu	7
F, K, Cr, Ni, Zn	6
B, Mg, Ti, Co, Br, Ag, Sn, I	
Au, Pb, U	5

[a]No ranking = 4

25×10^{10} US dollars, making it equivalent in value to the amount of hydrogen and carbon used and nearly six times that of the calcium used.

A means of measuring the relative importance of different elements in pure chemistry is to compare the relative numbers of compounds they form. Only approximate values for these can be obtained from Chemical Abstracts Service (CAS) statistics (Stobach, 1988). The reason that the values are approximate is because these statistics include theoretical species (on which calculations have been carried out) along with highly reactive species, and also list different isotopomers separately. Such comparison showed that iron occurred in around 2.3% of all known compounds in 1987, when the total number listed was around 8427300.

Table 1.1 (adapted from Nelson, 1991) summarises the overall importance of the more important elements based on factors including abundance in the atmosphere, lithosphere, biosphere and hydrosphere; the elements essential to life, their industrial importance in energy production, natural materials, manufactured materials and annual consumption, and their importance in pure chemistry. Iron has an overall ranking of eight; this finding brings out the fundamental importance of iron to man and is perhaps in itself a justification for this book.

1.3 Iron minerals and ores

Iron is the sixth most abundant element in the universe and the most abundant metal. It has been calculated that there are large amounts of iron on the moon's surface (0.5%) in the lunar soil. Iron is also known to occur in Martian soils. In the earth's crust it is the fourth most common element and the second most common metal after aluminium. The core of the earth is thought to be predominantly an iron nickel alloy from its known density, and from the velocity of sound propagation through it.

The reason for the high cosmic abundance of iron can be traced to the fact that its nuclei are especially stable. $_{26}Fe^{56}$ is at the peak of the nuclear binding-energy curve, and is believed to originate in the final seconds or minutes preceding the explosion of supernovae when temperatures in such stars are above $\sim 3 \times 10^9$ K. At such temperatures many kinds of nuclear reactions can occur allowing interconversions between various nuclei and free protons and neutrons establishing a statistical equilibrium. This explanation has been suggested to account for the observed abundance of elements from $_{22}$Ti to $_{29}$Cu in the cosmos. Thus having the most stable nucleus means that iron is readily formed by fusion when older stars burn their helium to produce heavier elements.

As meteorites contain nickel as well as iron, meteoritic iron does not rust easily in moist air. Meteoritic dust consisting mainly of iron continuously falls on the earth from space. Its presence is easily detectable on the surface of the snows and ice of the polar regions.

From the earth's integrated density and magnetic fields, and using echo soundings (earthquakes and nuclear explosions) from a large number of reflecting barriers, it is possible to make estimates of iron content, though these become increasingly less reliable as they progress deeper into the earth. About two-thirds of the earth's mass is in the mantle, and here the main constituents are the silicates rich in magnesium and iron, but apparently with an increasing proportion of the heavier iron atoms with greater depth into the earth. It is possible to simulate conditions there by subjecting heated Fe/Mg minerals to very high pressures. Increased temperature favours the endothermic side of the chemical equilibria present, but the more significant effect is the tremendously increased pressure which preferentially stabilises the more compact forms. In the deep mantle (Mao and Bell, 1978; Mao *et al.*, 1979; Navrotsky, 1981; Jamieson *et al.*, 1981), minerals like pyroxene $(Mg,Fe)SiO_3$, and olivine $(Mg,Fe)_2SiO_3$ (neither stable above 1×10^{-5} atm, becoming spinel AB_2X_4, β-phase) together with stishovite occur, when pyroxene is the starting composition. However, the materials do not simply separate out, even at quite high pressures (presumably modelling great depths in the mantle), perovskite (see chapter 3, Fig. 3.1) and magnesiowüstite, phases coexist, with iron preferentially distributed more into the latter. The following equations summarise important aspects of current knowledge to at least 700000 atm or a depth of 1600 km.

$(Mg,Fe)SiO_3$ composition (pyroxene):

$$\text{pyroxene} \xrightarrow{100\,000\ \text{atm}} (Mg,Fe)_2SiO_4 + \text{stishovite, } SiO_2$$

$(Mg,Fe)_2SiO_4$ composition (olivine, perovskite):

$$\text{olivine} \xrightarrow{100\,000\ \text{atm}} \text{spinel} \xrightarrow[-650\ \text{atm}]{190\,000\text{–}225\,000\ \text{atm}} \text{perovskite}$$

$$+ \text{magnesiowüstite } (Mg,Fe)_2O + \text{stishovite}$$

The inner and outer cores, which account for most of the other one-third of the earth's mass, probably consists of iron alloyed with some nickel.

On the surface of the earth iron occurs native (uncombined with other elements) in very small quantities, but with other elements it is found in a number of ores. The metal is so widely distributed that there are economic deposits of ores in all the continents. Due to the poor solubility of its oxides it is less abundant in the ocean, at about 2.3×10^{-8} mol/l or 1.3×10^{-3} mg/l, mainly as hydrated $[Fe(OH)_2]^+$ and $[Fe(OH)_4]^-$ (Schwochau, 1984). Among the major ores (Evans, 1972) are:

Magnetite, Fe_3O_4. This is described as iron-black with a metallic lustre. Deposits of this ore can confuse the unwary traveller by their action on his compass (see later). Magnetite is not found to any extent in the British Isles but occurs in Lapland, Sweden, Siberia, Germany, India and North America. The iron content of *c.* 72% makes magnetite the richest iron-containing ore. Mt. Whaleback in Western Australia attracts compass needles from great distances and was long known as a hazard to aircraft that navigate with the aid of magnetic readings. The mountains was only 'discovered' to be made largely of magnetite, Fe_3O_4, in the 1960s when the law on the distribution of profits between government and private enterprise was changed to favour the latter. Mt. Whaleback is the world's closest thing to an iron mountain. It has become *the* largest iron ore producer and a railway constructed by a Japanese consortium has the purpose of bringing most of it piecemeal to port for shipping to Japanese steel mills.

Haematite, Fe_2O_3. This ore varies from steel-grey to iron black, it is known as specular iron or kidney ore from its common external appearance. It is also found in a crystalline red form.

Siderite is iron carbonate, $FeCO_3$. It occurs in various shades of brown. Other iron carbonates are chalybite and spathic iron ore. These occur in Alpine regions and in Hungary. *Limonite* is a hydrated oxide of iron, $2Fe_2O_3 \cdot 3H_2O$, it is often referred to as brown haematite.

Iron pyrites is an iron sulphide, FeS_2. This mineral is also known as 'fool's gold', the name arising from its brassy colour and lustre that have misled many an unwary prospector. Although iron pyrites is common, it is not used as an ore because of the difficulty in eliminating sulphur.

The production, purification and metallurgy of iron are in themselves subjects for many books; they are briefly covered in chapter 2 of this book.

1.4 Pure iron and its main chemical and physical properties

Chemically pure iron is obtained by reduction of pure iron oxide using hydrogen. The iron oxide is prepared by thermal decomposition of a ferrous complex such as oxalate, carbonate or nitrate. Other ways of preparing pure iron include electro decomposition from aqueous solutions of iron salts, or by thermal decomposition of iron carbonyl. Some of the physical properties of iron are summarised in Table 1.2.

Table 1.2 Some properties of iron

Property	
Atomic number	26
Number of naturally occurring isotopes	4
Total number of isotopes reported	10
Atomic weight	55.847
Electronic configuration	$[Ar]3d^6 4s^2$
Electronegativity	1.83[a]
Metal radius (12-coordinate) (Å)	1.26
Effective ionic radius (Å)	
II	0.61 (low spin) 0.78 (high spin)
III	0.55 (low spin) 0.645 (high spin)
Melting point (K)	1808
Boiling point (K)	3023
Density (293 K, g cm^{-3})	7.874
Electrical resistivity (293 K/Ω cm)	9.71×10^{-8}
Molar volume (cm^3)	7.09
Coefficient of linear thermal expansion (K^{-1})	12.3×10^{-6}

[a]From A. L. Allred (1961) *J. Inorg. Nucl. Chem.*, **17**, 215.

Pure iron is a silvery-white, soft metal that rusts rapidly in moist air. As the pure element iron is of little use, it is strengthened by the addition of small amounts of carbon and of other transition metals. It is in the great variety of steels that iron has its most widespread uses (see also chapter 2). Carbon steels contain between 0.1 and 1.5% carbon and may also contain small amounts of other metals; in the latter case they are known as alloy steels. Different metals convey different properties to steel. Examples include vanadium for springiness, chromium to improve hardness and corrosion properties, while manganese aids resistance to wear. Nickel toughens steel for use in armour plating, whereas molybdenum and tungsten promote heat resistance. Stainless steels resist corrosion and find widespread use, particularly in hospital and medical laboratory equipment and in cutlery. Common stainless steel contains 18% chromium and 8% nickel.

Rusting of iron (Evans, 1979) is due to the formation of a hydrated oxide $Fe(OH)_3$ or FeO(OH). The process is an electrochemical one which necessitates the presence of water, oxygen and an electrolyte; without any one of these significant rusting does not occur. In moist air iron rusts rapidly, the hydrous oxide gives no protection as it flakes off exposing fresh metal surfaces (for greater details on rusting, corrosion and protection see chapter 2).

If iron is very finely divided it is pyrophobic. This fact should be borne in mind if an iron compound is being reduced to the metal state in such a way that very small iron particles may be formed.

Iron combines very vigorously with fluorine and chlorine (on mild heating) to give the iron(III) halides. It will react readily with many other non-metals including the other halogens, boron, carbon and silicon, phosphorus and sulphur. The carbide and silicide phases are of major importance in the metallurgy of iron (see chapter 2).

Iron is attacked by dilute mineral acids. In the absence of air and in non-oxidising acids iron(II) (ferrous) is the oxidation state of iron that is formed. However, if air is present or warm dilute nitric acid is used, some of the iron is present as iron(III) (ferric). Air/oxygen-free water and dilute air/oxygen-free hydroxides have little effect on iron, though hot concentrated sodium hydroxide will attack it.

In addition to the cheapness and tensile strength of its various alloys, iron is valued for its distinctive magnetic properties. Ferromagnetism derives its name from the magnetic properties of α-iron, the normal bcc phase at room temperature. In this the free electron gas in the metallic conduction band is formed largely from the d_{xy}, d_{xy}, and d_{xy} orbitals. The other orbitals retain between them an average of two unpaired electrons for each atom, this therefore acts as a local magnet on the atomic scale. Once oriented by an external magnetic field, a fraction of these individual atomic magnets remain preferentially ordered with the parallel moments each adding to a macroscopic magnetic field. Other pure elements which share this property are cobalt and nickel, both these are successively less ferromagnetic. Soft iron and mild steels (0.1–0.5% C) are easily magnetised, but this magnetisation is easily lost or reversed, a property which makes it valuable as a core in electric transformers and motors. Others, e.g. cobalt steels, are magnetised less readily but retain the magnetisation better, making them valuable as permanent magnets, though for sheer power, alnico, various rare earth cobalt alloys and NdFeB alloys have supplanted simple magnetic steels.

Iron is ferromagnetic up to its Curie temperature of 1041 K, where it becomes simply paramagnetic.

At temperatures up to 1183 K, iron has a body centred lattice, however, from 1183 to 1663 K it is cubic close packed, then above 1663 K it is again body centred.

The outer electronic configuration of iron in the metal (0) oxidation state is $4s^2 3d^6$. The unpaired electrons that give rise to the magnetic properties of metallic iron and indeed those found in the iron oxide magnetite are of great importance in the understanding of the chemistry of iron. Depending on the oxidation state and the nature of the surrounding ligands, their number can vary from 0 to 5. The number of unpaired electrons present gives information on the spin state of the iron, and this in turn gives information on the nature of the bonding around the iron. From this knowledge bonding strengths of ligands present may be judged. Techniques that measure the number or give information on the spin state of iron include e.s.r. and Mössbauer spectroscopy. These are discussed in detail in chapter 5.

Mössbauer spectroscopy can also give information on the oxidation state, coordination number and relative amount of s electron density (and amount of d electron shielding) at the iron nucleus in a compound.

Table 1.3 Isotopes of iron

Isotope	% natural abundance	Atomic mass	Lifetime	Mode of decay	Use
$_{26}Fe^{52}$			8.2 hours	β^+, EC	Tracer
$_{26}Fe^{53}$			8.5 minutes	β^+, EC	—
$_{26}Fe^{54}$	5.8	53.9396			—
$_{26}Fe^{55}$			2.6 years	EC	Tracer (medical)
$_{26}Fe^{56}$	91.8	55.9349			—
$_{26}Fe^{57}$	2.1	56.9354			Mössbauer and NMR spectroscopy
$_{26}Fe^{58}$	0.3	57.9333			—
$_{26}Fe^{59}$			45.1 ± 0.5 days	β^-	Tracer (medical)
$_{26}Fe^{60}$			3×10^5 year	β^-	—
$_{26}Fe^{61}$			6.0 minutes	β^-	—

The spin states of the various oxidation states of iron are discussed later in this chapter, and in detail in chapter 3.

1.5 Isotopes of iron

The four naturally occurring stable isotopes of iron are tabulated in Table 1.3 along with the six known radioactive isotopes. As stated in section 1.3, $_{26}Fe^{56}$ is at the peak of the nuclear binding curve and it accounts for 91.8% of the naturally occurring iron on the earth. $_{26}Fe^{57}$ is the isotope used in NMR and Mössbauer spectroscopies (see chapter 5).

1.6 Ionisation energies

Ionisation energies have been found for all ten outer electrons (see Table 1.4). These are given in kJ mol^{-1} but can be converted to eV by dividing by 96486 (for mJ mol^{-1} divide by 1000).

Table 1.4 Iron ionisation energies (kJ mol^{-1})

1.	M	→	M^+	=	759.3
2.	M^+	→	M^{2+}	=	1561
3.	M^{2+}	→	M^{3+}	=	2957
4.	M^{3+}	→	M^{4+}	=	5290
5.	M^{4+}	→	M^{5+}	=	7240
6.	M^{5+}	→	M^{6+}	=	9600
7.	M^{6+}	→	M^{7+}	=	12100
8.	M^{7+}	→	M^{8+}	=	14575
9.	M^{8+}	→	M^{9+}	=	22678
10.	M^{9+}	→	M^{10+}	=	25290

1.7 Oxidation states

The oxidation states known for iron compounds range from $-$II(d^{10}) to $+$VIII(d^0), examples are given in Table 1.5. This table also presents examples of the kinds of complexes that form their stereochemistry, and their spin state where established.

Of the oxidation states VII(d^1) is unknown and several are not well established, the $-$I(d^9) and V(d^3) states are rare. I(d^7) is not common and there is only one report of VIII(d^0). The common oxidation states of iron are II(d^6) and III(d^5); the ferrous and ferric states respectively. IV(d^4) the ferryl state is becoming more recognised, particularly in biological systems (see chapter 6). The more common nomenclature and that used in this book are iron(II), iron(III) and iron(IV), respectively.

It is apparent that in assignment of oxidation state, the way electrons are counted becomes of paramount importance. Low oxidation states are usually assigned to complexes where the iron has gained electrons over the number of its 8 original valence electrons. However, this is not always satisfactory, and it would be better to have physical evidence that any 'gained' electrons do indeed reside on or near the iron atom in the complex. Such evidence may be forthcoming from techniques such as Mössbauer spectroscopy (see chapter 5) and ^{57}Fe NMR (Houlton *et al.*, 1991).

The low oxidation states of iron such as I, 0, $-$I and $-$II are represented mainly by carbonyls, nitrosyls and phosphines, their derivatives and mixtures thereof. These are often considered under organometallic or carbonyl compounds in inorganic chemistry textbooks (Cotton and Wilkinson, 1988; Greenwood and Earnshaw, 1984). In this book, however, brief comments, examples and references will be given under the relevant oxidation state. Each oxidation state will be discussed in brief here, but many are discussed again in succeeding chapters in this book.

1.7.1 *Iron($-$II), d^{10}*

There are very few examples of iron in this oxidation state. The pale yellow compound [$Na_2Fe(CO)_4$] and other [$Fe(CO)_4$]$^{2-}$ complexes are familiar because they are of use in the synthesis of aldehydes and ketones (Collman, 1975; Colquohoun *et al.*, 1984). [$Fe(CO)_4$]$^{2-}$ is found to have essentially undistorted tetrahedral symmetry in [Na(crypt)]$_2$[$Fe(CO)_4$] (Teller *et al.*, 1977), but distorts in the presence of other cations. All the low oxidation states of iron are stabilised by π-acceptor ligands such as phosphines and carbon monoxide.

1.7.2 *Iron($-$I), d^9*

This is one of the rarest oxidation states of iron. Depending on how electrons are counted, it is possible to assign complexes such as $[Fe_2(CO)_8]^{2-}$ (Chin *et al.*, 1974) to this state. Such compounds contain a metal–metal bond.

1.7.3 *Iron(0), d^8*

There are a large number of iron(0) complexes, though current opinion best regards most of these as organometallic compounds (Shriver and Whitmire, 1982) (see chapter 4). $[Fe(CO)_5]$ (Beagley *et al.*, 1969), $[Fe(PF_3)_5]$ (Kruck and Prash, 1964) and mixed $[Fe(CO)_3(L)_2]$ (Clifford and Mukherjee, 1963) species ($L = A(C_6H_5)_3$, $A = P$, As or Sb) are fairly assigned to this oxidation state. (See chapter 4 for organometallic compounds, many of these can be assigned to the iron(0) oxidation state.)

1.7.4 *Iron(I), d^7*

This is not a common oxidation state of iron, yet complexes of iron(I) have been prepared and characterised. Amongst these are $[Fe(H_2O)_5NO]^{2+}$ (Griffith *et al.*, 1958) in which (NO) is present as (NO^+) coordinated to iron. Reduction of iron(II) phthalocyanine [Fe(pc)] electrochemically, or using sodium in tetrahydrofuran, yields $[Fe(pc)]^-$, though more highly reduced (pc) species probably contain iron in the same oxidation state with additional electrons being accommodated on the macrocyclic ring (Clack and Yandle, 1972). Many attempts to make iron(I) complexes have failed due to disproportionation reactions forming more stable iron(0) and iron(II) complexes.

1.7.5 *Iron(II), d^6*

This is one of the most common oxidation states found in iron chemistry. Iron(II) occurs in a variety of coordination geometries and spin-states (see Table 1.5). Iron(II) will form salts with practically all anions. In the anhydrous state, unless the anion is coloured, these are white. If they are obtained by evaporating aqueous solutions of the correct composition, then they are invariably green or blue-green, examples include $FeCl_2.4H_2O$, $FeSO_4.7H_2O$, $Fe(ClO_4)_2.6H_2O$, $FeC_2O_4.\frac{3}{2}H_2O$ and $Fe(SCN)_2.3H_2O$. The sulphate and perchlorate have been shown to contain octahedral $[Fe(H_2O)_6]^{2+}$ ions.

Many iron(II) salts darken in the presence of air (due to oxidation); examples are the carbonate and hydroxide which may be precipitated from aqueous solutions of ferrous salts. If excess CO_3^{2-} is added to an aqueous solution of iron(II) in the absence of air, the slightly soluble $[Fe(HCO_3)_2]$ salt is formed. When this is present in natural underground waters, it leads first to the production of $FeCO_3$ on exposure to air (when such waters reach the surface of the earth), then to iron(III) oxide that causes the widespread brown deposits found in many streams.

Pure aqueous solutions of iron(II) that contain no other complexing agents

Table 1.5 Oxidation states, spin states and stereochemistry of iron in its compounds

Oxidation state	Spin state (where established)	Coordination number	Geometry	Examples	Reference
Fe(-II), d^{10}	—	4	Tetrahedral	$[Fe(CO)_4]^{2-}$	Pensak and McKinney (1979); Mosback and Poulsen (1974)
Fe(-I), d^9	—		5-coordinate (including Fe–Fe bond)	$[Fe_2(CO)_8{}^{2-}]$	Chin *et al.* (1974)
Fe(0), d^8	low	5	Trigonal bipyramidal	$[Fe(CO)_5]$	Beagley *et al.* (1969)
				$[Fe(PF_3)_5]$	Kruck and Prasch (1964)
Fe(I), d^7	high	6	Octahedral	$[Fe(H_2O)_5NO]^{2+}$	Griffith *et al.* (1958)
				$[Fe(CO)_2(PMe_3)_3I]^+$	Pankowski and Bigorgne (1977)
Fe(II), d^6	high	4	Tetrahedral	$[FeCl_4]^{2-}$	Lauher and Ibers (1975)
				$[FeCl_2(dppe)]$	Barclay *et al.* (1988)
	intermediate	4	Square planar	[Fe(TPP)]	Lang *et al.* (1978)
				[Fe(PPIX)]	Medhi and Silver (1989)
					Lukas and Silver (1983)
					Silver *et al.* (1987)
				[Fe(pc)]	Kerner *et al.* (1976)
					Dale *et al.* (1968)
	high	5	Five coordinate*	$[FeCl_2(bipy)]$	Charron and Reiff (1986)
		5	Square pyramidal	$[Fe(ClO_4)(OAsMe_3)_4]ClO_4$	Brodie *et al.* (1968)
	high	5	Square pyramidal	[Fe(PPIX)OH]	Silver *et al.* (1984)
				$[Fe(TP_{piv}P)Cl]^-$	Schappacher *et al.* (1985)
	high	4 or 6	Square planar or octahedral	$[FeI_2(16[ane]S_4)]$	Hills *et al.* (1989)

	high	6	Octahedral	$[Fe(H_2O)_6]^{2+}$	Kermarrec (1964)
	high	6	Octahedral	$[FeCl_2(opdp)_2]$	Barclay *et al.* (1988)
	low	6	Octahedral	$[Fe(MeCN)_2(dppe)_2]I_2$	Barclay *et al.* (1988)
	low	6	Octahedral	$[FeCl_2(dmpe)_2]$	Barclay *et al.* (1988)
	low	6	Pseudo octahedral	$[Fe(phen)_3]^{2+}$	Johansson *et al.* (1978)
	low	—	Sandwich structures	$[(\eta^5\text{-}C_5H_5)_2Fe]$	Seiler and Dunitz (1979)
					Roberts *et al.* (1988)
				$[(\eta^5\text{-}C_5H_5)(\eta^6\text{-}C_6H_6)Fe][PF_6]$	Houlton *et al.* (1987)
				$[(\eta^6\text{-}C_6H_6)_2Fe][PF_6]_2$	Roberts *et al.* (1981)
		8	Dodecahedral (D_{2h})	$[Fe(1,8\text{-naphthyridine})_4]ClO_4$	Singh *et al.* (1971)
Fe(III), d^5		3	Trigonal	$[Fe(N(SiMe_3)_2)_3]$	Buerger and Wannagut (1963)
	high	4	Tetrahedral	$[FeCl_4]^-$	Paulus and Schäfer (1978)
		5	Square pyramidal	$[FeCl(dtc)_2]$	Martin and White (1967)
		5	Trigonal bipyramidal	$[FeF_3]^{2-}$	Creaser and Creighton (1979)
	high	6	Octahedral	$[Fe(C_2O_4)_3]^{3-}$	Gerloch *et al.* (1969)
	low	6	Sandwich	$[Fe(CN)_6]^{3-}$	Vannerberg (1972)
	low		Sandwich	$[(\eta^5\text{-}C_5H_5)_2Fe]^+$	Paulus and Schäfer (1978)
	high	7	Close to pentagonal	$[FeEDTA(H_2O)]^-$	Lind *et al.* (1964)
		8	Dodecahedral	$[Fe(NO_3)_4]^-$	King *et al.* (1971)
Fe(IV), d^4		4	Tetrahedral	$[Fe(I\text{-norbornyl})_4]$	Bower and Tennent (1972)
		6	Octahedral	$[Fe(diars)_2Cl_2]^{2+}$	Hazeldean *et al.* (1966)
FeV (d^3)		4	Tetrahedral	FeO_4^{3-} (only in solid oxides)	Levason and McAuliffe (1974)
Fe(VI), d^2		4	Tetrahedral	FeO_4^{2-}	Audelte and Quail (1972)
Fe(VIII), d^0			Not established	Fe in high pH alkaline media	Kopelev *et al.* (1990)

*proposed structure possibly Tbp

are a pale turquoise colour due to the presence of the hexaquo iron(II) ion, $[Fe(H_2O)_6]^{2+}$. This ion is easily converted to iron(III) by the presence of molecular oxygen in acid solution.

$$2Fe^{2+} + \tfrac{1}{2}O_2 + 2H^+ = 2Fe^{3+} + H_2O \quad E^0 = 0.46\ V$$

This oxidation is more favourable in basic solution

$$\tfrac{1}{2}Fe_2O_3.3H_2O + e = [Fe(OH)_2]_s + [OH]^- \quad E^0 = -0.56\ V$$

(s = solid fresh precipitate)

The oxidation of iron(II) to iron(III) is thought to involve an ion FeO_2^+ in which the oxygen may be initially bound. Reaction kinetics have been explained in terms of a binuclear species being a transient intermediate formed by attack of Fe^{2+} on FeO_2^{2+}. The overall reaction that has been suggested is:

$$FeO_2^{2+} + Fe(H_2O)^{2+} \rightarrow Fe(OOH)^{2+} + Fe(OH)^{2+}$$

This leads to the hydroperoxo ion which rapidly decomposes to generate iron(III) and HO_2^-, the latter can oxidise more iron(II) or decompose to O_2. The hydroperoxo ion has been detected in reactions of iron(III) with H_2O_2.

Iron(II) forms a vast number of complexes (see chapters 3 and 4), many of these are octahedral or distorted octahedral. The effect of complexing agents (Cotton and Wilkinson, 1988) on the relative stabilities in aqueous solution of the oxidation states is readily seen in the following examples:

$$[Fe(CN)_6]^{3-} + e \rightarrow [Fe(CN)_6]^{4-} \quad E^0 = 0.36\ V$$
$$[Fe(H_2O)_6]^{3+} + e \rightarrow [Fe(H_2O)_6]^{2+} \quad E^0 = 0.77\ V$$
$$[Fe(phen)_3]^{3+} + e \rightarrow [Fe(phen)_3]^{2+} \quad E^0 = 1.12\ V$$

In the presence of chelating ligands many low-spin iron(II) complexes that are stable in aqueous solution are known. A useful example involves the ethylenediemine ligand (en):

$$[Fe(H_2O)_6]^{2+} + en = [Fe(en)(H_2O)_4]^{2+} + 2H_2O \quad K = 10^{4.3}$$
$$[Fe(en)(H_2O)_4]^{2+} + en = [Fe(en)_2(H_2O)_2]^{2+} + 2H_2O \quad K = 10^{3.3}$$
$$[Fe(en)_2(H_2O)_2]^{2+} + en = [Fe(en)_3]^{2+} + 2H_2O \quad K = 10^2$$

Other low-spin iron(II) octahedral complexes include those formed by biby, phen and other strongly binding nitrogenous ligands. Iron(II) cyanides are discussed in chapter 3.

Low spin six-coordinate iron(II) complexes are usually intensely coloured (often purple or red). The colours arise from strong ligand to metal charge transfer bands (Sanders and Day, 1970). Such bands will obscure d–d transitions.

$Fe(phen)_3^{2+}$ is used as the radox indicator ‘ferroin’. This works by a sharp colour change which takes place when strong oxidising agents are added to it in solution (Walden *et al.*, 1931). The reaction is

$$\underset{\text{red}}{[\text{Fe(phen)}]^{2+}} \rightleftharpoons \underset{\text{blue}}{[\text{Fe(phen)}_3]^{3+}} + \text{e}$$

A property displayed by low spin iron(II) octahedral complexes which can be understood in terms of their crystal field stabilisation energies is their stability and kinetic inertness (Basolo and Pearson, 1958). This has resulted in a considerable body of work on their mechanisms of reaction and preparation (Shriver and Whitmere, 1982). However, it must be stated that this is not as extensive as that for the isoelectronic Co(III) complexes.

Iron(II) forms a number (which is constantly increased in the literature) of tetrahedral complexes. The $[FeX_4]^{2-}$ (X = halide) ions exist as salts with a range of cations (Lauher and Ibers, 1975). A number of neutral $[FeX_2L_2]$ or $[FeX_2L']$ (L = sulphur or phosphorus ligands; L′ = similar bidentate ligand) and cationic, $[FeL''_4]^{2+}$ (L″ = $(Me_2N)_3PO$ or Ph_3PO) complexes have been characterised (Edwards *et al.*, 1967; Birchall and Morris, 1972; Barclay *et al.*, 1988). In addition, tetrahedral iron sulphur complexes have been reported as models for iron sulphur proteins (see chapters 6 and 7). In fact, iron(II) carries out a number of important roles in biological systems (see chapters 6–8). One of the most important iron(II) complexes is the oxygen-carrying protein molecule found in human blood, haemoglobin. The molecule consists of protein surrounding 4 haem units. The haem found in haemoglobin is protoporphyrin(IX) iron(II), it carries oxygen in an axial position with the other axial site being occupied by the imidazole ring of a histidine residue (Figure 1.1).

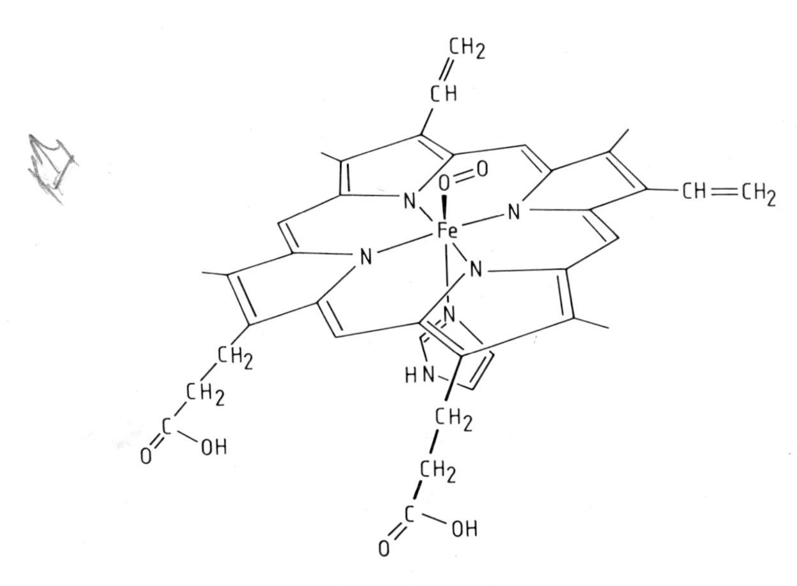

Figure 1.1

For more details on how haemoglobin operates and other biological molecules see chapter 6.

The porphyrin in Figure 1.1 is itself a tetradentate nitrogen donor macrocycle and will form square planar complexes with iron as will phthalocyanine, which is related to it and a number of synthetic macrocycles, in the absence of additional ligands. The magnetic properties and Mössbauer spectra of some of these complexes suggest they are in the intermediate spin (S = 1) ground state (Mehdi and Silver, 1989; Long *et al.*, 1978; Lukas and Silver, 1987; Kerner *et al.*, 1976; Dale *et al.*, 1968). When such complexes gain two-axial ligands (such as nitrogen donors or carbonyl ligands), they become six-coordinate and low spin; with one strong axial ligand they become five-coordinate and low spin (Scheidt and Gouterman, 1982). If the ligands are weaker binding (such as oxygen ligands) or sterically hindered ligands, five- or six-coordinate high spin complexes result (Scheidt and Gouterman, 1982). An 1987; Kerner *et al.*, 1976; Dale *et al.*, 1968). When such complexes gain (Caron *et al.*, 1979).

The obvious conclusion that can be drawn from the known iron(II) porphyrin compounds with the benefit of hindsight is that structured ligands (such as porphyrins) can be used to control stereochemistry.

Five-coordinate iron(II) complexes that are not based on such macrocyclic ligands can also be easily prepared from tripod ligands; these may be high-spin, low-spin or indeed exhibit spin-crossover as a function of temperature.

It is beyond the aim of this chapter to go into all the possible structures and spin states found for iron(II) in great depth. However, some points are briefly discussed below.

Iron(II) forms octahedral complexes with a large number of ligands (see chapter 3). These allow the iron to exhibit high spin, low spin and spin–spin crossover properties (see section 1.8.1). The spin crossover may be manipulated by choice of ligand, temperature or even by the use of energy in the visible region. Spin states and spin crossover are discussed in more detail in chapter 3 and the reader is referred there and also to inorganic chemistry textbooks (Cotton and Wilkinson, 1988; Greenwood and Earnshaw, 1984).

There are some compounds known for iron(II) that have high coordination numbers. Seven-coordination has been established in several iron(II) macrocyclic complexes (Nelson *et al.*, 1966), while eight-coordinate iron(II) complexes have also been characterised. An example of the latter is found with 1,8-naphthyridine. The $[Fe(naph)_4]^{2+}$ ion has a distorted dodecahedral environment (Clearfield *et al.*, 1970).

1.7.6 Iron(III), d^5

In common with iron(II) the majority of iron(III) complexes are octahedral or distorted octahedral, but as is apparent in Table 1.5, examples of three-, four-, five-, seven- and eight-coordinate complexes are known.

Iron(III) forms strong complexes with a wide range of oxygen ligands, and these have been intensely studied because of their biological significance (in siderophores utilised by many bacteria, in other aspects of bio-inorganic chemistry, in models for such compounds and in the pharmaceutical industry) (see chapters 6, 7 and 8). Iron(III) has a tendency to form oxo-bridged structures (Table 1.6 contains examples of these). More extensive oxygen-bridged networks can be thought of as building up to the iron oxides (referred to earlier in this chapter and more extensively in chapter 3), or as models for the core of ferritin (see more details in chapter 7).

In compounds containing linear Fe–O–Fe entities, there is π-bonding across the bridges (Table 1.6). Indeed even in those with bond angles less than 180°, there is evidence both for π-bonding and antiferromagnetic coupling of the electron spins on each iron(III) (Murray, 1974). The length of the Fe–O–Fe bond is iron coordination number dependent (Table 1.6).

Three iron(III) ions form a triangle around an oxygen atom in the basic carboxylates of general formula $[Fe_3O(O_2CR)_6L_3]_x$ (here R = Me, CMe_3, amino acid residue, etc., L = e.g. H_2O, MeOH, py, DMF and x = anion) (Dziobkowski *et al.*, 1981; Catterick *et al.*, 1977). The M_3O unit is usually planar because of π-bonding. Larger oxygen centred iron polyhedra have also been reported such as (μ_4-O) in $[(PPh_3)_2][Fe_3Mn(CO)_{12}(\mu_4\text{-}O)]$ (Schauer and Shriver, 1987). More complicated polymeric oxo species have also been studied such as $Fe_4O_6(PhCO_2)_{15}(OH)_6$ (Gorun *et al.*, 1987).

Many references to complicated iron oxygen bridged structures are to be found in chapter 7 as model compounds for various metallo enzymes and iron storage and transfer proteins.

Aspects of iron(III) sulphur chemistry are covered extensively in chapters 3, 6 and 7. It is worthwhile to note here that a number of iron(III) dithiocarbonate complexes of the general formula $[Fe(S_2CNR_2)_3]$ that have a trigonally distorted octahedral geometry have temperature-dependent high-spin/low-spin crossover properties (Ganguli and Marath, 1978). See section 1.8.1 for more on this topic, especially for iron(II) compounds.

The magnetic properties of monomeric iron(III) complexes are very varied and extremely ligand-dependent. Weak field ligands that produce high-spin complexes have magnetic moments close to the spin-free values, whereas very strong field ligands produce low-spin complexes which manifest higher than expected room temperature magnetic moments because of the presence of large orbital contributions (Nelson *et al.*, 1987).

In common with low-spin iron(II) compounds low-spin iron(III) materials are much more kinetically inert than high-spin complexes.

Iron(III) forms salts with all common anions except for those such as iodide which have reducing abilities. Such salts can be crystallised in pale-violet hydrated forms that, like low pH solutions in which iron(III) is surrounded by water ligands, presumably contain $[Fe(H_2O)_6]^{3+}$ cations. Such solutions (if the pH is raised) turn yellow due to hydrolysis (Flynn, 1984; Mann, 1987).

Table 1.6 Examples of structures containing μ-oxo bonds

Complex	Fe–O–Fe angle (degrees)	Fe–O distance (Å)	Comments	Reference
Singly bridged				
$[Cl_3Fe–O–FeCl_3]^{2-}$	162.4(9)	1.757(2)	Cations are ferrocenium; Chlorides are eclipsed	Bellen *et al.* (1986)
$[Cl_3Fe–O–FeCl_3]^{2-}$	155.6(7)	1.755(3)	Cations are pyridinium; Chlorides are eclipsed	Drew *et al.* (1978)
$[Cl_3Fe–O–FeCl_3]^{2-}$	180.0	1.740(1)	Cations are tetraphenylphosphonium; Chlorides are staggered	Dehnicke *et al.* (1983)
$[Cl_3Fe–O–FeCl_3]^{2-}$	161.6(9)	1.76(2)	Cations are tris (1,10-phenanthroline) iron(II); Chlorides are staggered	Reiff *et al.* (1983)
$[Cl_3Fe–O–FeCl_3]^{2-}$	180.0	—	Cations are $\{[(C_6H_5)_3P]_2CSe\}_2^{2+}$	Schmidbauer *et al.* (1983)
$[Br_3Fe–O–FeBr_3]^{2-}$	159.8(4)	1.752(2)	Cations are ferrocenium; Bromides are eclipsed	Evans *et al.* (1992)
$[L^*FeOFeCl_3]^+$	149.8(3)	1.751(4)	Anion is chloride	Gomez-Romero *et al.* (1986)
$[SalenFeOFeSalen].py_2$	139.1(9)	1.820(16)	Complex is neutral	Gerlock *et al.* (1969)
$[Fe(TPP)]_2O$	176.1(2)	1.759(1)	Complex is neutral	Swepston and Ibers (1985)
$[Fe(TPC)_2O$	180.0	1.747(5)	Complex is neutral	Strauss *et al.* (1987)
Triply bridged				
$[BPz_3)(OA_c)Fe]_2O$	123.6(1)	1.784(6)	Complex is neutral	Armstrong *et al.* (1984)

*L is a pentadentate η^5 ligand

In the absence of coordinating ligands which would cause further complications, the following equilibria are considered to be important:

$$[Fe(H_2O)_6]^{3+} \rightleftharpoons [Fe(H_2O)_5(OH)]^{2+} + H^+$$

$$[Fe(H_2O)_5(OH)]^{2+} \rightleftharpoons [Fe(H_2O)_4(OH)_2]^+ + H^+$$

and

$$2[Fe(H_2O)_6]^{3+} \rightleftharpoons [(Fe(H_2O)_4(OH))_2]^{4+} + 2H^+ + 2H_2O$$

The dimer in the last equation is (Morrison *et al.*, 1978):

$$[(H_2O)_4Fe\begin{matrix} \diagup OH \diagdown \\ \diagdown OH \diagup \end{matrix}Fe(H_2O)_4]^{4+}$$

If the pH is raised above 2–3, further condensation occurs, colloidal gels begin to form and eventually a reddish-brown precipitate of hydrous iron(III) oxide is formed. It is this precipitate that is responsible for the low solubility of iron(III) at pH 7, and for this reason iron(II) was originally used as a source of soluble iron for pharmaceutical compositions. That was before suitable ligands capable of solubilising iron(III) at pH 7 were developed (see chapters 6 and 8).

As stated in the presence of complexing anions such as chloride, the hydrolysis of iron(III) is more complicated. Chloro, aqua and hydroxy species are formed as well as $[FeC1_4]^-$ at high chloride concentration (Schlogel and Jones, 1983; Giubileo *et al.*, 1983).

The colours of iron(III) solutions are of interest as the d^5 electron configuration might be expected to give rise to weak spin forbidden bands as found for manganese(II) (which is also d^5). However, the presence of the extra positive charge on iron(III) gives it a greater polarising ability with coordinated ligands, producing intense charge transfer absorption bands for all the hydrolysed species. The bands are though in the ultraviolet region of the electromagnetic spectrum tail into the visible region giving the yellow colours and obscuring weak d-d bands. Other ligands bring these intense bands further into the visible region of the spectrum (see examples in Hider *et al.*, 1981 for instance).

1.7.7 *Iron(IV),* d^4

Only a small number of iron(IV) compounds are well characterised. However, transient FeO species in which the iron valence state is IV appear to be involved in many oxidation systems found in the biochemistry of iron (see chapter 6), particularly in compounds of the type (porphyrin)Fe=O (Schappacher *et al.*, 1986).

Iron(IV) can be stabilised in the solid state as in a perovskite-type oxide (such as $SrFeO_3$) (Takano and Takeda, 1983; Gibb, 1986), $[Fe(1\text{-norbonyl})_4]$,

(Bower and Tennent, 1972) probably owes its stability to the steric bulk of the ligand.

Cationic complexes can be made by chemical or electrochemical oxidation of related iron(III) complexes, e.g.$[Fe(S_2CNR_2)_3]^+$ (Crisponi *et al.*, 1986). However, in $[Fe(biby)_3]^{4+}$ the electron is removed from a ligand orbital (Gaudiello *et al.*, 1984). A neutral iron(IV) complex of *o*-phenylenebis-(dimethylphosphine) has also been prepared (Harbron *et al.*, 1986).

1.7.8 *Iron(V)*, d^3

This oxidation state is very rare for iron but some evidence has been found in Mössbauer studies. Among these are studies on the oxides $Ca_{1-x}Sr_xFeO_3$ and $Sr_{1-y}Ba_yFeO_3$. Mössbauer spectroscopic studies have revealed that these oxides contain iron(III) and iron(V) ions and show a transition between the average valence state and the mixed valence state (Takano *et al.*, 1982). Similar studies on $Sr_3Fe_2O_{6.5}$ have also found evidence for iron(V) (Al-Rawwas *et al.*, 1992).

1.7.9 *Iron(VI)*, d^2

Hypochlorite oxidation of iron(III) nitrate in strong alkali solution has been found to yield alkali metal salts of iron(VI) (Audette and Quail, 1972). An example of such a salt is K_2FeO_4 which is isomorphous with K_2CrO_4 and $K_2M_4O_4$ (Hoppe *et al.*, 1982).

1.7.10 *Iron(VIII)*, d^0

To date the only evidence for this oxidation state has been found in frozen solution Mössbauer studies on iron in high pH alkaline media (Kopelev *et al.*, 1990).

1.8 Spin states

The possible spin states of iron for the various oxidation states are a useful handle to probe iron chemistry. Some examples of spin state are given in Table 1.5, and more details and a brief introduction based on simple crystal field theory are to be found in chapter 3. A more exhaustive introduction to spin states is beyond the scope of this book but standard inorganic textbooks (Greenwood and Earnshaw, 1984; Cotton and Wilkinson, 1988) give good introductions to their origin and implications.

The range of spin states that are found in iron chemistry covers every integral and half-integral value of *S* from 0 to 5/2, that is every value possible for a d block element. Table 1.7 summarises these spin states but does not cover more

Table 1.7 Electronic spin states of iron

Spin quantum number (S)	Description of spin state	Electronic configuration
0 (diamagnetic)	Low spin Fe(II)	$t_{2g}{}^6$
1/2 (1 unpaired e^-)	Low spin Fe(III)	$t_{2g}{}^5$
	Low spin Fe(I)	$t_{2g}{}^6e_g{}^1$
1 (2 unpaired e^-)	Intermediate spin Fe(II) in square planar geometry	$d_{z^2}{}^2d_{zx}{}^2d_{xy}{}^1d_{yz}{}^1$
	Low spin Fe(IV)	$t_{2g}{}^4$
	Tetrahedral Fe(VI)	e^2
3/2 (3 unpaired e^-)	High spin Fe(I)	$t_{2g}{}^5e_g{}^2$
	Fe(III) in distorted square pyramidal arrangement	$d_{x^2-y^2}{}^2d_{yz}{}^1d_{xz}{}^1d_{z^2}{}^1$
2 (4 unpaired e^-)	High spin Fe(II)	$t_{2g}{}^4\ e_g{}^2$
	High spin Fe(IV)	$t_{2g}{}^3\ e_g{}^1$
5/2 (5 unpaired e^-)	High spin Fe(III)	$t_{2g}{}^3\ e_g{}^2$

complicated spin admixed systems, and oversimplifies some intermediate spin states. In addition, the table does not cover oxidation states O, – I, V, part of VI and VIII which have only one spin state each. These have respectively 2, 1, 3, 2 and 0 unpaired electrons.

Information on the nature of the spin state of an iron compound or complex can often be gained from magnetic measurements and ^{57}Fe Mössbauer spectroscopy (chapter 5). Such information aids in the understanding of the nature of the bonding type and strengths of bonding of ligands as well as the chemistry of the iron oxidation state. Indeed, understanding the forces that control spin state is a goal in the quest for new magnetic materials, both organometallic and inorganic as well as for molecular magnetics. Compounds and materials that can exhibit spin crossover may be sensitive to change in temperature, pressure or light (wavelength). Slight chemical modification may cause significant changes in their behaviour (see the following section and chapter 3 for more details of spin equilibria).

1.8.1 Spin crossover

As stated above, control of spin state is important and can be very sensitive in some systems. The first report of an iron(II) spin crossover system was by Madega and König in 1963, but the phenomenon had been predicted by Orgel in 1956.

There are now over 50 systems known for iron(II) chemistry. The phenomenon can usually be observed when the ground state is $^1A_1$ and the $^5T_2$ state is close enough in energy to be thermally populated (Toftund, 1989). Amongst a number of reviews of the subject that have appeared since 1968

(Martin and White, 1968; Sacconi, 1971; Goodwin, 1976; Gütlich, 1981; König *et al.*, 1985) the last two give a very comprehensive coverage of the literature. The most recent review (Toftund, 1989), though concentrating on Danish work, covers the design of new spin equilibria systems including ligand design, anion and solvent effects, characterisation of spin equilibria in the solid state and in solution, and dynamic aspects of spin change reactions.

The light induced spin-state trapping (LIESST) (Hauser *et al.*, 1986) observed at very low temperatures in some stystems, opens the possibility of the fabrication of advanced optical materials. If this phenomenon can be achieved at room temperature, materials for optical data storage might be obtained. However, the construction of optical components based on the non-linear optical properties of the materials offers a more easily achieved application (i.e. an optical switch (Toftund, 1989)).

1.9 Magnetics

The presence or absence of unpaired electrons in the various spin states of the oxidation states of iron makes the use of techniques that measure these, attractive probes for iron chemistry. Such techniques and the theory on which they are based are to be found in elementary textbooks and the user of this book is referred to them (Sharpe, 1986; Mackay and Mackay, 1973; Huheey, 1978; Atkins, 1983; Castellan, 1983).

It is worth pointing out here that values of magnetic moments based only on electron spin require small corrections for the presence of orbital angular momentum, again this is well covered in undergraduate textbooks. However, such a simple approach is not able to explain all bi- and polynuclear complexes where spin–spin coupling may be present, originating either from direct orbital overlap or mediated by bridging ligands. Also spin-crossover compounds can give strange results and thus single magnetic measurements or even multi measurements over temperature ranges must be interpreted with extreme caution. Complementary techniques such as Mössbauer spectroscopy and e.s.r. often make such interpretation easier and the combination of techniques often enables a more useful understanding of the data.

1.10 Electronic structure

The electronic structure of iron is dependent on valence state, geometry and ligand strength. The overall bonding ideas that have been developed to deal with the electronic structure of transition elements are often called *ligand field theory* (LFT). This title encompasses at one extremity crystal field theory (CFT) and at the other all manner of molecular orbital treatments. Such theories are to be found in elementary undergraduate texts referred to in section 1.9 (see also chapter 3). Spectroscopic properties are also discussed in these texts. For some complexes (those with relatively high symmetry and only moderate covalent bonding in the metal to ligand bonds) the electronic absorption spectra can be easily understood (Lever, 1984).

1.11 Organometallic chemistry of iron

Chapter 4 of this book is devoted to the subject of organo-iron chemistry. As defined in that chapter, this title include all compounds that contain direct metal to carbon bonds. The main classes are the carbonyls and the metallocenes and in both of these it has transpired that the iron compounds have had a significant role.

Such iron–carbon bonds can be considered to be of three basic kinds. The simplest would be single covalent FeC (σ) bonds, however, these are unstable in the absence of good donor or π-bonding ligands binding to the iron centre. This explains why 'homoleptic' alkyls and aryls of the type $Fe(CH_3)_n$ or $Fe(C_6H_5)_n$ have never been isolated, but compounds such as the methyl-iron complex (Figure 1.2a) and the phenyl iron complex (Figure 1.2b) (Guilard *et al.*, 1985) are stable at room temperature. The former molecule (Figure 1.2a) (Piper and Wilkinson, 1956; Jordan and Norton, 1979) is an illustration of stabilisation by the π-bonding ligands carbon monoxide and cyclopentadienyl, while in Figure 1.2b the chelating macrocyclic nitrogen donor ligand provides the stabilisation.

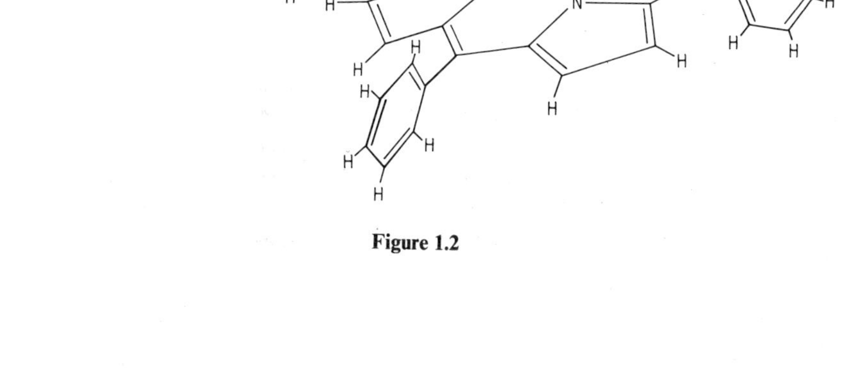

Figure 1.2

The methly-iron complex (Figure 1.2a) also exemplifies the two other Fe–C bonding modes. These are the 'end-on' π-bonding of the carbonyl groups and the 'side-on' π-bonding of the cyclopentaldienyl group.

The bonding in carbon monoxide and similar ligands such as the isocyanides, RNC, and the cyanide ion, CN^-, can be considered as a combination of σ-donation of electrons from the filled carbon sp-orbital to the metal and concominant back-donation of electrons from a filled d-type orbital of the metal into the antibonding (π^*) orbitals on carbon as illustrated in the bonding picture (Figure 1.3a). This back-donation allows the iron centre to distribute away excess electron density, but at the same time increases the strength of the Fe–C bond. The 'side-on' mode is illustrated (Figure 1.3b) for the simplest case, ethylene: here the forward donation involves the π-electrons of the unsaturated hydrocarbon ligand, while back-donation is again from d-type into π^* orbitals; again the result is relatively strong bonded though this is ligand-dependent. Some modern ideas on such bonding in cyclopentadienyl and aryl iron sandwich compounds have recently been discussed (Houlton *et al.*, 1990, 1991; Silver, 1990).

More extended π-systems bond similarly and organometallic complexes are classified according to the number of metal carbon interactions; this is also the number of electrons contributed by the *neutral* hydrocarbon ligand and is termed the 'hapticity' and denoted by η^n. Thus, alkyls are η^1-, carbonyl and alkene complexes η^2-, and cyclopentadienyls η^5-complexes. A summary of the principal types of carbon ligands and their hapticities are presented in Table 1.8

For iron, hexahapto is the highest known hapticity, but for earlier transition metals (e.g. Ti, Zr) and for actinides (Th, U) η^7 and η^8 ligands in the form of

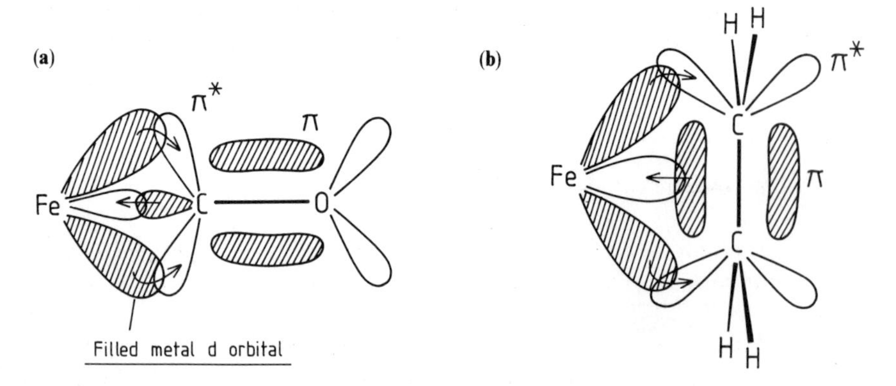

Figure 1.3

Table 1.8 Hapacity classification of organo ligands found in iron compounds

Numeric description	Hapticity	Ligand types
Monohapto	η^1	Alkyls, aryls, acyls
Dihapto	η^2	Alkenes, alkynes, CO, CS, CNR, CN, carbenes
Trihapto	η^3	π-Allyls
Tetrahapto	η^4	Conjugated dienes including cyclobutadienes, trimethylenemethane
Pentahapto	η^5	Dienyls including cyclopentadienyls, 'heterocyclopentadienyls' (e.g. pyrrolyl, phospholyl)
Hexahapto	η^6	Trienes, benzenoid aromatics

the cyclic planar tropylium (=cycloheptatrienyl) and cyclooctatetraene yield stable complexes.

It should also be noted that nearly all such π-complexes of iron satisfy the 'inert gas rule' or 'effective atomic number (EAN) rule', i.e. the metal atom acquires a share in sufficient electrons to complete the outer shell of 18 electrons (i.e. the number for the next inert gas, krypton).

This 'inert gas rule' has proved useful, particularly in counting electrons in iron carbonyl cluster compounds to ascertain when metal–metal bonding is taking place and how many such bonds are present. Using this rule, the iron in $[Fe(CO)_5]$ is found to be in oxidation state (0) as it obeys the rule whereas $[Fe_2(CO)_9]$ and $[Fe_3(CO)_{12}]$ only fit the rule if metal–metal bonding is assumed to occur. The former (which has 17 electrons per iron atom) contains one metal–metal bond) and the latter (which has 16 electrons per iron atom) contains three such bonds.

The final section of chapter 4 demonstrates that from the now vast organometallic chemistry of iron, a number of possible industrial uses are emerging. In addition, some ferrocene, ferrocenium and bisarene iron(II) complexes have been tested for possible cytotoxic applications for cancer therapy (Houlton *et al.*, 1991).

1.12 The biochemical importance of iron

As stated in section 1.2, iron is the most important element involved in the living process, being indispensable to all members of the plant and animal kingdom. Iron deficiency in plants is manifest by stunted growth. This is usually due to either the soil (growing medium) being itself deficient in iron or due to high alkalinity rendering the iron too insoluble to be accessible to the plants. The problem is usually alleviated by either adding simple iron salts or an iron complex which is stable enough to prevent the precipitation of the iron (see Sigel, 1978).

An adult human body contains around 4 g of iron, about three-quarters of which is in the form of haemoglobin. The remainder is found in a wide variety of environments; of these many are discussed in chapter 6 and models used to help the understanding of their chemistry are reviewed in chapter 7.

Iron has the status of being the first minor element to be listed as essential. In 1681, the physician T. Sydenham used iron 'steeped in cold Rhenish wine' to treat anaemia. Modern iron treatments for this and other iron related diseases are discussed in chapter 8.

1.13 The future of iron chemistry and biological chemistry

The purpose of this book is to give an overview of the main areas of interest in iron chemistry, and direct the reader to prime literature for in-depth studies. All the areas covered in the following chapters are being actively researched. A glance at chemical abstracts indexes under iron, iron compounds or iron chemistry quickly shows the thousands of scientific papers appearing every year in these areas (see section 1.1.), yet there is still much to be investigated. Although iron, as the metal, in alloy steels and in its compounds has been put to many industrial uses already, its abundance and relative cheapness together with the understanding of its catalytic roles in living systems will ensure that still more uses will emerge for this versatile element in the future. For scientists of today and tomorrow, this challenge will be both stimulating and rewarding.

References

Al-Rawwas, A.D., Thomas, M.F., Dann, S.E. and Weller, M.T. (1992) A Mössbauer study of $Sr_3Fe_2O_{6.5}$. Poster presented at the 33rd meeting of the Mössbauer Discussion Group of the Royal Society of Chemistry, Handbook p. 21.

Atkins, P.W. (1983) *Molecular Quantum Mechanics*, Second Edition, Oxford University Press, Oxford.

Audette, R.J. and Quail, J.W. (1972) Potassium, rubidium, cesium and barium ferrates(VI). Preparations infrared spectra and magnetic susceptibilities. *Inorg. Chem.*, **11**, 1904–1908.

Armstrong, W.H., Spool, A., Papaefthymiou, G.C., Frenkal, R., Lippard, S.J. (1984) Assembly and characterization of an accurate model for the diiron center in hemerythrin. *J. Am. Chem. Soc.*, **106**, 3653–3667.

Barclay, J.E., Leigh, G.J., Houlton, A. and Silver, J. (1988) Mössbauer and preparative studies of some iron(II) complexes of diphosphines, *J. Chem. Soc., Dalton Trans.*, 2865–2870.

Basolo, F. and Pearson, R.G. (1958) *Mechanisms of Inorganic Reactions*, John Wiley, New York, pp. 108–115.

Birchall, T. and Morris, M.F. (1972) Mössbauer and infrared spectra of tetrahedral complexes of iron(II) halides with thioamides and related ligands, *Can. J. Chem.*, **50**, 211–222.

Beagley, B., Cruicksank, D.W.J., Pinder, P.H., Robiette, A.G. and Sheldrick, G.M. (1969) The molecular structure of $Fe(CO)_5$ in the gas phase. *Acta Cryst.*, **B25**, 737–744.

Bower, B.K., and Tennent, H.G. (1972) Transition metal bicyclo [2.2.1] hept-1-yls, *J. Amer. Chem. Soc.*, **94**, 2512–2514.

Brodie, A.M., Hunter, S.H., Rodley, G.A. and Wilkins, C.J. (1968) High-spin five coordinated $[ML_5]^{2+}$ and $[ML_4(ClO_4)]^+$ complexes with trimethylphosphine oxide and trimethylarsine oxide. *Inorg. Chim. Acta*, **2**, 195–201.

Buerger, H. and Wannagut, U. (1963) Silylamide derivatives of iron and cobalt. *Monatsh. Chem.*, **94**, 1007–1012.

Bullen, G.J., Howlin, B.J., Silver, J., Fitzsimmons, B.W., Sayer, I. and Larkworthy, L.F. (1986) The crystal and molecular structure of ferrocenium μ-oxo-bis-[trichloroferrate(III)]; a product of reaction of ferrocene with iron(III)-chloride. *J. Chem. Soc., Dalton Trans.*, 1937–1940.

Caron, C., Mitschler, A., Riviere, G., Richard, L., Schappacher, M. and Weiss, R. (1979) Models for the reduced states of cytochrome P-450 and chloroperoxodase, structures of a pentacoordinate high spin iron(II) mercaptide mesoprophyrin derivative and its carbonyl adduct. *J. Am. Chem. Soc.*, **101**, 7401–7402.

Castellan, G.W. (1983) *Physical Chemistry*, Third Edition, Addison-Wesley Publishing Co., Reading, Mass.

Catterick, J., Thornton, P. and Fitzsimmons, B.W. (1977) Synthesis, magnetic properties and Mössbauer spectra of polynuclear iron carboxylates. *J. Chem. Soc., Dalton Trans.*, 1420–1425.

Charron, F.F. Jr. and Reiff, W.M. (1986) The second isomer of dichloro(2,2′-bipyridine) iron(II); syntheses and spectroscopic and magnetic characterizations of three related dichloro-(α-diimine) iron(II) complexes containing five coordinate, high-spin iron(II). *Inorg. Chem.*, **25**, 2786–2790.

Childe, V.G. (1942) *What Happened in History*, Penguin Books, London, pp. 182–185.

Chin, H., Smith, M.B., Wilson, R.D. and Bau, R. (1974) Variations in molecular geometry along the isoelectronic series $[Co_2(CO)_8]$, $[FeCo(CO)_8]^-$ and $[Fe_2(CO)_8]^{2-}$. *J. Amer. Chem. Soc.*, **96**, 5285–5287.

Clack, D.W. and Yandle, J.R. (1972) Electronic spectra of the negative ions and some metal phthalocyanines. *Inorg. Chem.*, **11**, 1738–1742.

Clearfield, A., Singh, P. and Bernal, I. (1970) The crystal structure of eight-coordinate iron(II) complex: tetrakis-(1,8-naphthyridine)-iron(II) perchlorate. *J. Chem. Soc., Dalton Trans.*, 380–391.

Clifford, A.F. and Mukherjee, A.K. (1963) Iron carbonyl complexes of triphenylphosphine, triphenylarsine and triphenylstibine. *Inorg. Chem.*, **2**, 151–153.

Collman, J.P. (1975) Disodium tetracarbonylferrate, a transition metal analog of a Grinard reagent. *Acc. Chem. Res.*, **8**, 342–347.

Colquhoun, H.M., Holton, S., Thompson, D.J. and Twigg, M.V. (1984) *New Pathways for Organic Synthesis*, Plenum Press, New York.

Cotton, F.A. and Wilkinson, G. (1988) *Advanced Inorganic Chemistry*, Fifth Edition, Chapters 22, 25 and 26, John Wiley and Sons Inc., New York.

Creaser, C.S. and Creighton, J.A. (1979) 5-Coordinate halogeno-complexes and related ions: halogeno-complexes of Fe(III) and Ti(IV). *J. Inorg. Nucl. Chem.*, **41**, 469–472.

Crisponi, G., Deplano, P. and Trogu, E.F. (1986) Mechanism of the reaction of iron(III) dithiocarbamates with iodine, *J. Chem. Soc., Dalton Trans.*, 365–368.

Dale, B.W., Williams, R.J.P., Edwards, P.R. and Johnson, C.E. (1968) S = 1 Spin state of divalent iron(II). A Mössbauer-effect study of phthalocyanine iron(II). *J. Chem. Phys.* **49**, 3445–3449.

Dehnicke, K., Prinz, H., Massa, W., Pebler, J. and Schmidt, R. (1983) Reaction of trichloronitromethane with iron carbonyls. Crystal structure of $(PPh_4)_2[Fe_2OCl_6].2CH_2Cl_2$. *Z. Anorg. Allg. Chem.*, **499**, 20–30.

Drew, M.G.B., McKee, V. and Nelson, S.M. (1978) Crystal and molecular structure and some properties of pyridinium μ-oxo-bis[trichloroferrate(III)]-pyridine. *J. Chem. Soc., Dalton Trans.*, 1980–1984.

Dziobkowski, C.T., Wrobleski, J.T. and Brown, D.B. (1981) Magnetic properties and Mössbauer spectra of several iron(III) dicarboxylic acid complexes. *Inorg. Chem.*, **20**, 671–678 and refs. therein.

Edwards, P.R., Johnson, C.E. and Williams, R.J.P. (1967) Mössbauer spectra of some tetrahedral iron(II) compounds. *J. Chem. Phys.*, **47**, 2074–2082.

Evans, I.O. (1972), *Rocks, Minerals and Gemstones*, The Hamlyn Publishing Group Ltd., London, pp. 68–72.

Evans, U.R. (1979) *An Introduction to Metallic Corrosion*, Arnold, London, pp. 153.

Evans, P.J.M., Fitzsimmons, B.W., Marshall, W.G., Golder, A.J., Larkworthy, L.F., Povey, D.C.

and Smith, G.W. (1962) Ferrocenium-μ-oxo-bis[tribromoferrate(III)], a product of reaction of ferrocene with iron(III) bromide, *J. Chem. Soc., Dalton Trans.*, 1055–1068.
Flynn, C.M., Jr. (1984) Hydrolysis of inorganic iron(III) Salts. *Chem. Rev.*, **84**, 31–41.
Ganguli, P. and Marathe, V.R. (1978) Magnetic susceptibility and magnetic anisotropy studies in some ferric dithiocarbamates exhibiting spin-crossover phenomena. *Inorg. Chem.*, **17**, 543–550 and refs. therein.
Gardiello, J.G., Bradley, P.G., Norton, K.A., Woodruff, W.H. and Bard, A.J. (1984) Electrochemistry in liquid sulphur dioxide 5.Oxidation of bipyridine. *Inorg. Chem.*, **23**, 3–10.
Gerloch, M., Lewis, J. and Slade, R.C. (1969) Paramagnetic anistotropies and zero-field splitting of some highfield iron(III) complexes. *J. Chem. Soc., (A)*, 1422–1427.
Gerloch, M., McKenzie, E.D. and Towl, A.D.C. (1969) Crystal and molecular structure of μ-Oxo-Bis-[NN-ethylenebis(salicylideneinato)-iron(III) bispyridine, *J. Chem. Soc. A*, 2850–2858.
Gibb, T.C. (1986) Magnetic exchange interactions in perovskite solid solutions. Part 6. Iron-57 and tin-119 Mössbauer spectra of $SrFe_{1-x}Sn_x-O_{3-7}$ $(O \leqslant x \leqslant 0.7)$. *J. Chem. Soc., Dalton Trans.*, 1447–1451.
Giubileo, G., Magini, M., Licheri, G., Paschina, G., Piccaluga, G. and Pinna, G. (1983) On the structure of iron(III) chloride solutions. *Inorg. Chem.*, **22**, 1001–1002.
Gomez-Romero, P., Detotis, G.C. and Jameson, G.B. (1986) Structure and magnetic properties of an unsymmetrical (μ-Oxo)diiron(III) complex. *J. Am. Chem. Soc.*, **108**, 851–853.
Goodwin, H.A. (1976) Spin transitions in six-coordinate iron(II) complexes. *Coord. Chem. Rev.*, **18**, 293–325.
Gorun, S.M., Papefthymiou, G.C., Frenkel, R.B. and Lippard, S.J. (1987) Synthesis, structure and properties of an undecairon(III) oxo-hydroxo Aggregate: An approach to the polyiron core in ferritin. *J. Am. Chem. Soc.*, **109**, 3337–3348.
Greenwood, N.N. and Earnshaw, A. (1984) *Chemistry of the Elements*, Chapter 25, Pergamon Press, Oxford.
Griffith, W.P., Lewis, J. and Wilkinson, G. (1958) Some nitric oxide complexes of iron and copper. *J. Chem. Soc.*, 3993–3998.
Guilard, R., Boisselier-Cocolios, A., Tabard, A., Cocolios, P., Simonet, B. and Kadish, K.M. (1985) Electrochemistry and spectroelectrochemistry of η-bonded aryliron porphyrins. 3. Synthesis and characterization of high, low and variable spin-state five-coordinate η-bonded aryl- and perfluoroaryliron(III) Complexes. *Inorg. Chem.*, **24**, 2509–2520.
Gütlich, P. (1981) Spin crossover in iron(II) complexes. *Struct. Bonding (Berlin)*, **44**, 83–195.
Harbron, S.K., Higgins, S.J., Levason, W., Garner, C.D., Steel, A.T., Feiters, M.C. and Hasnain, S.S. (1989) Iron K-edge EXAFS data on $[Fe(o\text{-}C_6H_4(PMe_2)_2)_2C1_2][BF_4]_n$ $(n = 0\text{–}2)$. The structure of an iron(IV) complex of o-phenylenebis(dimethylphosphine), *J. Am. Chem. Soc.*, **108**, 526–528.
Hauser, A., Gütlich, P. and Spiering, H. (1986) High-spin$\rightarrow$low-spin relaxation kinetics and cooperative effects in the $[Fe(ptz)_6](BF_4)_2$ and $[Zn_{1-x}Fe_x(ptz)_6](BF_4)_2$ (ptz = 1-Propyltetrazole) spin-crossover systems. *Inorg. Chem.*, **25**, 4245–4248 and refs. therein.
Hazeldean, G.S.F., Nyholm, R.S. and Parish, R.V. (1966) Octahedral ditertiary arsine complexes of quadrivalent iron, *J. Chem. Soc. (A)*, 162–165.
Hider, R.C., Mohd-Nor, A.R., Silver, J., Morrison, I.E.G. and Rees, L.V.C. (1981) Model compounds for microbial iron-transport compounds. Part 1. Solution chemistry and Mössbauer study of iron(II) and iron(III) complexes from phenolic and catecholic systems. *J. Chem. Soc., Dalton Trans.*, 609–622.
Hills, A., Hughes, D.L., Jimenez-Tenorio, M., Leigh, G.J., Houlton, A. and Silver, J. (1989) Large Mössbauer quadrupole splittings in high-spin iron(II) complexes; the structure of di-iodo-1,5,9,13-tetrathiahexadecane iron(II) [for 1,5,9,13-Tetrathiohexadecaneiron(II) Di-iodide]. *J. Chem. Soc., Chem. Commun.*, 1774–1775.
Hoppe, M.L., Schlemper, E.O. and Murmann, R.K. (1982) Structure of dipotassium ferrate(VI). *Acta Crystallogr.*, **B38**, 2237–2239.
Houlton, A., Ofori-Okai, K.G., Roberts, R.M.G., Silver, J. and Wells, A.S. (1987) An iron-57 Mössbauer study of (η^6-Arene) (η^5-cyclopentadienyl) iron(II) salts. *J. Organometal. Chem.*, **326**, 217–228.
Houlton, A., Miller, J.R., Roberts, R.M.G. and Silver J. (1990) Studies of the bonding in iron(II) cyclopentadienyl and arene sandwich compounds. Part 1. An interpretation of the iron-57 Mössbauer data. *J. Chem. Soc., Dalton Trans.*, 2181–2184.

Houlton, A., Miller, J.R., Roberts, R.M.G. and Silver, J. (1991) Studies of the bonding in iron(II) cyclopentadienyl and arene sandwich compounds. Part 2. Correlations and interpretations of carbon-13 and iron-57 nuclear magnetic resonance and iron-57 Mössbauer data. *J. Chem. Soc., Dalton Trans.*, 467–470.

Houlton, A., Roberts, R.M.G. and Silver, J. (1991) Studies on the anti-tumour activity of some iron-sandwich compounds, *J. Organometal. Chem.*, **418**, 107–112.

Huheey, J.E. (1978) *Inorganic Chemistry, Principles of Structure and Reactivity*, Second Edition, Harper and Row, New York.

Jamieson, J.C., Manghani, M.H. and Ming, L.C. (1981) Chapter 18, in *Structure and Bonding in Crystals. Vol. II* (eds M.F. O'Keefe, and A. Navrotsky), Academic Press, New York.

Johansson, L., Molund, M. and Oskarsson, A. (1978) Compounds with Intermediate Spin(II). The crystal structure of tris(1,10-phenanthroline) iron(II) iodide dihydrate. *Inorg. Chim. Acta.*, **31**, 117–123.

Jordan, R.F. and Norton, J.R. (1979) ^{13}C NMR relaxation mechanisms in methyl-transition metal compounds. *J. Am. Chem. Soc.*, **101**, 4853–4858.

Kermarrec, M.Y. (1964) Absorption spectra of metallic cations with six molecules of water of coordination. *Hebd. Sceances Academ. Sci.*, **258**, 5836–5838.

Kerner, J.F., Dow, W. and Scheidt, W.R. (1976) Molecular stereochemistry of two intermediate spin complexes iron(II) phthalocyanine and manganese(II) phthalocyanine. *Inorg. Chem.*, **15**, 1685–1690.

King, T.J., Logan, N., Morris, A.J. and Wallwork, S.C. (1971), Eight-co-ordinate iron(III) complex: structure of tetraphenylarsonium tetranitratoferrate(II), *Chem. Commun.*, 554.

König, E., Ritter, G. and Kulshreshtha, S.K. (1985) The nature of spin-state transitions in solid complexes of iron(II) and the interpretation of some associated phenomena. *Chem. Rev.*, **85**, 219–234.

Kopelev, N.S., Perfil'ev, Yu. D. and Kiselev, Yu. M. (1990) The Mössbauer spectra of solutions of compounds of iron in higher oxidation states, *Russian J. Inorg. Chem*,, **35**, 1022–1025.

Kruck, T. and Prasch, A. (1964) Pentakis(trifluorophosphine) iron(O). *Angew. Chem. Internat. Edit.*, **3**, 754.

Lang, G., Spartalian, K., Reed, C.A. and Collman, J.P. (1978) Mössbauer effect study of the magnetic properties of $S = 1$ ferrous tetraphenylporphyrin. *J. Chem. Phys.*, **69**, 5424–5427.

Lauher, J.W. and Ibers, J.A. (1975) Structure of tetramethylammonium tetrachloroferrate(II) $[N(CH_3)_4]_2[FeCl_4]$. Comparison of iron(II) and iron(III) bond lengths in high-spin tetrahedral environments. *Inorg. Chem.*, **14**, 348–352.

Levason, W.A. and McAuliffe, C.A. (1974) Higher oxidation state chemistry of iron, cobalt and nickel, *Coord. Chem. Rev.*, **12**, 151–184.

Lever, A.B.P. (1984) *Inorganic Electronic Spectroscopy*, 2nd ed., Elsevier, Amsterdam.

Lind, M.D., Hoard, J.L., Hamor, M.J., Hamor, T.A. and Hoard, J.L. (1964) Stereochemistry of ethylenediaminetetraacetato complexes II. The structure of crystalline $Rb[Fe(OH_2)y].2H_2O$ III. The structure of crystalline $Li[Fe(OH_2)y]\ 2H_2O$. *Inorg. Chem.*, **3**, 34–43.

Lukas, B. and Silver, J. (1983) Mössbauer studies on protoporphyrin IX iron(II) solutions. *Inorg. Chim. Acta*, **80**, 107–113.

MacKay, K.M. and MacKay, R.A. (1973) *Introduction to Modern Inorganic Chemistry*, Second edition, Intern. Textbook, London.

Madega, K. and König, E. (1963) Complexes of iron(II) with 1,10-phenanthraline. *J. Inorg. Nucl. Chem.*, **25**, 377–385.

Mann, S. (1987) Biomineralisation of iron oxides. *Chem. Brit.*, **23**, 137–140.

Mao, H.K. and Bell, P.M. (1978) High-pressure physics sustained static generation of 1.36 to 1.72 megabars. *Science*, **200**, 1145–1147.

Mao, H.K., Bell, P.M. Yagi, T. (1979) Geophysical Laboratory Report, Carnegie Institution Yearbook.

Martin, R.L. and White, A.H. (1967) A novel series of five-coordinated iron(III) complexes with the square-pyramidal configuration and spin $S = 3/2$, *Inorg. Chem.*, **6**, 712–717.

Martin, R.L. and White, A.H. (1968) Nature of the transition between high-spin and low-spin octahedral complexes of the transition metals. *Transition Met. Chem.*, **4**, 113–198.

Medhi, O.K. and Silver, J. (1989) The Mössbauer Spectrum of an Intermediate ($S = 1$) Four-coordinated (protoporphyrinato IX) iron(II) complex in a frozen aqueous solution of cetyltrimethylammonium bromide, *J. Chem. Soc., Chem. Comm.*, 1199–1200.

Morrison, T.I., Reis, A.H., Knapp, G.S., Fradin, F.Y., Chen, H. and Klippert, E. (1978) Extended X-ray absorption fine structure studies of the hydrolytic polymerisation of iron(III). 1. Structural characterisation of μ-dihydroxo-octaaquodiiron(III) dimer. *J. Am. Chem. Soc.*, **100**, 3262–3264.

Mosback, H. and Poulsen, K.G. (1974) Mössbauer investigation of dinitrosyl iron compounds, *Acta Chem. Scand.*, **A28**, 157–161.

Murray, K.S. (1974) Binuclear oxo-bridged iron(III) complexes. *Coord. Chem. Rev.*, **12**, 1–35.

Navrotsky, A. (1981) Chapter 17, in *Structure and Bonding in Crystals, Vol. II*, (eds M.F.O'Keefe and A. Navrotsky), Academic Press, New York.

Nelson, S.M. (1988) Iron(III) and Higher States in *Comprehensive Coordination Chemistry*, Vol. 4, (eds Sir G. Wilkinson, R.D. Gillard and J.A. McCleverty), Pergamon Press, Oxford, pp. 217–276.

Nelson, P.G. (1991) Important elements. *J. Chem. Ed.*, **68**, 732–737.

Nelson, S.M. Bryan, P. and Bush, D.H. (1966) Seven-coordinate complexes of iron(III). *Chem. Commun.*, 641–642.

Nelson, S.M., Hawker, P.N. and Twigg, M.V. (1987) Iron(II) and lower states, in *Comprehensive Coordination Chemistry*, Vol. 4 (eds. Sir G. Wilkinson, R.D. Gillard and J.A. McCleverty), Pergamon Press, Oxford, pp. 1179–1388.

Orgel, L.E. (1956) Quelques problèmes de chimie minerales, 10ème Conseil de Chimie, Bruxelles, p. 289.

Paulaus, E.F. and Schäfer, L. (1978) The crystal and molecular structure of ferricinium tetrachloroferrate, *J. Organometal. Chem.*, **144**, 205–213.

Pankowski, M. and Bigorgne, M. (1977) Synthesis and iomerisation of the complexes of the halogeno carbonyl iron series: $[FeX(CO)_{5-n}L_n]^+$, $[FeX_2(CO)_{4-n}L_n]$ and $[FeX_3(CO)_3]^-$ $[L = PMe_3; n = 1,2,3; X = Cl, Br, I)$. *J. Organomet. Chem.*, **125**, 231–252.

Pensak, D.A. and McKinney, R.J. (1979) Application of molecular orbital theory to transition-metal complexes. Fully optimised geometries of first-row metal carbonyl compounds. *Inorg. Chem.*, **18**, 3407–3413.

Piper, T.S. and Wilkinson, G. (1956) Alkyl and aryl derivatives of π-cyclopentadienyl compounds of chromium molybdenum, tungsten and iron, *J. Inorg. Nucl. Chem.*, **3**, 104–124.

Reiff, W.M., Witter, E.H., Mottle, K., Brennan, T.F. and Garafalo, A.R. (1983) The ferric chloride α-di-imine system. Part V. X-ray structure determination of tris(1,10-phenanthroline) iron(II) μ-oxo-bis-(trichloroiron(III)). DMF. *Inorg. Chim. Acta*, **77**, L83–L88.

Roberts, R.M.G., Silver, J., Yamin, B.M., Drew, M.G.B. and Eberhardt, U. (1988) Crystal structures of *o*-, *m*- and *p*- nitrophenylferrocenes and their relevance to other sterically crowded phenylferrocenes. *J. Chem. Soc., Dalton Trans.*, 1549–1556.

Roberts, R.M.G., Silver, J., Wells, A.S. and Wilkinson, S.P. (1987) Synthesis and structure of η^6-(biphenyl)-η^5-(cyclopentadienyl)iron(II) hexafluorophosphates. *J. Organometal. Chem.*, **327**, 247–254.

Sacconi, I. (1971) Conformational and spin-state interconversions in transition-metal complexes. *Pure Appl. Chem.*, 161–191.

Sanders, N. and Day, P. (1970) The Spectra of complexes of conjugated ligands. Part IV. Zero differential overlap calculations on phenanthroline and its mono-complexes. *J. Chem. Soc. (A)*, 1190–1196.

Schappacher, M., Chottard, G. and Weiss, R. (1986) Resonance Raman observation of the $Fe^{IV} = O$ stretching vibration in models for the active site of horse radish peroxidase compound II. *J. Chem. Soc., Chem. Commun.*, 93–94.

Schappacher, M., Ricard, L., Weiss, R., Montiel-Montoya, R., Gonser, U., Bill, E. and Trautwein, A. (1983) Synthesis and spectroscopic properties of a five coordinate tetrafluorophenylthiolato ironII 'picket fence' porphyrin complex and its carbonyl and dioxygen adducts: synthetic analogs for the active site of cytochromes P450. *Inorg. Chim. Acta*, **78**, L9–12.

Schauer, C.K. and Shriver, D.F. (1987) Synthesis and structure of $[(PPh_3)_2N][Fe_3Mn(CO)_{12}(\mu_4\text{-}O)]$: A butterfly oxo-cluster. *Angew. Chem., Int. Ed. Engl.*, **26**, 255–256.

Scheidt, W.R. and Gouterman, M. (1982) *Ligands, Spin State and Geometry in Hemes and Related Metalloporphyrins in Iron Porphyrins. Part One.* (eds A.B.P. Lever and H.B. Gray), Addison-Wesley Publishing Co., Reading, Mass., USA.

Schlögel, R. and Jones, W. (1984) On the structure and properties of iron halide-graphite intercalates. A comparison of the reactivity of free and graphite intercalated iron(III) chloride. *J. Chem. Soc., Dalton Trans.*, 1283–1292.

Schmidbauer, H., Zybill, C.E. and Neugebauer, D. (1983) oxidative coupling of the selenolate functions in $(C_6H_5)_3PC(Se)P(C_6H_5)_3$: crystal structure of $[\{(C_6H_5)_3P\}_2(Se)_2]^{2+}[Fe_2OCl_6]^{2-}$. $4CH_2Cl_2$ containing a linear Fe-O-Fe axis. *Angew. Chem., Int. Ed. Eng.*, **22**, 145–161.
Schwochau, K. (1984) Extraction of metals from sea water. *Topics in Current Chemistry*, **124**, 91–133.
Seiler, P. and Dunitz, J.D. (1979) A new interpretation of the disordered crystal structure of ferrocene, *Acta Cryst.*, **B35**, 1068–1074;
Sharpe, A.G. (1992) *Inorganic Chemistry*, Second Edition, Longman, London.
Shriver, D.F. and Whitmire, K.H. (1982) Iron compounds without hydrocarbon ligands, in *Comprehensive Organometallic Chemistry*, Vol. 4, (eds G. Wilkinson, F.G.A. Stone and E.W. Abel) Pergamon, Oxford, 243–649.
Sigel, H. (ed.) (1978) Iron in Model and Natural Compounds, Vol. 7 of *Metal Ions in Biological Systems*, Marcel Dekker, New York.
Silver, J. (1990) Qualitative interpretation of Mössbauer data for some [1]ferrocenophanes; Fe-Pd dative bonding in $[S(C_5H_4)_2FePd(PPh_3)]$ and Fe-Hg and Fe-H^+ bonding in ferrocene *J. Chem. Soc., Dalton Trans.*, 3513–3516
Silver, J., Al-Jaff, G. and Al-Taies, J. (1987) Mössbauer spectroscopic studies on concentrated protoporphyrin IX iron(II) solution. *Inorg. Chim. Acta*, **135**, 151–153.
Silver, J., Lukas, B. and Al-Jaff, G. (1984) Mössbauer studies on protoporphyrin IX iron(II) frozen solutions containing ligands that cause the iron to be a five coordinate high-spin iron(II) environment, *Inorg. Chim. Acta*, **91**, 125–128.
Singh, P., Clearfield, A. and Bernal, I. (1971) The crystal structure of an octacoordinated iron(II) compound tetrakis (1,8-naphthyridine)Fe(II) perchlorate. *J. Coord. Chem.*, **1**, 29–37.
Stobach, R.E. (1988) Chemical abstracts, service chemical registry system. II. Substance-related statistics: update and additions. *J. Chem. Inf. Comp. Sci.*, **28**, 180–187.
Strauss, S.H., Pawlik, M.J., Skowyra, J., Kennedy, J.R., Anderson, O.P., Spartatian, K. and Dye, J.L. (1987) Comparison of the molecular and electronic structures of (μ-Oxo)-bis[(5,10,15,20-tetraphenylporphyrinato)iron(III)] and (μ-Oxo)-bis[(7,8-dihydro-5,10,15,20-tetraphenylporphyrinato)iron(III)]. *Inorg. Chem.*, **26**, 724–730.
Swepston, P.N. and Ibers, J.A. (1985) Redetermination of the structure of μ-oxo-bi[5,10,15,20-tetraphenylporphyrinato)iron(III)] at 122 K, $[Fe_2O(C_{44}H_{28}N_4)_2]$. *Acta Crystallogr.*, **C41**, 671–673.
Takano, M. and Takeda, Y. (1983) Electronic state of iron (Fe^{4+}) ions in perovskite-type oxides. *Bull. Inst. Chem. Res. Kyoto Univ.*, **61**, No. 5–6, 406–425.
Takano, M., Nakanishi, N., Takeda, Y. and Shinjo, T. (1982) On the electronic state of iron(4^+) ferrites, *Proc. ICF*, 3rd, 1980 (publ. 1982) 389–392.
Teller, R.G., Finke, R.G., Collman, J.P., Chin, H.B. and Ban, R. (1977) Dependence of the $[Fe(CO)_4]^{2-}$ geometry on counterion: Crystal structures of $K_2Fe(CO)_4$ and $[Na(crypt)]_2$ $[Fe(CO)_4]$ [crypt = $N(CH_2CH_2OCH_2.CH_2OCHH_2CH_2)_3N$]. *J. Amer. Chem. Soc.*, **99**, 1104–1111.
Toftlund, H. (1989) Spin equilibria in iron(II) complexes. *Coord. Chem. Rev.*, **94**, 67–108.
Vannerberg, N-G. (1972), The OD structures of $K_3Fe(CN)_6$ and $K_3Co(CN)_6$, *Acta. Chem. Scand.*, **26**, 2863–2876.
Walden, G.H., Jr., Hammett, L.P. and Chapman, R.P. (1931) A reversible oxidation indicator of high potential especially adapted to oxidimetric titrations. *J. Am. Chem. Soc.*, **53**, 3908.

2 Industrial chemistry of iron and its compounds

F.J. BERRY

2.1 History

The origin of the use of iron can be traced far back into the history of ancient civilisations. Certainly metallic iron was known in pre-dynastic Egypt before *c.* 3400 BC but the metal was exceedingly scarce and used only as beads for jewellery. Recent analysis of this iron has shown it to contain some nickel and speculation exists as to whether it was derived from meteoritic sources. Iron of seemingly terrestrial origin was also used in Mesopotamia during this time. The general use of iron in Egypt can be traced to a much later date, *c.* 1500 BC, and appears to have spread from the Hittites in Asia Minor who may have been the first smelters of iron in the third millenium BC. The great value of the smelting process resulted in it being kept very secret, and only when the Hittite empire fell around 1200 BC did the knowledge of smelting become dispersed and the Iron Age begin (see Childe, 1942).

Iron was also much used by the Assyrians about 600 BC and was also known in pre-classical Greece. For example, in the Mycenaen period described by Homer, iron was regarded as a precious metal as evidenced by its use as a prize given to Achilles. The Greeks also made extensive use of iron. The Etruscans worked the mines of Elba which were subsequently taken over by the Romans who also worked the mines of Spain and Noricum. Iron was known in India from at least 900 BC and has been used in China since about 500 BC, indeed, the Chinese appear to have used cast iron since *c.* 200 AD.

2.2 Production of iron

The successful extraction of iron from its ore by reducing the iron oxide with carbon marked a significant development in modern civilisation and the chemical implications of the processes and the conversion of iron into steel have received much attention (Kirk-Othmer, 1981). Although the Iron Age can be considered to originate from the time when primitive man discovered how to use the charcoal formed by burning wood to extract iron from iron ores and then use the iron to make tools and utensils, the Industrial Revolution owes it origin to Abraham Darby who in 1773 developed a process for producing carbon in the form of coke from coal and used this instead of charcoal which was becoming difficult to make because of a shortage of timber. The discovery had an impact of enormous proportions such that the consequent increase in the economics and scale of iron-making gave rise to the development of steam engines, railways, ships, bridges and iron-framed buildings.

The basic process in iron production, of reducing the iron ore, is achieved in the blast furnace (Figure 2.1). The use of small simplified blast furnaces dates from *c.* 1500 AD and the large modern versions are now capable of producing *c.* 10000 tonnes of iron each day. The furnace has an outer shell of steel plates and is lined with refractory bricks. The mouth is closed with a cup and cone through which the charge of ore (usually haematite), limestone and coke is introduced by lowering the cone. A double cup and cone are frequently used to prevent the escape of gas when the lower cone is opened. The gas is emitted through a pipe and is burnt in the Cowper stoves to heat the air blast to 700–800°C. The furnace below the boshes narrows to the hearth at the base. The hearth contains holes for a number of water-jacketed iron blowing pipes or tuyeres through which the air blast is forced from an annular pipe. The hearth also contains a hole lined with clay from which the iron is tapped into sand moulds on the ground. A slag notch at a higher level allows slag (mainly calcium silicate which floats on the molten iron and thereby prevents its oxidation) to be removed. The slag is used as a building material and in the manufacture of cement. The process of charging and removing the iron is a continuous one.

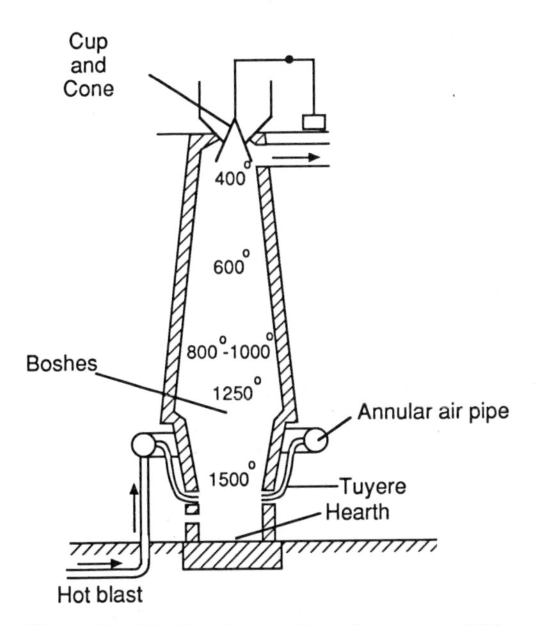

Figure 2.1 The blast furnace (from Partington, 1961).

2.3 Chemistry in the blast furnace

The basic principle of the process is that the coke burns with the evolution of intense heat such that temperatures approaching 2000°C are achieved near the hearth of the furnace with the iron ore being reduced to iron. The detailed chemistry of the blast furnace is complex and is probably still not fully understood. The speed at which the blast rises through the furnace (about 10 seconds) precludes the gas–solid reactions reaching equilibrium and adds to the difficulty in defining all the steps involved. However, it can be assumed that the oxygen of the blast combines with carbon at very high temperatures in the hearth (~2000°C) to produce carbon monoxide which rises through the furnace.

$$2C + O_2 \rightarrow 2CO$$

The temperature of the charge passing down the furnace increases as it moves from the mouth to the hearth. In the higher regions of the furnace some reduction of ferric oxide by carbon monoxide occurs but it is in the region above the boshes at 700–1200°C that the bulk of the ferric oxide and other iron oxides are reduced by carbon monoxide to iron.

$$Fe_2O_3 + 3CO \rightleftharpoons 2Fe + 3CO_2$$

The reaction is reversible and the escaping gases contain both carbon monoxide and carbon dioxide. In the higher zone of the furnace the limestone is also decomposed. Some carbon dioxide is reduced to carbon monoxide in the lower region of the furnace at *c.* 700°C.

$$CaCO_3 \rightarrow CaO + CO_2$$

$$CO_2 + C \rightarrow 2CO$$

In regions near the centre of the furnace finely divided carbon is deposited which, together with the carbon of the charge, completes the reduction.

$$2CO \rightarrow CO_2 + C$$

$$Fe_2O_3 + 3C \rightarrow 2Fe + 3CO.$$

Phosphorus, arising from the reduction of phosphates in the ore, is absorbed by the iron at higher temperatures. Some silicon is formed from the reduction of silica by carbon in the presence of iron and this alloys with the iron.

$$Ca_3(PO_4)_2 + 3SiO_2 + 5C \rightarrow 3CaSiO_3 + 2P + 5CO$$

$$SiO_2 + 2C(+Fe) \rightarrow 2CO + Si(+Fe)$$

Manganese is also formed by reduction of manganese compounds in the ore.

$$Mn_2O_3 + 3C \rightarrow 2Mn + 3CO$$

The molten iron formed in the white heat region of the furnace, which contains carbon, silicon, manganese, phosphorus and some sulphur, is tapped from the hearth into the sand moulds to form cast iron or pig iron.

2.4 Commercial iron

The three important varieties of commercial iron are (i) cast iron or pig iron, (ii) malleable iron or wrought iron, and (iii) steel.

2.4.1 Cast iron or pig iron

Cast iron or pig iron contains 2.2 to 4.5% of carbon together with some silicon, manganese, sulphur and phosphorus. The formation of pig iron by rapid cooling of a melt containing a low silicon content but a high proportion of manganese gives white pig iron in which all the carbon is in the form of the iron carbide of composition Fe_3C which is called cementite. White pig iron is brittle and coarsely crystalline and dissolves nearly completely in dilute hydrochloric acid liberating a mixture of hydrogen and hydrocarbons. If, on the other hand, the molten iron contains at least 2.5% of silicon and is slowly cooled, most of the carbon separates in the form of fine laminae of graphite and the iron adopts a softer and finer texture. This variety of cast iron is known as grey pig iron and dissolves in hydrochloric acid with the evolution of hydrogen to leave a black residue of graphite.

2.4.2 Malleable or wrought iron

Malleable or wrought iron is nearly pure iron containing *c.* 0.12 to 0.25% carbon and less than 0.5% of total impurities in the form of carbon, sulphur, phosphorus and silicon. Malleable iron is obtained from cast iron by the puddling process invented by Henry Cort of Lancaster in 1784. The cast iron is fused in a reverberatory furnace in which the hearth is lined with haematite which oxidises the carbon.

$$3C + Fe_2O_3 \rightarrow 2Fe + 3CO$$

The carbon monoxide bubbles through the molten iron. Any sulphur, phosphorus and silicon are oxidised and pass into the slag. When the melt adopts the consistency of paste it is formed into lumps or blooms which are beaten to squeeze out the slag. The iron welds together at bright red heat to give a coherent mass.

Malleable iron is tough and fibrous and, since it welds when two or more pieces are hammered at red heat, can be used in the fabrication of shapes.

Table 2.1 Carbon content of some commercially important steels.

% Carbon	Name of steel
0.15–0.3	Mild steel
0.3 –0.6	Medium steel
0.6 –0.8	High carbon steel
0.8 –1.4	Tool steel

2.4.3 *Steel*

The most commercially important form of iron for industrial exploitation is steel. All forms of steel are composed of iron with less than 2% carbon. Steel is ductile and can be rolled or machined into shape. The carbon content of some commercially important steels is shown in Table 2.1. The hardness and strength increase with increasing carbon content. The most common forms of steel are mild steel, which is sometimes called soft steel or low-carbon steel, and hard steel which is sometimes called high carbon steel. Steel differs from iron in that it acquires a 'temper' by heating and quenching. This means that it becomes soft when heated and slowly cooled.

Steel can be made by the now obsolete process of increasing the carbon content of wrought iron, or by removing some of the carbon and all of the impurities from cast iron. Several processes have been developed for this. The Siemans-Martin or open hearth process developed in 1864 and involving the external heating of molten pig iron and the oxidation of impurities by air is now obsolete. Three processes are now in use.

2.4.3.1 The Bessemer process. This process was patented in 1855 by the Frenchman Bessemer and involves the use of the Bessemer converter (Figure 2.2). The furnace is lined with silica and, while horizontal, is filled with molten pig iron. It is then tilted to vertical and compressed air blasted through holes in the bottom of the furnace and through the molten metal. This causes the silicon and manganese content of pig iron to burn and to form a slag composed of silica, iron silicate, and manganese oxide. These exothermic reactions induce the oxidation of carbon to carbon monoxide and carbon dioxide. When the carbon content is sufficiently low the converter is tipped and the molten steel is poured into moulds to give ingots which can be rolled or forged. The process takes *c.* 20 minutes to produce a 6 tonne ingot of steel.

A variation of the process to handle phosphorus-containing iron made from phosphorus rich iron ores and which damage the silica lining of the Bessemer process, was developed by Thomas and Gilchrist in 1879. In this process the converter is lined with a basic material such as calcined dolomite or limestone. Limestone, $CaCO_3$, or lime, CaO, are added as the slag forms.

The Bessemer processes have dominated the world-wide manufacture of steel and were still used in England up to the 1960s.

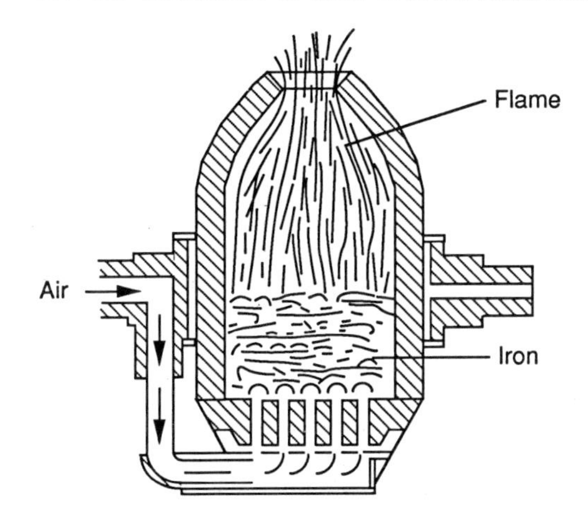

Figure 2.2 The Bessemer converter (from Partington, 1961).

2.4.3.2 Siemans electric arc furnace. In 1878 Sir William Siemans patented an electric arc furnace for the production of steel. Heat is provided by having an electric arc just above the metal or by passing an electric current through the metal. This process continues to be used to produce steel alloys and high quality steel. For example, stainless steel containing 12–15% of nickel, and steel for cutlery containing 20% chromium and 10% nickel are both made by this method, as is high speed cutting steel containing 18% tungsten and 5% chromium and steel containing manganese. Some of these steels have a high tensile strength and others, which are very tough, are used for rock-working machinery and excavators.

2.4.3.3 Basic oxygen process. The main process for steel production is now the basic oxygen process which was developed in Australia in 1952 as the Kaldo and LD processes. The process was designed to avoid the uptake of nitrogen by the molten metal which occurred during the Bessemer process and which caused the steel to become brittle and, when the surfaces became nitrided, difficult to weld. The new process used pure oxygen as opposed to air to oxidise the pig iron. However, this led to the generation of excessive heat which tended to melt the bottom of the Bessemer converter. Hence a new system was developed. In the LD process strong convection currents are set up in the melt to obtain effective reaction whilst in the Kaldo process this is achieved by rotating the converter to ensure good mixing. In the basic oxygen process the furnace is charged with a mixture of molten pig iron and lime and oxygen is blown onto the surface of the liquid metal at great speed through water-cooled tubes. The oxygen penetrates the melt and oxidises the impurities with the generation of heat which keeps the contents of the furnace molten despite the rise in melting point as the impurities are removed.

Table 2.2 World production of steel (1988)

Country	% Total world production
USSR	21
Japan	14
USA	12
China	7
West Germany	5
Brazil	3
Italy	3
UK	2
France	2
South Korea	2
Poland	2
Canada	2
Czechoslovakia	2
Romania	2

Total production = 776 million tonnes

Hence the process requires no external heating. The furnace is eventually tipped and the molten steel is poured either into moulds to give steel castings or into ingots which can be rolled.

Commercially the process using oxygen is very advantageous in that it gives faster conversion and hence greater production. It also enables the handling of larger quantities of steel and the formation of purer products free from surface nitrides.

The worldwide extent of steel production is summarised in Table 2.2.

2.5 Pickling of steel

The steel produced by the processes described above is often formed with an oxide layer which adheres to the surface. The oxide layer presents itself as a scale formed during the heat-treatment processes or as rust as a result of the corrosive action of water. The rust consists largely of Fe_2O_3 while the scale is composed of three iron oxides: wustite FeO, magnetite Fe_3O_4 and heamatite Fe_2O_3 in a ratio which depends on the steel composition, annealing conditions, temperature of rolling and rate of cooling after rolling. The process known as 'pickling' is designed to remove this oxide layer which adheres to the steel surface. The process involves the treatment of the steel with acid solutions.

The pickleability of a steel and the acid consumption for pickling depend on many factors including the degree of adherence of the oxide scale, the composition of the steel, the type and composition of the pickling solution, and the pickling conditions. The type of pickling acid selected is determined by the nature of the required quality of surface and by economic factors. The most important pickling acids for iron and steel are sulphuric and hydrochloric acid. Phosphoric nitric and hydrofluoric acids are used for special purposes and stainless steel.

Any Fe^{3+} species which dissolve during the pickling process are reduced by metallic iron so that the pickling acid contains primarily Fe^{2+} salts. The quantity of iron dissolved during pickling is called chemical pickling loss and accounts for 0.2 to 1.2 wt% of the iron or steel being pickled. The removal of the iron salts from the pickling solution enables the unconsumed pickling solution to be used again.

2.6 Main uses of steel

The largest use of mild steel which is malleable and amenable to bending or machining is in ship building, the production of girders, and the manufacture of car bodies. Thin sheets of mild steel electroplated with a very thin protective layer of tin are used for packaging food and other materials such as tinned food. Several iron-containing alloys are produced in large quantities particularly ferrosilicon, ferrochrome, ferromanganese and ferronickel.

Iron has also found extensive use as a catalyst in the Haber–Bosch process for making ammonia.

2.7 Corrosion of iron and steel

It is well known that the interactions between solid surfaces and a gaseous environment are very complex. This is very well illustrated by the atmospheric corrosion of iron and steel which, despite investigations for at least two centuries, remain only understood in terms of simplified laboratory simulations using the simplest and cleanest of environments. Our understanding of the chemistry involved in the corrosion of steel in real atmospheric environments containing a wider variety and greater concentration of reactive trace constituents is very primitive. However, the situation is helped by the fact that although iron, the steels with $< 1\%$ carbon, steels with varying amounts of manganese sulphur, phosphorus and silicon, and the weathering steels (iron with $< 1\%$ copper, manganese, silicon, chromium or nickel) differ in composition, their interactions with corrosive atmospheres do show similarities. For example, even though corrosion rates and surface morphologies may vary, the reactive atmospheric species and many of the corrosion products have been found to be the same. The whole subject of corrosion mechanisms in iron and low carbon steels exposed to the atmosphere has been recently reviewed by Graedel and Frankenthal (1990).

2.7.1 Formation of rust

The formation of rust on iron and steel is a chemically complex process which can be considered to involve at least three stages. The first stage involves the formation of a 1–4 nm thin oxide/hydroxide film which is stable and passivating in the absence of atmospheric impurities and/or high relative humidities or liquid water. After exposure to near-neutral aqueous environments the oxide/hydroxide film changes into either of the two types of green precipitate known as green rust I and green rust II. In the third stage the green rusts are transferred into the fragile brown layer of iron oxides and hydroxides.

It is important to appreciate that in most real rust-forming environments constituents such as sulphides, sulphates and chlorides are present together with oxygen and water. It is clear that these other constituents influence the kinetics of the three stages of rust formation which vary according to the environments. In general, the first stage occurs on a time scale of milliseconds to seconds and the second stage occurs in 2–3 hours. The third stage, which is often evidenced by the formation of lepidocrocite (γ-FeOOH), may occur after two weeks or less and is followed in a few days by the formation of magnetite, Fe_3O_4. The corrosion rates for the carbon steels become quasi-linear in the first stage and encompass the ranges summarised by Mattson (1985) and shown in Table 2.3. Weathering steels generally corrode more slowly than carbon steels and under urban conditions may be of the order of 4–10 μm/year.

2.7.2 Atmospheric corrosion

With this background it is possible to say a little more about the multistep atmospheric corrosion of iron and low alloy steels. It seems that during the first few hours or days of exposure a moderately protective surface layer of oxides and hydroxides forms on the surface of the metal. The presence of any atmospheric hydrogen peroxide enhances the passivating properties of the layer. The subsequent adsorption of water and deposition of corrosive gases and chemically complex particulate matter produces an electrolytic solution on the metal into which Fe^{2+} can enter either by oxidation of the metal or by

Table 2.3 Corrosion rates in carbon steels

Environment	Rate of corrosion μm/year
Rural	4–65
Urban	23–71
Industrial	26–175
Marine	26–104

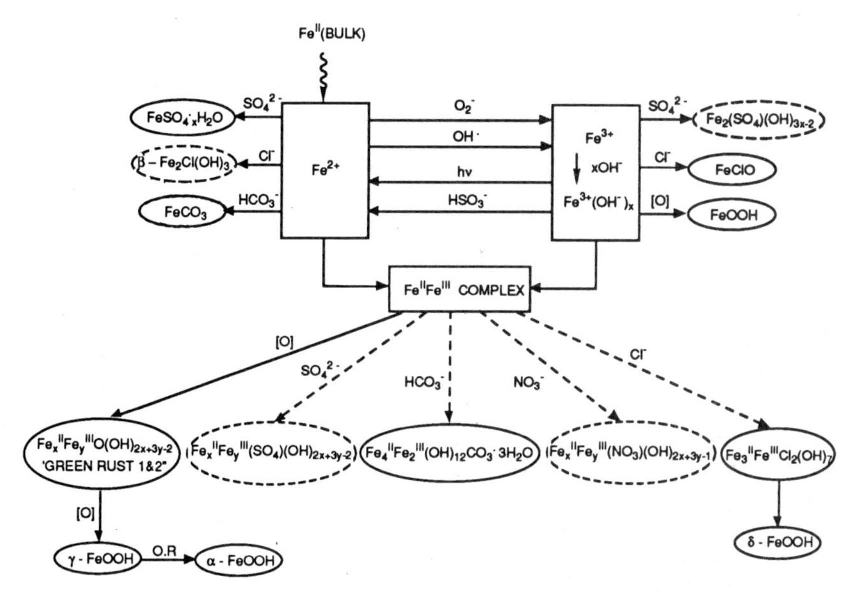

Figure 2.3 A schematic diagram of the formation of intermediates and final products in the atmospheric corrosion of iron and low alloy steels, from Graedel and Frankenthal (1990). Because water layers and droplets exposed to the atmosphere are acidic, the diagram applies within the approximate pH range 3–6. Species shown within rectangular boxes are present as solution constituents and those in oval boxes are present as precipitates. Dotted ovals or lines indicate constituents or reactions that are currently unconfirmed by laboratory studies. D, R indicates dissolution and reprecipitation respectively.

reduction of Fe^{3+}. The subsequent chemistry has been summarised by Graedel and Frankenthal (1990) and is described schematically in Figure 2.3 where specific oxidation processes and the effects of trace anions in solution are considered. The transformations involving the oxidation and reduction processes that link Fe^{2+} and Fe^{3+} are important. The available anions such as hydroxysulphate and hydroxychloride in the surface film compete for the iron cations producing compounds that are either soluble or less protective than any oxyhydroxides which would occur without them. The amenability of iron to be transformed between Fe^{2+} and Fe^{3+} results in a tendency to form mixed $Fe^{2+/3+}$ complexes. Although some of these products remain on the surface a significant proportion are lost.

In situations where the humidity exceeds *c*. 50%, multilayers of water vapour are adsorbed especially on any small particles. Any indoor photon irradiation may be sufficient to liberate Fe^{2+} ions although it is unlikely that the aqueous surface layers under indoor conditions will become highly acidic. Under some circumstances sulphate or chloride aerosols may interact with the surface to produce rust components.

In outside atmospheres the presence of airborne corrodents must also be considered. Atmospheric sulphur levels are now lower as a result of the use of low-sulphur coal and the more efficient means of sulphur emission control. In contrast, atmospheric concentrations of NO_x and nitric acid have increased as a result of an increased rate of high temperature combustion. Concentrations of atmospheric chlorine are also likely to increase as a consequence of increased combustion of chlorine-containing polymers.

All these types of reactions occur most readily in pits and crevices in the metal surface. The pH of the moisture on the surface, which is largely controlled by atmospheric sulphur dioxide and NO_x, is crucial because it controls the dissolution of the passive oxyhydroxide surface.

2.7.3 Stainless steel

Stainless steels are different from the low carbon steels in that they have *c.* 2 nm of a chromium-rich passive film which strongly inhibits atmospheric attack. Any corrosion which does occur tends to be localised in pits which form below the crevices at flaws in the passive layer. Only under extreme conditions such as highly acidic fogs or in marine locations does the chemical degradation of stainless steel begin to resemble that of the low carbon steels.

2.8 Prevention of corrosion

The foregoing has shown that the origin of corrosion lies in the inherent instability of many metals such that their ionic solutions and compounds including hydroxides, oxides, carbonates, sulphides and sulphates are more stable than the free metal under the atmospheric conditions to which the metals are exposed. The free energy change for the conversion of many metals to a compound is highly negative and there is therefore a great driving force for the conversion of an elemental metal or an oxide to a salt. Thus the prevention of corrosion involves the protection of the metal from the environment to which it is to be exposed. The objective of most protective systems is to reduce the rate of corrosion to a value that is tolerable or which will enable the material to survive its desired lifetime. Only rarely is it possible to eliminate corrosion completely. The subject of corrosion protection has recently been reviewed by Leidheiser (1987).

2.8.1 Cathodic protection

Since corrosion is an electrochemical process the corrosion rate of a corroding metal is determined by the kinetics of the anodic and cathodic parts of the corrosion reaction. If the rates of either of these reactions can be changed, then the corrosion rate can be reduced. A widely used method for taking advantage of the corrosion kinetics is known as cathodic protection and involves the application of a small potential between an inert electrode and the metal to be protected. The potential of the metal is made more negative and the rate of the anodic reaction is reduced.

2.8.2 Barrier protection

Another method of corrosion protection is based on the concept of coating the metal with a barrier which resists penetration by the atmospheric agents which cause corrosion.

2.8.2.1 Anodic oxides. The formation of anodic oxides on the surface of stainless steel and other iron-containing alloys to give a passive oxide layer which resists attack is a very satisfactory method of producing a protective barrier. This can be done by application of an anodic potential which is sufficient to maintain a barrier oxide on the surface.

2.8.2.2 Ceramic coatings. Ceramic coatings are also used to protect steel from corrosion. The inorganic coatings such as silica which can be used to protect stainless steel from tarnishing serve as a barrier between the steel and the atmosphere and also maintain an alkaline environment at the interface between steel and the ceramic in which the corrosion rate is low.

2.8.2.3 Corrosion inhibitors. The purpose of corrosion inhibitors is largely to form a thin surface coating which acts as a barrier. Organic inhibitors based on hydrocarbons containing a reactive group have been developed. In these inhibitors, the reactive group interacts with the metal surface while the hydrocarbon entity remains in contact with the environment. The salts or esters of cinnamic acid which are called cinnamates have been used to protect the surface of steel against corrosion in neutral or slightly alkaline media. The method involves the inhibitor changing the corrosion potential of the metal.

2.8.2.4 Organic coatings. Organic coatings are widely used to protect the surface of steel from atmospheric corrosion. The coating is designed to act as a barrier to water, oxygen and dissolved salts and prevent the cathodic reaction $H_2O + \frac{1}{2}O_2 + 2e^- \rightarrow 2OH^-$ from occurring beneath the coating. The protective barrier properties may be lost if water collects as thin layers or blisters at the interface since such aqueous regions facilitate the electrochemical corrosion reactions by charge transfer through the coating. The need to have a good barrier to water means that an additional defence in the form of a corrosion inhibitor such as lead chromate primer is required. If an aqueous phase does develop then the chromate begins to dissolve and deactivates any regions of the surface which become in contact with the water.

2.8.2.5 Conversion coatings. Organic coatings are only effective if they protect the entire metal surface and, in many cases where they are used commercially, such as on automobiles which may suffer damage during use, it is important that any damaged area does not result in a complete loss of protection to the neighbouring region. Hence conversion coatings composed of phosphates, chromates or oxides are sometimes applied before the organic coating is used. These coatings which are therefore situated between the metal and the organic coating are known as conversion coatings because they convert some of the base metal to a coating in which ions of the base metal are a component.

Other methods of corrosion protection are less dependent on chemical principles and are more dependent upon design. For example, appliances may be designed so that rainwater is not allowed to collect in recesses and thereby provide regions where the metal can support for a long period of time the electrochemistry of the processes responsible for corrosion.

2.9 Uses of iron compounds

There are numerous uses of iron compounds as detailed in Ullmann's Encyclopedia (1989) and only a few are outlined here.

2.9.1 Iron sulphates

Iron(II) sulphate, $FeSO_4.7H_2O$, is commonly used for the preparation of other iron-containing compounds. It is used to a lesser extent in writing inks, wood preservatives, iron-containing pigments such as Prussian blue and Turnbull's blue, as an etchant for aluminium and for process engraving and lithography. Large amounts of ferrous sulphate are also used for the clarification of phosphate-containing community effluents as well as decontamination of effluents containing cyanides and chromates. It can also be used to eliminate chlorine in waste gases. Iron(II) sulphate can be used as an additive to reduce the content of water-soluble chromates in cement. It is also used to combat a disease of vines called chlorosis and to treat alkaline soils and kill moss.

Iron(III) sulphate $Fe_2(SO_4)_3$, is used to prepare alums and iron oxide pigments and as a coagulant for the treatment of liquid effluents. Ammonium ferric sulphate is used in tanning. A mixture of iron(III) sulphate, iron(III) chloride and calcium hydroxide can be used to reduce the volume of sludge from effluent treatment plants by at least 35%.

2.9.2 Iron chlorides

Iron(II) chloride is used as a reducing agent and for the production of other iron compounds. It is used as a solution to precipitate, flocculate, and reduce contaminants such as chromates in wastewater. Pure ferrous chloride solution is used to prepare acicular goethite, α-FeOOH, and lepidocrocite, γ-FeOOH, which are further processed to γ-Fe_2O_3 magnetic pigments.

Iron(III) chloride is used as a chlorinating agent for aliphatic hydrocarbons and for aromatic compounds. It acts as a catalyst in Friedel-Crafts synthesis and condensation reactions. Since solutions of ferric chloride dissolve metals such as copper and zinc without evolving hydrogen, they are used for the etching or surface treatment of metals in, for example, the manufacture of electronic printed circuits, copperplate printing, and textile printing rolls.

The largest use of iron(III) chloride is in the form of dilute solutions which gives iron(III) hydroxide which induces flocculation and precipitation of heavy metals, phosphates and sulphides in municipal and industrial water treatment. Sludge conditioning with iron(III) chloride and lime improves the dewatering of filter sludge so that drier sludges better suited to disposal or incineration are obtained.

2.9.3 *Iron pentacarbonyl*

The compound iron pentacarbonyl, $Fe(CO)_5$, has found widespread use not only as a product in its own right but also for the preparation of other commercially important iron compounds. The largest use of iron pentacarbonyl is for the production of carbonyl iron powder. This is achieved by decomposition of $Fe(CO)_5$ in a cavity decomposer to give spherical iron particles of 1–10 μm diameter which can be used for making the magnetic cores of electronic components. The powder is also used in metallurgy for making pure iron and iron alloy components by pressing and metal injection moulding techniques. It is also used for the production of iron-containing pharmaceuticals and is incorporated into rubber and plastic which are used for microwave attenuation.

The conversion of iron pentacarbonyl to iron oxide is an important clean method of making iron oxide. The only by-product of the conversion is carbon dioxide and no water pollution by salts occurs. Hence, finely divided red iron oxide can be obtained by the atomisation of iron pentacarbonyl and burning it in an excess of air. By variation of the temperature and residence time in the reactor, transparent or highly transparent red pigments can be obtained with surface areas between 160 and 10 m^2/g. The larger particles adopt the haematite structure. The iron oxide is used to make high quality ferrites and other ceramic materials. The pure oxide is used as a lightfast u.v.-blocking red pigment for paints, wood varnishes, fibre dyeing and colour printing.

The air oxidation of iron pentacarbonyl to magnetite, Fe_3O_4, is possible at high temperature. Mixed oxides in which iron is a major component can be made by atomisation of iron pentacarbonyl in the presence of the other metal. For example, iron chromium oxide which is used as a catalyst or as a brown pigment is made by atomising the carbonyl with aqueous solutions of chromic acid which gives the metal oxide at high temperatures.

Polycrystalline iron whiskers with high strength and suitable for making composites can be prepared from the decomposition of iron pentacarbonyl in a magnetic field.

Pigments ranging from bright yellow to brilliant red can be prepared in a fluidised bed of aluminium powder which is coated with iron oxide when iron pentacarbonyl vapour and air are admitted. Other pigments can be made on a mica base. These pigments are lightfast and corrosion-resistant and are suitable for use in automobile paints.

Carriers for the toner in photocopies can be made by coating fine spherical particles of iron. Suspensions of finely divided iron or its alloys prepared from the decomposition of iron pentacarbonyl in solution can be used to produce magnetic liquids. Pyrite films can be obtained from iron pentacarbonyl and sulphur or hydrogen sulphide by chemical vapour deposition. The films are photoactive and can be used to make solar cells.

2.9.4 Iron-containing pharmaceuticals

Dietary iron is essential for animal tissue growth (see chapters 6, 7 and 8). Although a sufficient supply of iron is usually found in the human diet the level of absorption of the element from food is low and the supply of iron can therefore sometimes become critical. Iron-deficiency anaemia is commonly encountered in pregnant women and some diseases such as rheumatoid arthritis result in poor distribution of iron to the body which can lead to chronic anaemia (see chapter 8).

More than thirty-four iron-containing preparations are currently used for the treatment of iron-deficiency anaemia. Oral iron treatments to achieve the daily dose of 100–200 mg are usually used. Iron(II) salts are more commonly used than iron(III) salts because the former are more soluble in the pH range 3–7. The main compounds used in iron supplements are iron(II) fumarate, iron(II) gluconate, iron(II) sulphate, iron(II) succinate, iron(III) chloride solution, and iron(III) ammonium citrate. Iron dextran and iron sorbital can be used as injectable haematinics. A number of oral preparations contain ascorbic acid (vitamin C) to stabilise the iron(II) state. Folic acid is used in conjunction with the iron salt for the prevention of anaemia in pregnant women. Iron(II) complexes have a major disadvantage in that they are sensitive to oxidation in aqueous environments and some tablets are coated. The dissociation in the gut still enables oxidation which can cause irritation and gastrointestinal side-effects. Hence there is currently interest in developing uncharged soluble iron(III) complexes and some success has been achieved with iron(III) maltol (see chapter 8).

The occurrence in humans of severe tissue iron overload as a result of treatment for anaemia is also known. This is currently treated by injections of disferrioxamine mesylate which is a good chelating agent for iron(III), however this treatment has severe disadvantages which are discussed in chapter 8.

2.9.5 Other uses of iron compounds

There are a multitude of other uses for the large number of iron compounds currently known. Compounds such as iron(II) carbonate-, citrates-, iron(III) ammonium citrate, iron-halides and iron fumarates are used as iron additives in foods, mainly for animals. Other components, such as iron acetates, are used in dyeing, while iron(III) nitrate is used for the preparation of iron-containing catalysts. The green compound iron(II) phthalocyaine is used as a catalyst for a variety of chemical and electrochemical redox reactions.

References

Childe, V.G. (1942) *What Happened in History*, Penguin, London, pp. 182–185.

Graedel, T.E. and Frankenthal, R.P. (1990) Corrosion mechanisms for iron and low alloy steels exposed to the atmosphere. *J. Electrochem. Soc.* **137**, 2385–2394.

Kirk-Othmer Encyclopedia of Chemical Technology (1989), 3rd edn., Vol. 13, Interscience, New York, pp. 735–763.

Leidheiser, H. (1987) Fundamentals of corrosion protection in aqueous solutions, in *Metals Handbook*, Ninth Edn., Vol. 13, Corrosion ASM International, Ohio, pp. 377–379.

Mattson, E. (1988) Corrosion rates in steels. *Chemtech.* **15**, 234–239.

Partington, J.R. (1961) *A Text Book of Inorg. Chem.*, Macmillan, London.

Ullmann's Encyclopedia of Industrial Chemistry (1989), Fifth Edn., Vol. A14, VCH, Weinheim, pp. 591–610.

3 Inorganic chemistry of iron

E. SINN

3.1 Introduction

As in every sense we live by iron (see chapter 1), so its physical and chemical properties are nothing if not important. However, no less important is the chemistry of its compounds (Cotton and Wilkinson, 1989; Greenwood and Earnshaw, 1986; Nelson *et al.*, 1987; Nicholls, 1973; Porterfield, 1984; Wells, 1975), which range from apparently simple compounds like rust (see chapter 2) to the complex biological macromolecules like oxygen-carrying haemoglobins and electron-carrying cytochromes (see chapter 6). Thus it is also of major importance in biology and, with the possible exception of a few bacteria, no life exists without it.

3.1.1 *Iron chemistry: solid state, coordination, and organometallic*

The inorganic chemistry ranges from simple *infinite lattice* structures with atoms arranged in a 3-D matrix of points but usually without molecular boundaries, to *coordination* compounds. We think of coordination compounds as discrete molecules, in which the iron atom has several, most commonly four or six, non-metallic atoms attached by covalent bonds. If one of these donor atoms is carbon, we generally consider the compound *organometallic* (and exclude it from this chapter). However, these divisions are artificial, and we can change the chemistry of a compound more dramatically by pairing all the d-electrons than by introducing a Fe–C bond to make it organometallic. Organometallics and coordination complexes are now increasingly made as precursors for infinite lattice, or solid state, compounds: often systematic removal of ligand fragments gives a residual lattice compound with similar structure.

Solid state is the fastest growing branch of inorganic chemistry in the wake of the publicity from the high T_c superconductors. The role of iron in the superconductor has so far been negative: T_c and critical current fall with increased iron doping. However, the familiar perovskite structure of $YBa_2Cu_3O_7$, now seen on posters and in books everywhere, is one of the typical solid state structures shared by many iron compounds such as Fe_2SiO_4. Figure 3.1 shows the idealized perovskite structure.

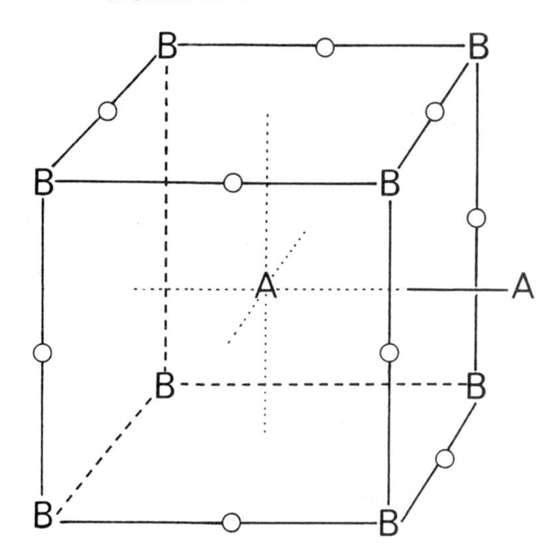

Figure 3.1 Perovskite ABO_3, e.g. A at cage centre (= Ca, La, ...), B at corners (= Ti, Fe, ...), O at mid-edge. Non-ideal perovskites (e.g. $YB_2Cu_3O_7$ and Fe_2SiO_4) may have vacant sites. Idealized perovskite structure, e.g. Fe_2SiO_4.

Before discussing the inorganic chemistry of iron some of the concepts outlined in chapter 1 need to be explored in greater depth.

3.1.2 Expected spin states and magnetism

As stated in chapter 1, the normal oxidation states are 2 and 3, but we consider the full range -2 to 8, all of which are met occasionally. The MO diagram for even a simple *octahedral* complex of iron looks complicated (Figure 3.2), but most of the time we can obtain the information we need by just considering the circled region. There are five orbitals, t_{2g} with threefold degeneracy, and e_g with twofold degeneracy. This is qualitatively equivalent to the results of the simpler crystal field (CF) theory, which ignores orbital overlap with the ligand orbitals altogether and uses just the electrostatic repulsion of the metal d-orbitals and the negative charges of the ligands (Figure 3.3).

The d_{z^2} and $d_{x^2-y^2}$ orbitals (e_g) point at the corners of an octahedron, or exactly at the ligands of an octahedral complex. The d_{xy}, d_{xz} and d_{yz} orbitals (t_{2g}) point away from the ligands. Electrons in the t_{2g} orbitals are less repelled by the ligand charges, so these orbitals are lowest. The two approaches will give different values for the splitting energy Δ which is normally determined empirically.

For the species Fe^{IV} (d^4), Fe^{III} (d^5), Fe^{II} (d^6) and Fe^{I} (d^7) there are two choices, high spin or low spin, depending on how Δ compares with the

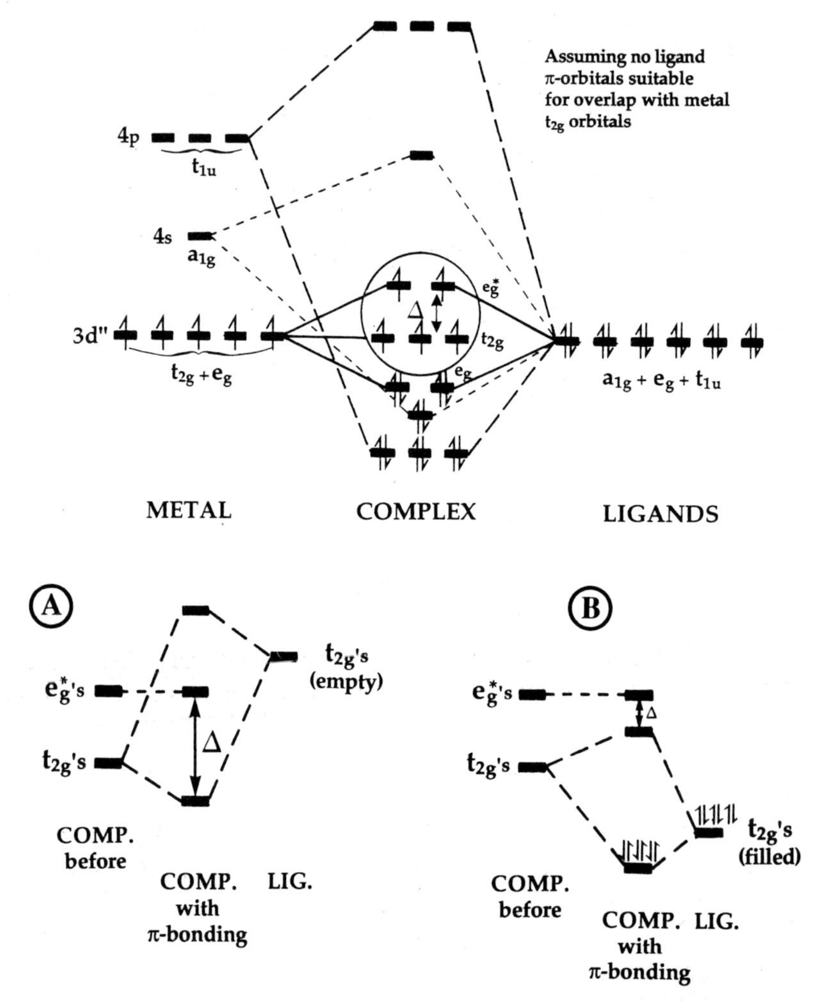

Figure 3.2 Upper: Mo diagram for a simple iron complex. Region inside the circle corresponds to orbitals taken in crystal field calculations. Lower: effect of ligand π-bonding with metal t_{2g} orbitals.

electron-pairing energy Π. If it takes more energy (Π) to put two electrons into the same (lower) orbital, than it takes (Δ) to raise one electron into an upper orbital, the complex has the highest possible number of unpaired electrons or the highest value of spin ('high spin'). The configurations are tabulated here for easy reference, where CFSE is crystal field stabilization energy.

High spin: Low spin:

$Fe^{I}, t_{2g}^{5}e_{g}^{2}, {}^{4}T_{1g}; \dfrac{CFSE}{\Delta} = -\dfrac{4}{5}$ $\quad$ $Fe^{I}, t_{2g}^{6}e_{g}^{1}, {}^{4}T_{1g}; \dfrac{CFSE}{\Delta} = -\dfrac{9}{5}$

$$Fe^{II}, t_{2g}^4 e_g^2, {}^5T_{2g}; \frac{CFSE}{\Delta} = -\frac{2}{5} \qquad Fe^{II}, t_{2g}^6, {}^1A_{1g}; \frac{CFSE}{\Delta} = -\frac{12}{5}$$

$$Fe^{III}, t_{2g}^3 e_g^2, {}^6A_{1g}; \frac{CFSE}{\Delta} = -\frac{0}{5} \qquad Fe^{III}, t_{2g}^5, {}^2T_{2g}; \frac{CFSE}{\Delta} = -\frac{10}{5}$$

$$Fe^{IV}, t_{2g}^3 e_g^1, {}^5E_g; \frac{CFSE}{\Delta} = -\frac{3}{5} \qquad Fe^{IV}, t_{2g}^4 \, {}^3T_{1g}; \frac{CFSE}{\Delta} = -\frac{8}{5}$$

The other oxidation states have only a single spin state to choose from:

$$Fe^0, t_{2g}^6\ e_g^2, {}^3A_{2g}; \frac{CFSE}{\Delta} = -\frac{6}{5}$$

$$Fe^{-I}, t_{2g}^6\ e_g^3, {}^2E_{2g}; \frac{CFSE}{\Delta} = -\frac{3}{5}$$

$$Fe^V, t_{2g}^3, {}^4A_{1g}; \frac{CFSE}{\Delta} = -\frac{6}{5}$$

$$Fe^{VI}, t_{2g}^2, {}^3T_{1g}; \frac{CFSE}{\Delta} = -\frac{2}{5}$$

$$Fe^{VII}, t_{2g}^1\ {}^3T_{2g}, \frac{CFSE}{\Delta} = -\frac{2}{5}$$

Fe^{-II} and $Fe^{VIII}$are the simplest to consider. With empty and filled d-shells respectively, they would have no CFSE, no magnetic properties, and no

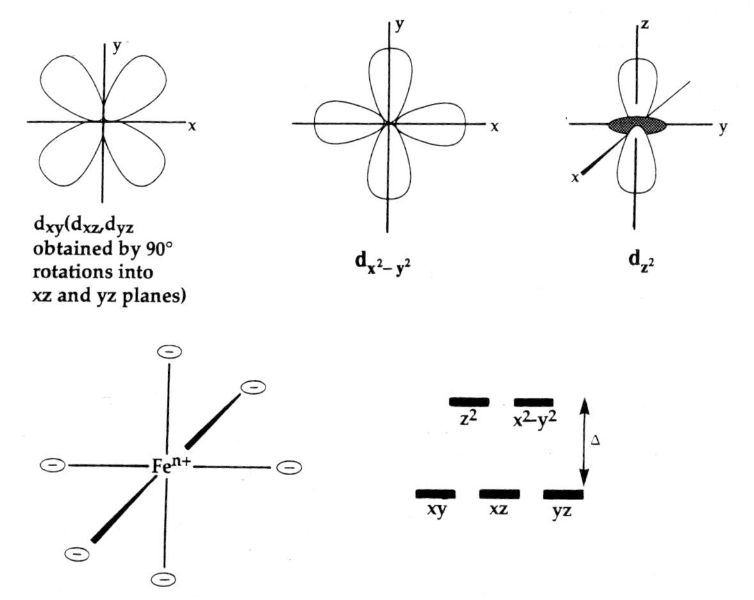

Figure 3.3 Octahedral configuration – electrostatic repulsion of the metal *d*-orbitals and the negative charges of the ligands.

ligand field spectra, as in non-transition metal complexes. There are some other special cases among the normal oxidation states of 2 and 3: low spin Fe^{II}, d^6, is also diamagnetic. This makes CN complexes and pyrite magnetically uninteresting, unlike high spin Fe^{II} and Fe^{III}, and low spin Fe^{III}, which are often characterized by their magnetic properties.

High spin Fe^{III}, having no CFSE, resembles the non-transition metals in having no preferred geometry. This enhances its ability to replace non-transition metal ions of similar size in minerals and complexes. An example is in ferrites, $M^{II}Fe_2^{III}O_4$, where the spinel structure is generally preferred, except that when M^{II} is d^{6-9}, the inverse spinel structure is stabilized with all the M^{II} in octahedral sites (CFSE) and the Fe^{III}, having no geometry preference, distributed equally over octahedral and tetrahedral sites. This enables strong magnetic coupling between the Fe^{III} in the octahedral and tetrahedral sites: hence $MnFe_2O_4$, $CoFe_2O_4$ and $NiFe_2O_4$ have moments near the 6BM, 5BM and 3BM values expected for the high spin Mn, Co and Ni respectively, while $MgFe_2O_4$ has a moment approaching the zero value for Mg.

Tetrahedral complexes can be treated in a similar manner to octahedral. Both octahedral and tetrahedral are cubic systems: in octahedral (Figure 3.3) the ligands are placed at the centres of the six faces of a cube. In tetrahedral the ligands are diagonally across the cube faces from each other (Figure 3.4). Now the d-electron repulsion by ligands works in exactly the opposite sense, giving rise to exactly the inverse picture for the orbital energies. The t_2 orbitals point closer to the ligands, and are therefore more unfavourable places for electrons to occupy than are the e_g. The subscript g is dropped for even parity, which applies to undistorted octahedral, but not to tetrahedral (no centre of inversion). By visualizing an energy in the inverse direction we can see that the ground state for octahedral d^n is the same as that for tetrahedral d^{10-n}. The state symbols and CFSE formulas also apply as given, except that low spin tetrahedral complexes tend not to occur: this would require a rigid ligand with donor atoms locked in tetrahedral geometry and a large value of Δ_t. Otherwise, any four-coordinated complex with Δ large enough to cause spin pairing would also distort to planar, which offers much larger CFSE,

Π does not vary much within a series of iron complexes, but the size of Δ, the 'strength of the ligand', varies widely. One way it can be estimated is from d–d transitions, or ligand field spectra (hence the name, spectrochemical series):

$$I^- < Br^- < Cl^- < NO_3^- < F^- < OH^- < C_2O_4^{2-} < H_2O$$
$$< NH_3 < en \ll \text{bpy, phen, terpy} < CO, CN^-$$

This list gives an approximate indication of relative ligand field strength, but other factors, such as different geometries, also affect the actual crystal

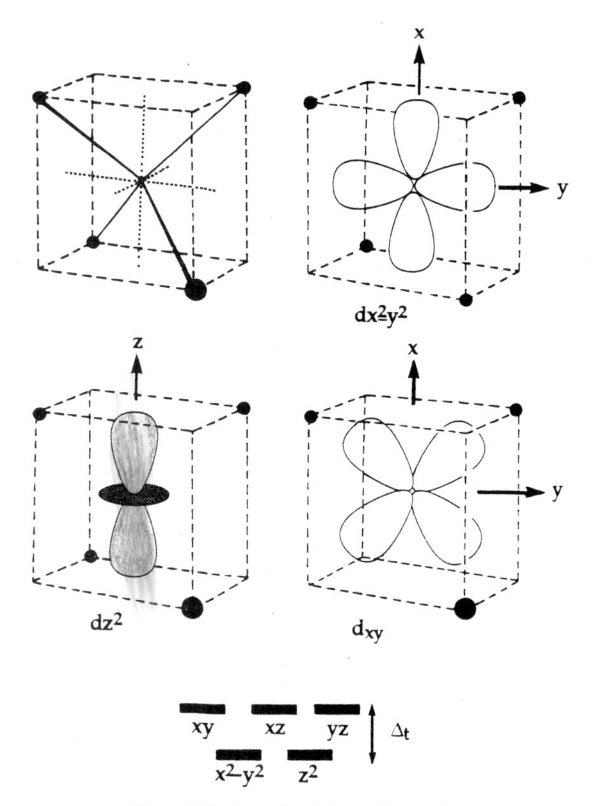

Figure 3.4 Tetrahedral configuration.

field experienced. In general, ligands before en in the list do not produce low spin compounds with iron, while those listed to the right of it do. Having the largest CFSE, low spin Fe^{II}, d^6 is especially stabilized, and is therefore easier to produce than low spin Fe^{III}.

The states listed as A and E have temperature independent magnetic moments which conform reasonably well to the spin only values, $g\sqrt{S(S+1)}$ BM (i.e. $\sqrt{3}$, $\sqrt{8}$, $\sqrt{15}$, $\sqrt{24}$, $\sqrt{35}$ for 1, 2, 3, 4, 5 unpaired electrons respectively). The moment values for T ground states, derived from the Hamiltonian $H = v\,\mathbf{L.S} + (\beta\,\mathbf{L} + 2\mathbf{S})$ on the appropriate T wave functions, $|M_S\,M_L\rangle$, are given on Figure 3.5. For A and E states, the quadrupole splitting is often small enough to produce only a single line in the Mössbauer spectrum. For real T states, quadrupole splitting invariably leads to a doublet, the magnitude of which depends on bond strength as well as degree of distortion from cubic symmetry (Golding *et al.*, 1969).

For distortion from regular cubic geometry, we add to H a term δl_s^2, which allows for tetragonal or trigonal distortion (but with very similar results

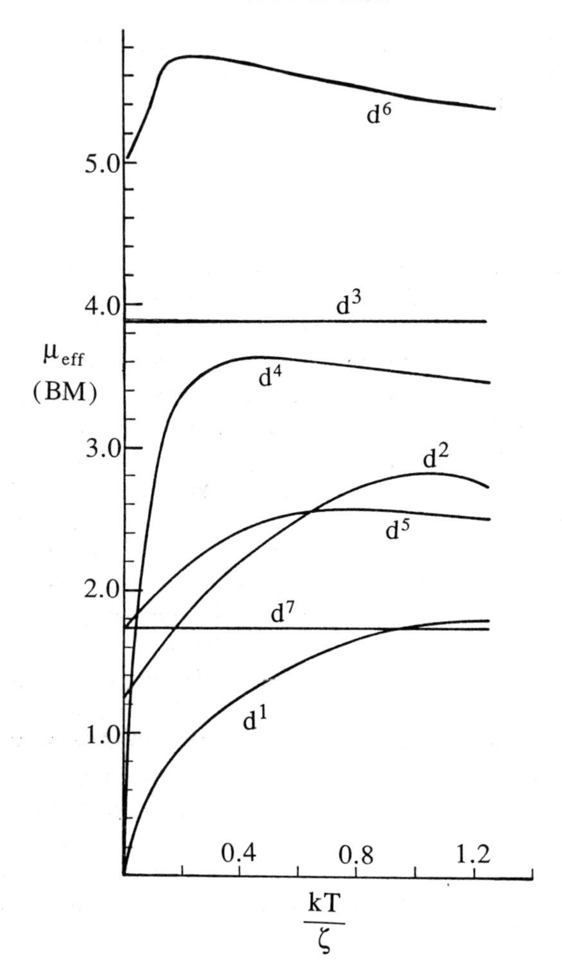

Figure 3.5 Effective magnetic moment μ_{eff} vs kT/ζ for the T states in the d^6, d^2, low spin d^4, low spin d^5, and high spin d^6 configurations. The temperature independent behaviour of A and E states (d^3 and low spin d^7) are shown for comparison. ζ is the one-electron spin-orbit coupling constant for the state ($= 2S\lambda$).

for other kinds of distortion) (Golding, 1972). This greatly reduces the temperature dependence of the moments and brings them closer to the spin-only values. Distortion from octahedral symmetry splits $^5T_2$ into $^5B_2$ and ^5E states. This was treated by König and Chakravarty (1967), but their calculation is wrong in that all terms based on the matrix elements $\langle M_S M_L|H|M_S\ M_L\rangle$, for which $(M_S + M_L) = 0$, matrix elements are doubled. The published numerical data (König *et al.*, 1967) based on this give the correct qualitative trends but are quantitatively wrong. The calculations can be corrected by eliminating the excess terms resulting from double weighting the zero–L wavefunctions; this has the consequence of having several terms in both the numerator and denominator wrongly weighted by a factor of two (Sinn, 1967).

3.2 Lattice structures

The range of lattice structures is wide but there are many similarities between them, and efforts to find more general correlations between different structures may lead to insights into the chemistry and physics of these compounds.

The point in seeking unexpected similarities between various lattice structures is that we may see how solid state reactions best proceed – with minimum movement of atoms to provide the minimum energy path. Further, we get a better idea of properties, e.g. olivine, mentioned above, has 2D layers of metal atoms in the configuration of a metallic phase (Vegas *et al.*, 1990, 1991; A. Vegas, personal communication), and it happens to be a 2-D conductor. This would be more than coincidental yet the two observations have not been related. The layers of metal atoms found in rutile structures should also have a profound influence on the properties.

3.2.1 Borides

The range of possible compounds may be almost limitless. This is because the building blocks in solid borides, as indeed in elemental boron, tend to be clusters like B_{12} rather than individual atoms of boron. An extreme example, characterized by single crystal X-ray work, is the discrete compound FeB_{49}, which is best described as Fe in β-rhombohedral boron (Callmer and Lundström, 1976). Simpler stoichiometry borides are FeB and Fe_2B (Bjurstroem, 1933).

3.2.2 Carbides

The interest in many solid state compounds derives from their relevance as additives in steel making. Carbon is the most common additive and forms a low melting azeotrope with iron which on cooling yields cast iron with at least 4% carbon by weight. A smaller amount of carbon ($< 1.5\%$) turns iron into common steel. Depending on the previous heat treatment, the carbon may be present as specks of graphite, as solid solution with carbon atoms placed interstitially in the bcc lattice, or as cementite, Fe_3C (Fruchart *et al.*, 1964; Herbstein and Smuts, 1964). Cementite is unusual in having carbon in a very distorted trigonal prism of six iron atoms at distances ranging from 1.89 Å–2.15 Å, with an additional two irons further away at 2.31 Å. Although Fe_3X stoichiometries are common, e.g. Fe_3P, Fe_3Ga, Fe_3Ge, Fe_3Ni, Fe_3O, Fe_3Pd, Fe_3Pt, Fe_3Si, Fe_3Sn, Fe_3Ti, Fe_3Zn, they tend not to emulate the irregular cementite structure. Other iron carbides are $Fe_4C_{0.63}$, an ordered phase of iron–carbide martensite (Nagakura and Toyoshima, 1979) and Fe_7C_3. (Bouchard, 1967).

3.2.3 *Nitrides*

Nitrogen is an additive for case-hardening steels. The phase diagram is complicated, and discrete compounds with iron include $Fe_{2.7}N$, (Fe_3N with defects) and Fe_2N, two ε-phase hexagonal iron nitrides, the ε-phase hexagonal nitrides $Fe_{24}N_{10}$ and Fe_2N (Jack, 1952; Pinsker and Kaverin, 1954). Each of the following has been prepared in two different forms, austentite (γ) and martenstite (α), with two distinct structures, body centered tetragonal (I_4/mmm) and face centered cubic (Fm3m): FeN_{0324}, FeN_{0499}, FeN_{056}, FeN_{0589}, FeN_{076}, FeN_{088}, FeN_{0897}, FeN_{0935}, FeN_{0939}, FeN_{095} (Jack, 1951). Nitrogen also forms ζ-Fe_2N (orthorhombic P222), ε-Fe_2N (hexagonal close packed, with one third of the octahedral sites occupied), Fe_3N (hexag P3-1M) and Fe_4N (γ, orthorhombic P222) (Jack, 1948).

3.2.4 *Phosphides*

Structures here include Fe_3P which is isomorphous with Pd_3As, Z_3Sb, Hf_3As, Zr_3Sb. Also Fe_2P, FeP, FeP_2 (Aronsson *et. al.*, 1965). Hexagonal Fe_2P forms from the reaction of red phosphorus with metallic iron (Fujii *et al.*, 1977, 1979). Such classical solid state compounds are now under intensive scrutiny, in case some may have been overlooked as ‘new, advanced’ materials. Fe_2P has two types of iron, Fe_I in a nearly perfect tetrahedron of 4 P atoms, and Fe_{II} in a pyramid of five phosphorus atoms. It is a ferromagnet with a Curie temp of 210 K, and magnetization per molecule of 2.94 μ_β. The transition from ferromagnetic at this point is assumed to be accompanied by a first order transition due to magnetoelastic effect. The magnetic anisotropy is larger than in other 3-d transition metal compounds.

There are three distinct FeP_4 structures (Jeitschko and Braun, 1978; Sugitan *et al.*, 1978). FeP has the orthorhombic MnP structure, also with interesting structural and magnetic properties (Evain *et al.*, 1987). FeP_2 is orthorhombic and similar to PtP_2 (Selte and Kjekshus, 1972; Dahl, 1969).

3.2.5 *Arsenides, etc.*

As expected (Boller and Parthe, 1963; Smith, 1973) many arsenides are found with similar stoichiometry to that of phosphides, but the greater size of the As atom tends to alter the structures: Thus Fe_2As is isomorphous with Cu_2Sb C38, Cr_2As, UAs_2, $ThSb_2$, USb_2, Mn_2Sb, $ThBi_2$ and UBi_2. Like FeP, Fe_2As has the Fe atoms close enough for strong magnetic coupling, but in this case it is antiferromagnetic, with a Neel temperature of 351 K.

FeAs is isomorphous with MnP (B31), VAs, CrAs, MoAs, RuAs, CoAs, RhAs, RuSb and RhSb.

$FeAs_2$ is isomorphous with FeS_2 (marcosite, C18, below), as well as with $FeSb_2$, $RuAs_2$, $OsAs_2$, $CrSb_2$, $RuSb_2$, $OsSb_2$ and $RuSb_2$. $FeAs_2$ and $FeSb_2$, with 19 valence electrons, are semiconductors (2 more electrons would be required for occupation of the conduction band and metallic conductivity). They have band gaps of 0.2 eV and 0.05 eV respectively. As $FeAs_2$ has the FeS_2 structure it is not surprising that elements like sulphur are often substituted for As. The structure of $FeAs_2$ is similar to that of FeAsS, FeAsSe, FeAsTe, FeSbS, FeSbSe and FeSbTe (the arsenopyrites, FeAsS, or $CoSb_2$ strtucture), as well as to RuAsS, RuAsSe, RuAsTe, RuSbS, RuSbSe, RuSbTe, OsAsS, OsAsSe, OsAsTe, OsSbS, OsSbSe, OsSbTe, RhAsSb, and IrAsSb.

3.2.6 Antimonides

Fe_3Sb_2 is isomorphous with Ni_2In ($B8_2$), Pd_5Sb_2 and Pd_5Bi_2.

FeSb is isomorphous with NiAs (B8), TiSb, VSb, CoSb, IrSb, NiSb, PdSb, PtSb, NiBi and PtBi. It has a hexagonal cell with each Fe having six As neighbours at the corners of an octahedron, and each Sb atom having six Fe atoms at the corner of a trigonal prism. The Fe atoms are close enough here to be considered bonding. The structure can also be described as a hexagonal-close-packed array of Sb atoms with Fe atoms in octahedral holes. If some of these holes are unfilled, the stoichiometry can very widely from 1:1. The same thing applies if holes other than the octahedral ones are occupied: if Fe atoms were placed in all the nearby five-coordinated holes the structure would become Fe_3Sb_2.

$FeSb_2$ is isomorphous with FeS_2 (marcosite), and $FeAs_2$ above and is important because of potential semiconductor applications (Hulliger, 1968; Johnston *et al.*, 1965).

FeTiSb and FeVSb have the MgAgAs structure, related to antifluorite, and are isomorphous with MgLiAs, MgAgAs, MgLiSb, MgNiAs, MgCuSb, MnNiSb, MnCuSb, CoLiSb, CoVAs, CdCuSb, NiTiSb, NiVSb, MgLiBi, MgNiBi, MgCuBi and Li_5TiAs_3.

3.2.7 Sulphides

Triolite, FeS, has six-coordinated iron and six-coordinated S. Below 138°C, it deviates from the ideal NiAs structure as the lattice shrinks. Another FeS (mackinawite) has a structure similar to that of LiOH. Fe_3S_4 has the spinel (AB_2X_4) structure. Fe_2S_3 has been prepared in amorphous form (Stiller *et al.*, 1978).

The structure of pyrite, FeS_2, is widely distributed in nature ('pyrite structure' C2, Figure 3.6). It is found in $PdAs_2$, $PtAs_2$, $PdSb_2$, $PtSb_2$, $AuSb_2$, and $PtBi_2$. SiP_2 acquires this structure at high pressures. It resembles the rutile structure, with S replacing O but moving closer together to form S_2^{2-}. The structure

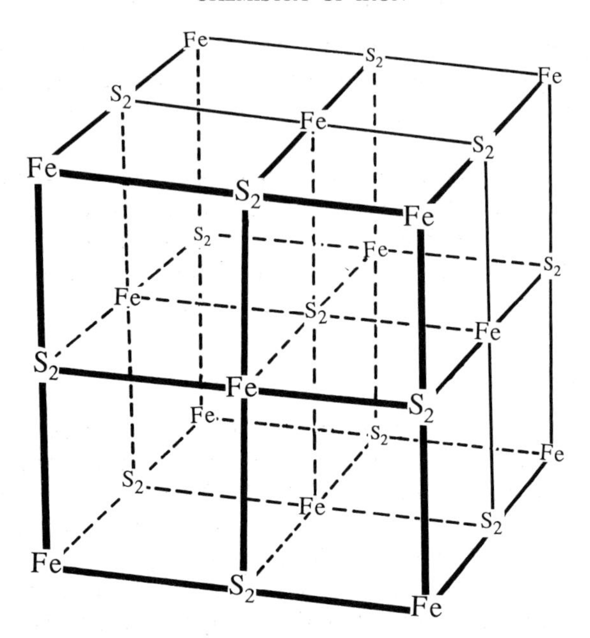

Figure 3.6 Pyrite, FeS_2 structure.

can also be derived from NaCl, with Fe sub 2+ at the Na^+ site and slanted S_2^{2-} at the C1 site. The structure of the isomeric marcosite, FeS_2, is very similar: in pyrite S–S is 2.171 Å, in marcosite it is 2.21 Å. In both pyrite and marcosite, the iron is six-coordinated by sulphur, while the S_2 unit is six-coordinated by iron.

CoAsS, NiP_2, and $PtAs_2$ have pyrite-like structures, as expected (counting 10 valence electrons, based on S_2^{2-}, Fe acquires the [Kr] configuration), but, interestingly, with similar structures, FeAsS (−1 electron by the same count) and $FeAs_2$ (−2e) are electron-deficient, while NiAsS, CoS_2 (1e), NiS_2 (2e) CuS_2 (high-P, 3e), and ZnS_2 (high-P, 4e) have an excess of electrons.

The ternary sulphide $CuFeS_2$ has the structure of zincblende, ZnS (B3), with Zn atoms replaced alternately by Cu and Fe in a regular manner. It is also isomorphous with AlAs, GaAs, InAs, AlSb, GaSb and InSb. These structures all lack a center of symmetry, so the 111 and −1 −1 −1 faces are non-equivalent, which should lead to interesting piezoelectric effects. The structures are open, with low coordination numbers of the elements. There is large separation of the metal or metalloid atoms compared to solid elemental states, but if high pressure is used to reduce the separations, they become metallic (Minomura and Drickamer, 1962).

Cubanite $CuFe_2S_3$ has edge-shared, regularly alternating FeS_4 and CuS_4 tetrahedra. The Fe–Fe separation of 2.81 Å is small enough to permit ferromagnetic interactions between the irons. Magnetoplumbite has slices of spinel-like (AB_2X_4) $PbFe_2O_4$ held together by Pb^{2+} ions. $FeSb_2S_4$ (berthierite) contains $(SbS_2)_n$ chains (Burger and Hahn, 1955. Burger and Niizeki, 1957). The quaternary $FePb_4Sb_6S_{14}$ (jamesonite) has SbS_3 rings linked to Sb_3S_7 units.

The iron nickel sulphides, pentlandite ($(Ni,Fe)_9S_8$), and violarite ($FeNi_2S_4$), are easily distinguished by their magnetic properties: pentlandite has a sharp maximum in the susceptibility near 120 K, indicative of strong long range ordering, while in violarite only a small smooth maximum is observed (P.P. Edwards, R. James and E. Sinn, unpublished).

3.2.8 *Iron–sulphur clusters*

Coordination chemistry has benefited greatly from the interest in small molecule iron biomolecules. FeS clusters in various proteins, and of both Fe and Mo in the most common nitrogenases have helped advance the multinuclear chemistry of iron as well as Mo (Burgess, 1990). For example, $[Fe(NO)(S_2MoS_2)_2]^{2-}$ is readily formed from Roussin esters $[Fe(SR)_2(NO)_4]$ (R = Me, Et, etc.) with tetrathiomolybdate, $[MoS_4]^{-2}$, a sulphate analogue. The ability to thiomolybdate into a range of FeMoS complexes makes them a favourite building block of chemists interested in modelling nitrogenase (Eldredge *et al.*, 1988; Coucouvanis *et al.*, 1988). It took a long time to accept that some nitrogenases contain V instead of Mo, e.g. in *A. chroococum* there is V with a V–Fe separation of ~2.75 Å (Garner *et al.*, 1989). This has spawned detailed exploration of FeVS chemistry. It is now clear that there are also nitrogenases which contain neither Mo nor V. The active site apparently contains an Fe_4S_4 centre (Chisnell *et al.*, 1988), which is one of the well-known, stable cluster fragments in FeS clusters as well as existing in FeS proteins.

3.2.9 *Oxides*

Compounds with oxygen ligands are very common, especially ones containing the FeO_6 core. Hematite, Fe_2O_3, a major ore (FeO_6 units), is an electrical semiconductor (Gharini *et al.*, 1990), as well as being a ferrimagnet. It is readily reduced into magnetite, Fe_3O_4, (iron ferrite) a mixed oxidation compound (FeO_6 and FeO_4 units, *vide infra*) and another major ore. A third major ore is geothite, FeO(OH), a hydrated ferric oxide.

FeO(OH) again has a distorted octahedral structure of six oxygens around the metal (FeO_6 centres). The octahedra are linked into corrugated layers, and these in turn are linked by hydrogen-bonding. The FeOCl and other MOX structures are quite similar except for the absence of the H-bonding allowing a close approach between the layers.

FeO has the NaCl structure (and FeO_6 units) and is a semiconductor. It can have a wide range of deviations from the 1:1 stoichiometry mainly by establishing vacancies at the Fe sites: $Fe_{1-x}O$. Each Fe^{2+} vacancy is compensated by oxidation of two nearby irons to Fe^{3+}. The result is a random distribution of mostly tetrahedral Fe^{3+} in the NaCl structure. Too great a concentration of the latter promotes transformation of the material to the more stable magnetite.

3.2.9.1 Rust. Iron is fairly reactive element (electron affinity, $Fe + e^- \rightarrow Fe^- = 0.16$ V; ionization potentials 7.87, 16.18, 30.65, 54.8V), and we see daily evidence that the elemental form is metastable under ambient conditions. The spontaneous corrosion in moist air, especially in the presence of electrolytes, to form the highly insoluble hydrated oxide, $Fe_2O_3.xH_2O$, is the most familiar reaction of iron. The product, rust, as well as being a very interesting substance, is an economic problem of monumental proportions: at least 1% of the world's economy is devoted to preventing rust, removing it or replacing rusted parts; in the US, the cost of corrosion and other metal failure has the same order of magnitude as the national debt (Cohen, 1987). As rust is porous, it does not inhibit further corrosion. The metal is rendered 'passive' in HNO_3 by the formation of an impervious film of highly insoluble oxide which resists further attack. However, hydrohalic acids quickly dissolve the metal, as well as the oxide film around passive iron.

The main route to rust is electrochemical, which is why iron rusts rapidly in shallow water (dissolved oxygen), in seawater (enhanced conductivity) or moist air. In dry air or air-free water it remains rust-free for extended periods. A good way to protect iron against rust is also electrochemical: another more reactive metal is placed in electrical contact with the iron. One way to do this is plating, as in zinc galvanizing. The zinc then reacts preferentially, protecting the iron from attack until all the zinc is oxidized. Any oxidation of the iron that did occur would be reversed by preferential oxidation of zinc.

$$2Fe^{3+} + Zn \rightarrow 2Fe^{2+} + Zn^{2+}; \; Fe^{2+} + Zn \rightarrow Fe + Zn^{2+}$$

Similarly lumps of magnesium are attached to the Alaska pipeline, and buried nearby, to make the magnesium a sacrificial electrode. Although the magnesium metal is relatively expensive, the cost is preferable to the risk of leaks in the pipes.

Aqueous iron(III) is readily oxidized at neutral pH to form insoluble hydrated oxides (Flynn, 1984):

$$Fe^{3+} \xrightarrow{pH \sim 1} > Fe(OH)^{2+} \xrightarrow{pH \sim 3} Fe_2(OH)_2^{4+} \xrightarrow{3 < pH < 10} FeO(OH) \cdot H_2O$$

The iron is then unavailable to biology unless it can be solubilized with mineral acids (e.g. stomach HCl), or with very strong ligands (enterobactin in *E. coli* or mycobactins in mycobacteria), or by reducing it to iron(II). Rust therefore has a chemistry which is as important to life as to technology (Lippard, 1986): the iron storage protein ferritin (Chasteen *et al.*, 1985) consists approximately of a protein sheath of adjustable size around a lump of rust (formulae such as $(FeOOH)_8(FeO.H_2PO_4)_n$ are suggested). The room temperature magnetic moment is about 3.8 BM per iron due to magnetic coupling between the FeO_6 sites through Fe–O–Fe bridges. The high insolubility of ferric oxide makes this a convenient form for iron storage. Reduction to iron(II) provides a convenient soluble form for taking the metal out when needed. Given stomach acids that can dissolve ferric oxide as the chloride, we can easily acquire the iron we need from many sources. Many bacteria find themselves in situations where the only source of iron may be poorly available, e.g. as ferric oxide in soil (K_{sp} for $Fe(OH)_3 \sim 10^{-38}$ so that a saturated solution would be about 10^{-18}M). Lacking stomach acid to solubilize the metal as the chloride, they can overcome the solubility problem with ligands of extremely high binding constants for Fe^{3+}, such as enterobactin ($K_{stab} = 10^{52}$ with Fe^{3+}), and other natural siderophores (Raymond *et al.*, 1984).

3.2.9.2 Mixed oxides. With lanthanide orthoferrites, or iron metal oxides, $FeLn^{III}O_3$ or $Fe_2M^{II}O_3$, properties can be varied by the choice of the lanthanide or the bivalent M^{II} (Mn, Ni, Zn). The same is true of the iron metal garnets $Fe_5Ln_3^{III}O_{12}$. The most interesting variants are those in which the lanthanides are paramagnetic, allowing tailored ferromagnetic properties. Other mixed iron metal oxides exist, e.g. $Fe_{12}PbO_{19}$, and a multitide of perovskites like $Ca_{1-x}M_xFe_xTi_{1-x}O_3$ (M = La, Eu) (Correia dos Santos and da Costa, 1991).

While $CuFeS_2$ and many analogs have the metals alternating systematically, in many oxide structures the type of ion on one site may be randomized through the lattice, as in the perovskites $Ca_{1-x}M_xFe_xTi_{1-x}O_3$; the structure (Figure 3.1) allows the FeO_6 units to be retained in all cases.

FeO_6 is also found in the phosphates $AFeP_2O_7$ (A = Na, K, Rb, Cs) (Millet and Mentzen, 1990). These have interesting ferromagnetic Fe_2O_9 clusters in $Fe_4(PO_4)_3(OH)_3$: the two Fe atoms belong to two different ferromagnet rows, so that the intradimer coupling is forced to be ferromagnetic to avoid spin frustration due to competition between the interdimer and intradimer interactions which would prefer to be, but cannot all be, simultaneously antiferromagnetic (Torardi *et al.*, 1989; Malaman *et al.*, 1991). A similar situation exists in a related compound, β-$Fe_2(PO_4)$ (Ijjaali *et al.*, 1990). There is a wide range of Fe^{II}, Fe^{III} and mixed valence phosphates, having antiferromagnetic and ferromagnetic interactions between various iron sites. The overall magnetism depends on which type of interaction dominates. Six-coordinated FeO_6 sites predominate, but a number of five-coordinated sites exist in some complexes (Gleitzer, 1991).

3.2.10 Fe–O–Fe

We have already met compounds with multiple Fe–O–Fe bridges, like those linking the FeO_6 units of rust and other minerals, and leading to strong magnetic coupling. The Fe–O–Fe linkage is also very often found as the sole link between two Fe^{III} species (Mockler *et al.*, 1983; Sinn *et al.*, 1983). In urease (Doi *et al.*, 1988), this linkage was first discovered from magnetic properties (Mockler *et al.*, 1983; Sinn *et al.*, 1983). A Fe–O–Fe bridge is implicated in ribonuclease reductase (Wilkins and Wilkins, 1987), and methane mono-oxygenase (Woodland *et al.*, 1986).

Fe–O_2 complexes, like oxyhaemoglobin, are dealt with in detail elsewhere in the book. The frequently quoted rise (up to 1 Å) of the metal atom above the heme plane in the 5-coordinated high spin deoxyhaemoglobin, is not necessarily due to the spin state change from the low spin oxy form – the raised geometry would occur anyway in a square pyramidal complex.

3.2.11 Halides

$FeCl_3$ is likewise a layer structure in the solid, each Cl bridging between a pair of octahedrally bonded Fe atoms; on vaporization, Fe_2Cl_6 with 2 of the Cl atoms bridging a pair of tetrahedrally coordinated Fe atoms. It forms layer complexes with graphite by intercalation between the graphite layers and also forms similar compounds with the BN analogue of graphite. At high temperatures, $FeCl_3$ species separate and on further heating decomposition occurs, first losing a single Cl. On the other hand, FeOF has the 3-D rutile structure. Like rutile (TiO_2) itself, this retains fragments of a metallic phase of the metal Fe or Ti. The hydrated form is $[Fe(H_2O)_4Cl_2]$ cations in an octahedral environment and Cl^- ions with two further H_2O molecules linked by hydrogen bonded to the cation. $FeBr_3$ is likewise a layer structure in the solid.

$FeCl_2$ also has an octahedron of Cl atoms, but in this case the chlorines must bridge more extensively in order to provide an octahedron. Consequently it is three-dimensionally linked in a rutile (TiO_2) structure.

As fluoride is a weak ligand, the FeF_6^{-3} ion is high spin. Fluoride bridging Fe–F–M leads to a wide range of magnetic properties due to magnetic coupling (Kahn, 1987). In Na_2CuFeF_7, a variant of the weberite family $Na_2M^{II}M^{III}F_7$, FeF_6 distorted octahedra are well separated from each other, but F-bridged to the Jahn-Teller distorted CuF_6 groups (Yakubovich *et al.*, 1990). Again there should be weak coupling through the bridges.

3.2.12 Lower dimensional structures

A newly emerging field is the synthesis of new types of materials for specialized applications: one simple strategy is to seek ways of incorporating new properties into known structures. Low-dimensional compounds of many kinds are known. An obvious example is liquid crystal compounds which have applications in electronics and displays; another example is layer compounds known for the intercalation chemistry that can be carried out to modify their properties.

Mesogenic (liquid crystal) materials were developed in the last few decades.

Metallomesogens are a more recent variant, modified so as to add in specific properties due to the metal, such as the d–d spectra or magnetism of transition metals. Figure 3.7 gives an example involving Fe.

An example of a 1-D structure is $KFeS_2$ which forms as steel blue fibrous crystals and has 1-D edge sharing $FeS_2FeS_2Fe\ldots$ tetrahedra. The chains are separated by K^+ ions. It is formed by fusion of iron, sulphur and K_2CO_3 and extraction with H_2O. It reacts with CuCl in NH_3 solution to form the brassy metallic $CuFeS_3$, which has the 3-D structure of zincblende and of the analogue Cu_2FeSnS_4.

$Ba_9Fe_{16}S_{32}$contains chains of FeS_4 tetrahedra joined via their shared edges (Hoggins and Steinfink, 1977). $Ba_7Fe_6S_{14}$ also contains FeS_4 tetrahedra. Units of three edge-sharing tetrahedra are connected to each other by a corner-sharing Fe–S–Fe linkage:

$$\ldots Fe\text{-}(\text{-}S(S)FeS_2FeS_2Fe(S)\text{-}S\text{-})\text{-}Fe\ldots$$

Molecular ferromagnets have also aroused considerable interest. Possible applications for them are in magneto-optics. $[Fe(C_5Me_5)]^+[TCNE]^-$, stacked in alternating layers, is an example (Kollmar *el al.*, 1991).

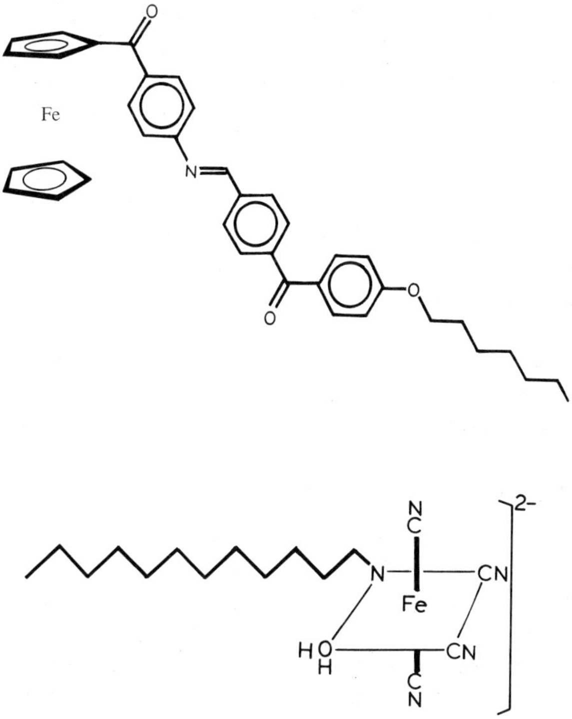

Figure 3.7 Iron-containing metallomesogens (Giroud-Godquin and Maitlis, 1991; Bruce, 1992).

3.2.12.1 FeOCl. We have already seen that $FeCl_3$, $FeBr_3$, FeOCl and FeO(OH) have layer structures. There is a great deal of information on how to modify the properties of graphite by intercalating many chemical species, including $FeCl_3$ between the sheets of graphite. The benefit of the transition metal in the intercalation host is that further properties can be added, such as different structures of the layers, electric and magnetic properties, and d–d spectra of the transition metal.

Intercalative polymerization of pyrrole in FeOCl produces a superior form of conductive polypyrrole (Kanatzidis *et al.*, 1987).

3.2.13 Cyanides and pseudo-halides

Here ligand π-bonding (back-bonding) must be considered. In using crystal field theory we have ignored the influences from outside the little circle in Figure 3.2: the t_2 and e orbitals may be involved in bonding. In particular, ligand orbitals of t_2 symmetry may overlap with the metal t_2-orbitals. High energy (and therefore empty) ligand orbitals will increase Δ while a low energy (full) one will decrease it. CN^- is an example of the former.

Thus ferrocyanide $Fe(CN)_6^{4-}$ with low spin Fe^{II}, is especially stabilized by CFSE, so much so that ferrous sulphate is an effective antidote to CN^- poisoning if taken quickly enough: $6CN^- + Fe^2 \rightarrow [Fe(CN)_6]^{4-}$. Ferrocyanides have other medical applications: the iron salt, Prussian Blue, administered orally is an effective antidote to Tl poisoning (Nelson *et al.*, 1987; Heydlauf, 1969). The CN^{-1} acts as a Fe–C$\equiv$N–M bridge linking the iron or other transition metals at the points of a simple cube in Prussian Blue and the transition metal derivatives of $[Fe(CN)_6]^{4-}$ (Figure 3.8). The very property that confers extra stability on ferrocyanides and makes it possible to form these and a wide range of derivatives makes the iron atom low spin and therefore magnetically uninteresting. Even when the other metals inserted have magnetic moments, the large separations prohibit strong magnetic coupling. However, there are a number of other interesting properties, e.g. $ALnFe(CN)_6$ (A = K, Rb, NH_4, Cs) often have zeolitic cavities (Hulliger and Vetsch, 1990).

Ferricyanides can be characterized by their magnetic properties, but the magnetism of the low spin Fe^{III} is necessarily relatively weak. Thiocyanates are weaker ligands and give similar but high spin complexes, which makes them more reactive than their low spin Fe^{II} analogues, e.g. $K_2[Fe_2(\mu\text{-NCS})(NCS)_6(CH_3OH)]$, and $[Fe_2(\mu\text{-NCS})_2(NCS)_4(H_2O)_4].4H_2O$, which presumably show antiferromagnetic interactions at low temperature, and $[Fe(NCS)_3(H_2O)_3].3H_2O$ (Böhland and Witzenhauser, 1991).

Much inorganic work on metallobiomolecules is dedicated to unravelling their properties by adding simple inorganic ligands like cyanides and azides (Saleem and Wilson, 1988). This may be especially diagnostic with iron compounds when it changes the spin state.

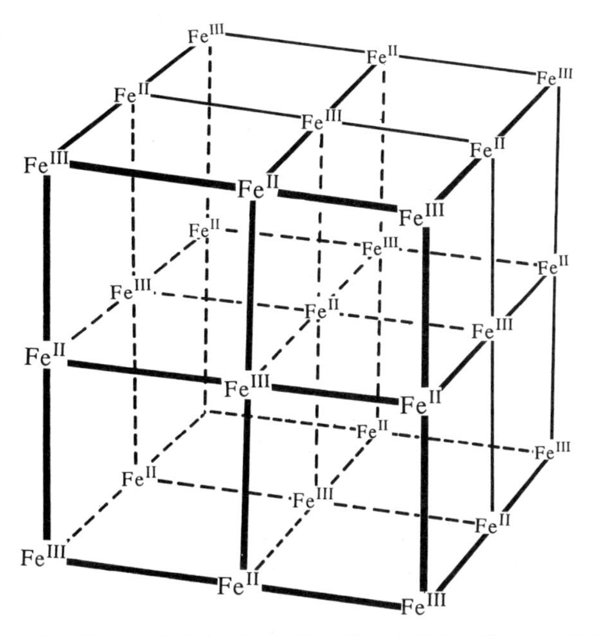

Figure 3.8 Prussian Blue (and derivatives). CN^- ligands link Fe^{II} and Fe^{III} ions (or other transition metals in derivatives) and the cell consists of eight smaller cubes, which are alternately empty or contain K^+ (or other cation).

3.2.14 Unusual oxidation states

As stated in chapter 1 the normal oxidation numbers of iron are 2 and 3, lower numbers being more common in organometallics, with formal charges as low as -2 and as high as $+6$ being cited, though mainly as chemical curiosities at present. Isolated molecules or ions of coordination or organometallic complexes are needed to stabilize these: 'solid state' lattices range almost exclusively from Fe^{II} to Fe^{III}. Negative formal charges are seen mainly in organometallics and in some ionic iron porphyrins, where the extensive delocalization allows a true charge closer to zero on the metal atom. Likewise, though Fe^{IV} is occasionally proposed for iron porphyrins, oxidation states other than 2 and 3 are generally too reactive for iron-containing biomolecules. Although covered in chapter 1, additional examples of unusual oxidation states are given here – Fe^{-II} in $Na_2[Fe(CO)_4]$ (Colquhoun *et al.*, 1984) and Fe^0 d^8 in $Fe(dpy)_3$ (Herzog and Weber, 1968).

- Examples of Fe^I d^7, are seen in the low spin square pyramidal $[Fe(NO)(S_2MoS_2)_2]^{2-}$ (Glidwell and Johnson, 1988), with $S = 1/2$, and in the entire series of square pyramidal $Fe(NO)(S_2CNR_2)^{2+}$ (Butcher and Sinn, 1980).

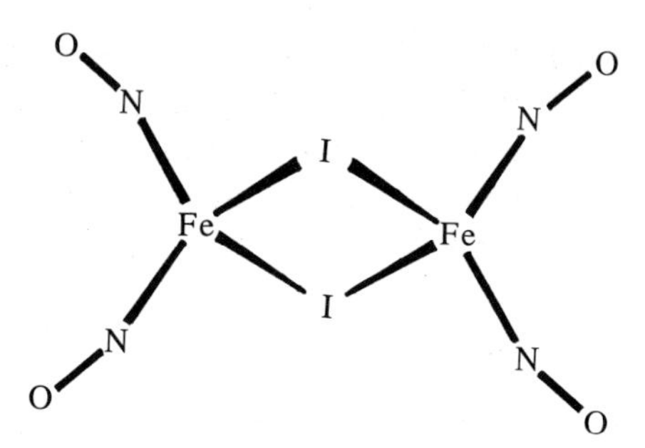

Figure 3.9 $[(ON)_2FeI_2Fe(NO)_2]$.

- Tetrahedral Fe^I is found in the S-bridged dimer (Thomas *et al.*, 1958) $[(ON)_2Fe(SC_2H_5)_3Fe(NO)_2]$ and the I-bridged (Dahl *et al.*, 1969) $[(ON)_2FeI_2Fe(NO)_2]$ (Figure 3.9) in which antiferromagnetic coupling pairs the electrons.
- Tetrahedral Fe^{IV} occurs in Ba_2FeO_4 which has the Ba_2TiO_4 structure (Scholder *et al.*, 1956).
- Octahedral Fe^{IV} occurs in $C_{15}H_{24}N_3S_6.Fe(ClO_4)$ (Martin *et al.*, 1974) and has Fe–S 2.300 Å.
  $C_{60}H_{40}As_2FeN_6S_6$ (Bernal, 1973), has Fe–S 2.266 Å. $[(C_7H_7\phi_3P]_2$ $[Fe(S_2CC(COOEt)_2)_3]$ (Coucouvanis *et al.*, 1977), has Fe–S 2.298 Å. Octahedral Fe^{IV} is indicated by a magnetic moment consistent with 2 unpaired electrons, i.e. low spin d^4, in $[Fe(diars)_2Cl_2](BF_4)_2$ and $[Fe(S_2CNR_2)_3]BF_4$ (Greenwood and Earnshaw, 1986; Pasek and Straub, 1977).
- Fe^V can be obtained by the electrochemical oxidation of $Fe(bpy)_3^{2+}$ in SO_2 (Garcia *et al.*, 1988).
- Tetrahedral Fe^{VI} has been prepared in K_2FeO_4 and $BaFeO_4$, which are isostructural with K_2SO_4 and $BaFeO_4$ respectively (Krebs, 1950).

3.3 High pressure effects

There has been a considerable amount of work on the effect of very high pressure on iron compounds, in order to find out what oxidation number is stable under extreme conditions. A reduction in oxidation state is often seen: high spin Fe^{III} is reduced to high spin Fe^{II} (see below), and low spin Fe^{II} in various $[Fe(CN)_6]^{4-}$ complexes is raised to high spin Fe^{III}. This has been attributed to decreased CN backbonding making it a weaker ligand. However, there is also the possibility that at least some of the observed changes are due to chemical reaction to form new compounds, rather than just compression of unchanged species (Sinn, 1974), since a number of the effects are irreversible and a number are sluggish to reverse (Fung and

Drickamer, 1969). $SrFe^{IV}O_{2.85}$ reduces to Fe^{III}, which is reversible. $BaFe^{VI}O_{4-x}$ reduces to Fe^{IV} (and some Fe^{III}) but this is not reversible (Sinn, 1974; Champion and Drickamer, 1969). In ferric dithiocarbamates, there is simply a change from high to low spin as pressure increases, but no sign of reduction to Fe^{II} (H.G. Drickamer and E. Sinn, unpublished work).

3.3.1 *Intermediate spin states*

With sufficient distortion, each of d^5 and d^6 can have intermediate spin states as the ground states (a split $^4T_1$ for d^5, and a split ^3T for d^6). The intermediate spin state, $S = B/2$, is found, for example in the halo-bisdithiocarbamato complexes of iron(III), $FeCl(dtc)_2$ (Sinn, 1970). The earlier literature reported the large quadrupole split Mössbauer data expected for the low symmetry $FeX(dtc)_2$ complexes, but misidentified the compounds as $Fe(dtc)_3$. This highlights how the compounds can form: the reaction is a peculiarity of the method of preparation of $Fe(dtc)_3$, which includes recrystallization from $CHCl_3$. Pure (redistilled) $CHCl_3$ reacts with air to form some phosgene, $COCl_2$, which in turn reacts with moisture to form HCl; the latter reaction produces new complexes: $Fe(dtc)_3 + HCl \rightarrow FeCl(dtc)_2Cl$. ($CHCl_3$ solvent is normally stabilized with a small admixture of ethanol (2%) to inhibit formation of phosgene.)

A typical $S = 3/2$ state species is Fe^{III} in tetrahedral $[Fe(S_2MoS_2)_2]^{3-}$ (Coucouvanis *et. al.*, 1980). To this can be added NO^+ to change the 4-coordinate Fe^{III} into formal Fe^I, low spin d^7, with $S = 1/2$, in the square pyramidal $[Fe(NO)(S_2MoS_2)_2]^{2-}$ (Glidwell and Johnson, 1988). The wide range of Fe–N–O bond angles observed in such complexes is often associated with the spin state, but prediction from spectroscopic data is difficult. For $Fe(NO)(dtc)_2$, there had been a variety of predictions, but crystal structures clearly show an essentially linear FeNO link. (Butcher and Sinn, 1980; Brewer *et al.*, 1983).

3.4 Spin equilibria

A high degree of distortion from octahedral symmetry is needed to produce intermediate spin states between the normal high and low spin states. However, even for relatively undistorted complexes, there is a region in which the high and low spin extremes can be in equilibrium. Spin crossover work had its beginning with iron. It was proposed by Cambi and Malatesta (1937) some 60 years ago that the ferric dithiocarbamates could explain magnetic moments that conformed neither to the high spin nor the low spin values, but seemed scattered randomly in between. Cambi's hypothesis was long ignored by most and scoffed at by others (Figgis and Lewis, 1960) until the correct measurements were made – these showed that the Fe(dtc) complexes were monomeric, so that no metal–metal coupling could explain reduction of moments from the HS values.

In the most crude description, the values Δ and Π are fixed in a particular complex and, when they are similar in value, thermal equilibrium results. However, high pressure measurements have shown that the low spin state has a smaller molar volume that the high spin (Ewald *et al.*, 1969; Sinn, 1974) which shows up as a shorter metal–ligand bond length (Butcher and Sinn, 1976; Cukauskas *et al.*, 1977; Greenaway and Sinn, 1978; Greenaway *et al.*, 1979; Healy and Sinn, 1975; Leipoldt and Coppens, 1973; Mitra *et al.*, 1976; Sim and Sinn, 1978, 1981; Sinn, 1976; Sinn *et al.*, 1978; Sim *et al.*, 1981; Timken *et al.*, 1985) brought about by the fact that the e_g orbitals are unoccupied – empty orbitals are essentially absent; they exist in our mathematical imagination, not for the molecule. The shorter metal–ligand bond length in low spin complexes implies a larger Δ ($\Delta \sim$ M–L^{-5}). Thus, when the two spin states are approximately equal in ground state energy, an activation barrier exists between the two, and a change in geometry is required as the molecule travels between the two states. The high spin molecules have higher degeneracy of magnetic states and this is one reason why they tend to be favoured at high temperatures. The requirement for crossover is that $\Delta_{LS} > \Pi > \Delta_{HS}$. When the barrier is minimum, the equilibrium should be fast, in the absence of external influences, i.e. in the gas phase. The best available approximation to this situation is in solution, where the best data have been recorded for a number of Co^{II}, Fe^{II} and Fe^{III} compounds by Dose *et al.* (1978).

As increasing external influences are applied, in the solid state, intermolecular forces and packing must be taken into account. Since the molecules in different spin states must differ in shape and magnetic properties, they effectively represent different phases and interchanges between the two require nucleation of one spin state in a host consisting of the other. In some systems, this may occur easily especially in rapidly interconverting molecules, as when there is high spin–orbit coupling. High rates mostly occur when there is minimal ligand constraint – maximal flexibility – and higher spin–orbit coupling (e.g. S-donor ligands, with higher spin–orbit coupling than O or N donors). For the faster interchanging complexes, we expect the macroscopic changes to be quicker as the probability of nucleating at a given site increases. Ways of increasing the number of nucleating sites (grinding) are known to increase the rate and so make the sharp transitions more gradual.

The magnetic properties can display an especially wide variety of behaviours near the $^5T_2$–$^1A_1$ crossover in iron(II) complexes (Cunningham *et al.*, 1972). The spin change in the solid material may be especially abrupt as in [$Fepy_2(NCS)_2$]. The magnetic properties of this and [$Fe(phen)_2(NCSe)_2$] were so surprising when first observed, that they were first interpreted as an antiferromagnetic dimer interaction (Baker and Bobonich, 1964); next it was treated as a smooth Boltzmann type equilibrium between spin states (König and Madeja, 1966), analogous to that in the ferric dithiocarbamates. However,

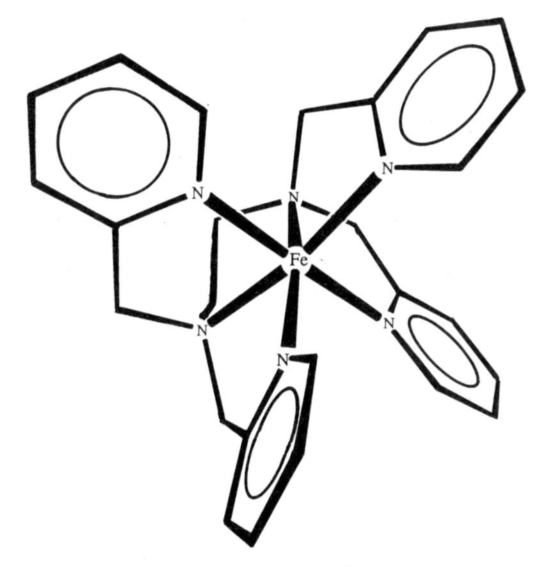

Figure 3.10 The first rapidly interconverting Fe^{II} crossover is observed in this ion.

the properties resembled neither and was clearly a phase transition from LS to HS (Harris and Sinn, 1968).

Usually Fe^{II} has gradual spin crossovers while fast ones are characteristic of Fe^{III}. However, there are exceptions: the complex in Figure 3.10 is Fe^{II}, newly discovered to be rapidly interconverting (Chang *et al.*, 1990).

Encapsulating or cage-like ligands (such as those based on tren) tend to produce spin crossovers, and in a constrained environment the conversion rate (usually) tends to be slow with Fe^{II} (Martin *et al.*, 1988). Fe-trenpyr$_3$ is low spin with Fe; but the Mn^{II} electronically analogous with Fe^{III}, and possibly with an analogous slow equilibrium existing in some porphyrin compounds (B. Landrum, personal communication).

In addition to temperature changes, the spin crossovers are sensitive to pressure, light, slight chemical modification and occluded solvent molecules in the lattice. The pressure effect has geological implications: high spin iron-containing minerals like olivine will convert to low spin in the earth's mantle, accompanied by phase change (lower density). The ability to induce the transitions optically (Poganiuch *et al.*, 1990; Conti *et al.*, 1989) means that the spin crossover materials may find application as light storage devices (Gütlich and Hauser, 1989).

3.4.1 Bond lengths in high and low spin states

The volume difference between high and low spin states means that the metal–ligand bond length increases steadily with increasing magnetic moment

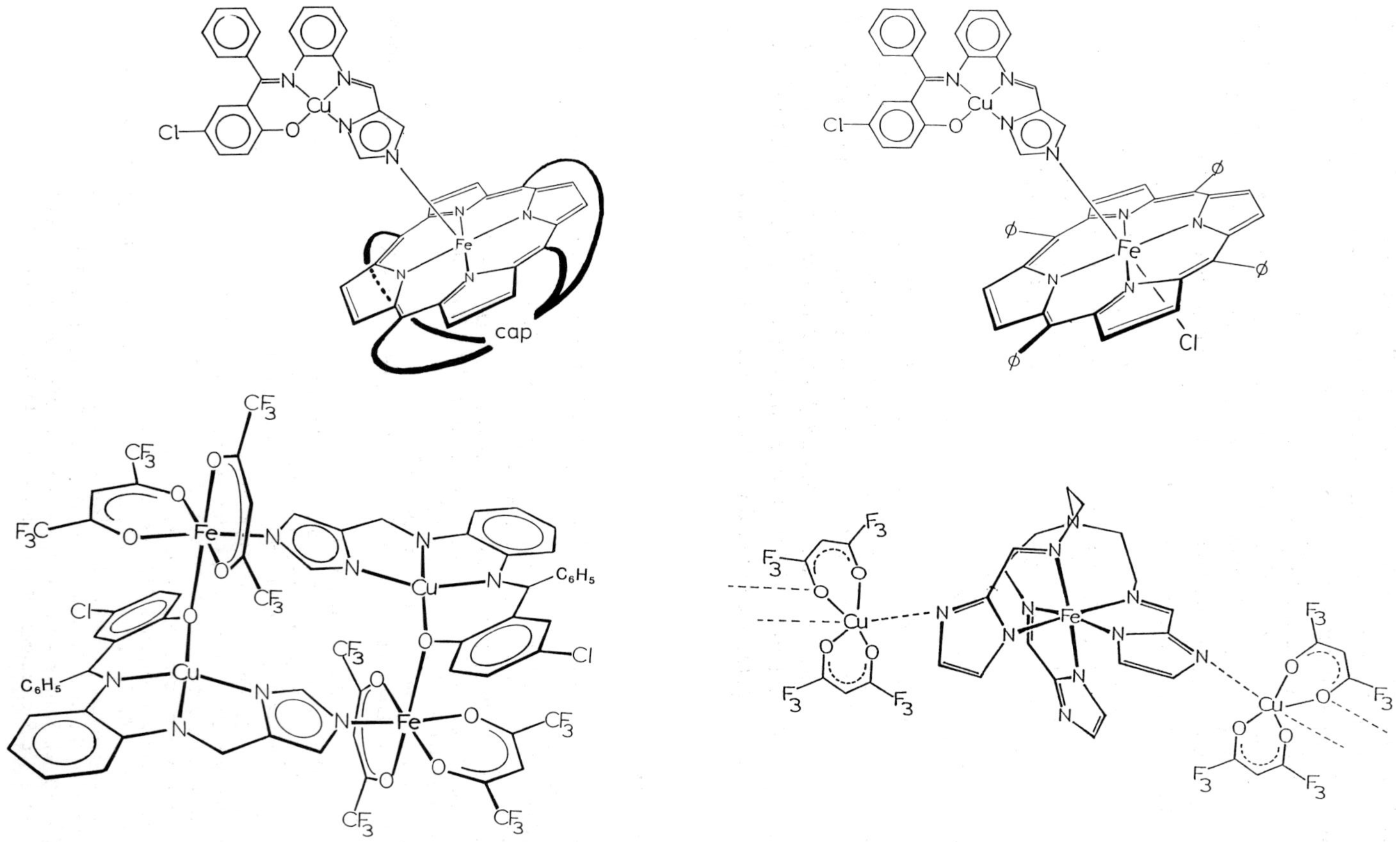

Figure 3.11 Some models for iron–copper metalloproteins.

in spin crossover complexes. The Fe–L bond length decreases significantly with the transformation from high spin to low spin. A range of measurements on ferric dithiocarbamates and some other iron(III) complexes (Healy and Sinn, 1975; Leipoldt and Coppens, 1973; Mitra *et al.*, 1976; Sim *et al.*, 1981; Sinn, 1974; Sinn *et al.*, 1978; Timken *et al.*, 1985) indicates a value of 0.12 Å. The analogous value is about 0.18 Å in Fe^{II} (Greenaway and Sinn, 1978; Greenaway *et al.*, 1979). This is interesting to compare with recently estimated sulphide crystal radii (Sharron, 1981):

6-coord.Fe^{IV} 0.59 Å
6-coord.Fe^{III} 0.595 Å, LS
6-coord.Fe^{III} 0.72 Å, HS
6-coord.Fe^{II} 0.55 Å, LS
6-coord.Fe^{III} 0.825 Å, HS
5-coord.Fe^{III} 0.62 Å, (LS?)
4-coord.Fe^{II} 0.555 Å, LS
4-coord.Fe^{II} 0.66 Å, HS

3.4.2 Fe and Cu bridging

There is considerable current interest in Fe–Cu bridging, engendered mainly by strong interactions between these metals in cytochrome oxidase. $Cu^{II}(en)_2[(Fe^{II}edta)_2O].2H_2O$ consists of bimetallic chains but with no magnetic coupling between the metals (Gomez-Romero *et al.*, 1988). Some examples of complexes exhibiting magnetic coupling not dealt with specifically in the bioinorganic section of this book, are shown in Figure 3.11.

References

Aronsson, B. Lundström, T. and Rundquist, S. (1965) *Borides, Silcides and Phosphides*, John Wiley and Sons, Inc. NY.
Baker, W.A. and Bobonich, H.M. (1964) *Inorg. Chem.*, **3**, 1184.
Bernal, I. (1973) *J. Cryst. Mol. Struct.*, **3**, 157.
Bjurstroem, T. (1933) *Arkiv. för Chemie, Mineralogi och Geologi*, **11**, 1.
Böhland, H. and Witzenhausen, U. (1991) *Z. anorg. allg. Chem.*, **600**, 131.
Boller, H. and Parthe, E. (1963) *Acta Cryst.*, **8**, 1439.
Bouchard, J.P. (1967) *Annales de Chemie*, 353.
Brewer, G.A., Butcher, R.J., Letafat, B. and Sinn, E. (1983) *Inorg. Chem.*, **22**, 371 and references cited.
Bruce, D.W. (1992) *Liq. Cryst.*, **11**, 127.
Burger, M.J. and Hahn, T. (1955) *Am. Miner.*, **40**, 226.
Burger, M.J. and Hahn, T. (1956) *Chem. Abs.*, **50**, 9239.
Burger, M.J. and Niizeki, N. (1957) *Z. Krist.*, **109**, 129, 161.
Burgess, B.K. (1990) *Chem. Rev.*, **90**, 1370.
Butcher, R.J. and Sinn, E. (1976) *J. Am. Chem. Soc.*, **98**, 2440, 5159.
Butcher, R.J. and Sinn, E. (1980) *Inorg. Chem.*, **19**, 3622.
Callmer, B. and Lundström, T. (1976) *J. Solid State Chem.*, **17**.
Cambi, L. and Malatesta, L. (1937) *Ber.*, **70**, 2067.
Chang, H.R., McCusker, J.K., Toftlund, H., Wilson, S.R., Trautwein, A.X., Winkler, H. and Hendrickson, D.N. (1990) *J. Am. Chem. Soc.*, 6814.
Champion, S.C. and Drickamer, H.G. (1969) *J. Chem. Phys.*, **51**, 4353.
Chasteen, N.D., Antanaitis, B.C. and Aisen, P. (1985) *J. Biol. Chem.*, **260**, 2926.
Chisnell, J.R. Premakumar, R. and Bishop, P.E. (1988) *J. Bacteriol.*, **170**, 27.

Cohen, M. (1987) in *Advanced Materials Research*, Pt 2, (eds P.A. Psaras and H.D. Langford), National Academy Press, Washington DC.
Colquhoun, H.M., Holton, J., Thompson, D.J. and Twigg, M.V. (1984) *New Pathways for Organic Synthesis*, Plenum, NY.
Conti, A.J., Xie, C.L. and Hendrickson, D.N. (1989) *J. Am. Chem. Soc.*, **111**, 1171.
Correia dos Santos, A. and da Costa, F.M. (1941) *Eur. J. Solid State Inorg. Chem.* **28**, 635.
Cotton, F.A. and Wilkinson, G. (1989) *Advanced Inorganic Chemistry* Interscience.
Coucouvanis, D., Hollander, F.J. and Pedelty, R. (1977) *Inorg. Chem.*, **16**, 2691.
Coucouvanis, D., Salifoglou, A., Kanatzidis, M.G., Dunham, W.R. and Simopoulos, A. (1988) *Inorg. Chem.*, **27**, 4066.
Coucouvanis, D., Simhon, E.D., Stremple, P. and Baenziger, N.C. (1980) *J. Am. Chem. Soc.*, **53**, 6646.
Cukauskas, E.J., Deaver, B.S. and Sinn, E. (1977) *J. Chem. Phys.*, **67**, 1257.
Cunningham, A.J., Fergusson, J.E., Powell, H.K.J., Sinn, E. and Wong, H. (1972) *J. Chem. Soc. Dalton Trans.*, 2155.
Dahl, E. (1969) *Acta Chem. Scand.*, **23**, 2677.
Dahl, L.F. de, Gil, E.R. and Feltham, R.D. (1969) *J. Am. Chem. Soc.*, **91**, 1653.
Doi, K., Antanaitis, B.C. and Aisen, P. (1988) *Structure and Bonding*, **70**, 1.
Dose, E.V., Hoselton, M.A., Sutin, N. and Tweedle, M.F. (1978) *J. Am. Chem. Soc.*, **100**, 1141.
Eldredge, P.A., Bryan, R.F., Sinn, E. and Averill, B.A. (1988) *J. Am. Chem. Soc.*, **110**, 5573.
Evain, M., Brec, R., Fichter, S. and Tributsch, M. (1987) *J. Sol. State Chem.*, **71**, 40.
Ewald, A.H., Martin, R.L., Sinn, E. and White, A.H. (1969) *Inorg. Chem.*, **8**, 1837.
Figgis, B.N. and Lewis, J. (1960) *Modern Coordination Chemistry* (eds J. Lewis and R.G. Wilkinson), Interscience, NY.
Flynn, C.H. Jr. (1984) *Chem. Rev.*, **84**, 31.
Fruchart, D., Chadouet, P., Fruchart, R., Roualt, A. and Senateur, J.P. (1984) *J. Sol. State Chem.*, **51**, 246.
Fujii, H. *et al.* (1977) *J. Phys. Soc. Japan*, **43**, 41.
Fung, S.C. and Drickamer, H.G. (1969) *J. Chem. Phys.*, **51**, 3305.
Fujii, H., Komura, S., Takeda, T., Okamoto, T., Ito, Y. and Akimitsu, J. (1979) *J. Phys. Soc. Japan*, **46**, 1616.
Garcia, E., Kwak, J. and Bard, A.J. (1988) *Inorg. Chem.*, **27**, 4377.
Garner, C.D., Arber, J.M., Hasnain, S.S., Dobson, B.R., Eady, R.R. and Smith, B.E. (1989) *Physica B*, **158**, 74.
Gerloch, M. (1983) *Magnetism and Ligand-field Analysis*, Cambridge University Press, Cambridge.
Giroud-Godovin, A.M. and Maitlis, P.M. (1991) *Angew. Chemie, Int. Ed. Engl.*, **30**, 375.
Gharini, E., Hbika, A., Dupre, B. and Gelitze, C. (1990) *Eur. J. Solid State Inorg. Chem.*, **27**, 647, and references cited.
Gleitzer, C. (1991) *Eur. J. Solid State Inorg. Chem.*, **28**, 77, and references cited.
Glidwell, C. and Johnson, I.L. (1988) *Polyhedron*, **7**, 1371.
Golding, R.M. (1972) *Applied Wave Mechanics*, van Nostrand.
Golding, R.M., Jackson, F. and Sinn, E. (1969) *Theor. Chim. Acta.*, **15**, 123.
Gomez-Romero, P., Jameson, G.B., Borras-Almenar, J.J., Escrva, E., Coronado, E. and Beltran, D. (1988) *J. Chem. Soc. Dalton*, 2747.
Greenaway, A.M., Schrock, A. and Sinn, E. (1979) *Inorg. Chem.*, **18**, 2692.
Greenaway, A.M. and Sinn, E. (1978) *J. Am. Chem. Soc.*, **100**, 8080.
Greenwood, N.N. and Earnshaw, A. (1986) *Chemistry of the Elements*, Pergamon.
Gütlich, P. and Hauser, A. (1989) *Pure Appl. Chem.*, **61**, 215.
Harris, C.M. and Sinn, E. (1968) *Inorg. Chim. Acta.*, **2**, 296.
Healy, P.C. and Sinn, E. (1975) *Inorg. Chem.*, 109.
Herbstein, F.H. and Smuts, J. (1964) *Acta Cryst.*, **17**, 1331.
Herzog, S. and Weber, A. (1968) *Z. Chem.*, **8**, 66.
Heydlauf, H. (1969) *Eur. J. Pharmacol.*, **6**, 340.
Hoggins, J.T. and Steinfink, H. (1977) *Acta Cryst.*, **B33**, 673.
Hulliger, F. (1968) *Structure and Bonding*, **4**, 83.
Hulliger, F. and Vetsch, H. (1990) *Eur. J. Solid State Inorg. Chem.*, **27**, 739.
Ijjaali, M., Malaman, B., Gleitzer, C., Warner, J.K., Hriljac, J.A. and Cheetham, A.K. (1990) *J. Solid State Chem.*, **86**, 195.

Jack, K.H. (1948) *Proc. Roy. Soc. London A*, **195**, 34 and references cited.
Jack, K.H. (1951) *Proc. Roy. Soc. London A*, **208**, 200.
Jack, K.H. (1952) *Acta Cryst.*, **5**, 404.
Jeitschko, W. and Braun, D.J. (1978) *Acta Cryst.*, **34**, 3196.
Johnston, W.D., Miller, R.C. and Damon, D.H. (1965) *J. Less-Common Metals*, **8**, 272.
Kahn, O. (1987) *Structure and Bonding*, **68**, 89.
Kanatzidis, M.G., Tonge, L.M., Marks, T.J., Marcy, M.O. and Kennewurf, C.R. (1987) *J. Am. Chem. Soc.*, **109**, 3797.
Kollmar, C., Couty, M. and Kahn, O. (1991) *J. Am. Chem. Soc.*, 7994.
König, E. and Chakravarty, A.S. (1967) *Theor. Chim. Acta*, **9**, 151.
König, E. Chakravarty, A.S. and Madeja, K. (1967) *Theor. Chim. Acta*, **9**, 171.
König, E. and Madeja, K. (1966) *Chem. Commun.*, 61.
Krebs, H. (1950) *Z. anorg. allg. Chem.*, **263**, 175.
Leipoldt, L.G. and Coppens, P. (1973) *Inorg. Chem.*, **12**, 2269.
Lippard, S.J. (1986) *Chem. in Britain*, 222.
Malaman, B., Ijjaali, M., Venturini, G., Gleitzer, C. and Soubeyroux, J.L. (1991) *Eur. J. Solid State Inorg. Chem.*, **28**, 519.
Martin, R.L., Rhode, N.M., Robertson, G.B. and Taylor, D. (1974) *J. Am. Chem. Soc.*, **96**, 3647.
Martin, L.L., Hagen, K.S., Hauser, A., Martin, R.L. and Sargeson, A.M. (1988) *J. Chem. Soc. Chem. Comm.*, 1313.
Millet, J.M.M. and Mentzen, B.F. (1990) *Eur. J. Solid State Inorg. Chem.*, **27**, 739.
Minomura, K.S. and Drickamer, H.G. (1962) *J. Phys. Chem. Solids*, **23**, 451.
Mitra, S., Raston, C.L. and White, A.H. (1976) *Australian J. Chem.*, **29**, 1899.
Mockler, G.M., O'Connor, C.J., deJersey, J., Zerner, B. and Sinn, E. (1983) *J. Am. Chem. Soc.*, **105**, 1891.
Nagakura, S. and Toyoshima, M. (1979) *Trans. Jap. Inst. Metals.*, **19**, 100.
Nelson, S.M., Hawker, P.N. and Twigg, M.V. (1987) *Comprehensive Coordination Chemistry, Vol 4, Middle Transition Elements*, (eds G. Wilkinson, R.D. Gillard and J.A. McCleverty).
Nicholls, D. (1973) Chemistry of Iron Ch. 40 in *Comprehensive Inorganic Chemistry*, (eds J.C. Bailar, Jr, H.J. Emeleus, R.S. Nyholm and A.F. Trotman-Dickenson), Pergamon Press, Oxford.
Pasek, E.A. and Straub, D.K. (1977) *Inorg. Chim. Acta*, **21**, 23.
Pinsker, Z.G. and Kaverin, S.V. (1954) *Doklady Akad Nauk SSSR*, **96**, 519.
Poganiuch, P., Decurtins, S. and Gütlich, P. (1990) *J. Am. Chem. Soc.*, **112**, 3270.
Porterfield, W.W. (1984) *Inorganic Chemistry, Unified Approach.* Addison-Wesley, Reading, Mass.
Raymond, K.N., Müller, G. and Matzanke, B.F. (1986) *Current Topics in Chemistry*, **123**, 649.
Saleem, M.M.M. and Wilson, M.T. (1988) *Inorg. Chim. Acta*, **153**, 93, 99 and references cited.
Scholder, R., Bunsen, H.V. and Zeiss, R.W. (1956) *Z. anorg. allg. Chem.*, **383**, 330.
Selte, K. and Kjekshus, A. (1972) *Acta Chem. Scand.*, **26**, 1276.
Shannon, R.D. (1981) Ch 16 in *Structure and Bonding in Crystals, Vol II*, (eds M.F. O'Keefe and A. Navrotsky), Acad. Press, NY.
Sim, P.G. and Sinn, E. (1978) *Inorg. Chem.*, **17**, 1288.
Sim, P.G. and Sinn, E. (1981) *J. Am. Chem. Soc.*, **103**, 241.
Sim, P.G., Sinn, E., Petty, R.H., Merrill, C.L. and Wilson, L.J. (1981) *Inorg. Chem.*, **20**, 1213.
Sinn, E. (1967) 37th A.N.Z.A.A.S. Congress, Melbourne, Australia, January, 1967.
Sinn, E. (1970) thesis.
Sinn, E. (1974) *Coord. Chem. Rev.*, **9**, 185.
Sinn, E. (1976) *Inorg. Chem.*, **15**, 1976.
Sinn, E., Sim, P.G., Dose, E.V., Tweedle, M.F. and Wilson, L.J. (1978) *J. Am. Chem. Soc.*, **100**, 3375.
Sinn, E., O'Connor, C.J., deJersey, J. and Zerner, B. (1983) *Inorg. Chim. Acta*, L13 and references cited.
Smith, J.D. (1973) The chemistry of arsenic, antimony and bismuth, Ch 21 of *Comprehensive Inorganic Chemistry*, (eds J.C. Bailar, Jr, H.J. Emeleus, R.S. Nyholm, and A.F. Trotman-Dickenson) Pergamon, Oxford.
Stiller, A.H., McCormick, B.J., Russel, P. and Montano, P.A. (1978) *J. Am. Chem. Soc.*, **100**, 2553.
Sugitan, M., Kinomura, N., Koizumi, M. and Kume, S. (1978) *J. Sol. State Chem.*, **26**, 195.
Timken, M.D., Hendrickson, D.N. and Sinn, E. (1985) *Inorg. Chem.*, 3947.

Thomas, J.T., Robertson, J.H. and Cox, E.G. (1958) *Acta Cryst.*, **11**, 599.
Torardi, C.C., Reiff, W.M. and Takacs, L. (1989) *J. Solid State Chem.*, **82**, 203.
Vegas, A., Romero, A. and Martinez-Ripol, M. (1990) *J. Solid State Chem.*, **fB 88**, 594.
Vegas, A., Romero, A. and Martinez-Ripol, M. (1991) *Acta Cryst.*, **B47**, 17.
Wells, A.F. (1975) *Structural Inorganic Chemistry*, Clarendon Press, Oxford.
Wilkins, P.C. and Wilkins, R.G. (1987) *Coord. Chem. Rev.*, **79**, 195.
Woodland, M.P., Daulat, S.P., Cammack, R. and Dalton, H. (1986) *Biochim. Biophys. Acta*, **873**, 237.
Yakubovich, O.V., Urusov, V.S., Frenzen, G., Massa, W. and Babel, D. (1990) *Eur. J. Solid State Inorg. Chem.*, **27**, 467.

4 Organo-iron compounds

P.L. PAUSON

4.1 Introduction

Organo-iron compounds have an important place in organotransition metal chemistry, both historically and in respect of their variety and utility.

The strict definition of organometallic is that the term denotes all compounds possessing a direct metal–carbon bond. It thus includes the cyanides which are treated in the preceding chapter, but it should be noted that ferricyanide-derived pigments of the Prussian Blue (or Berlin Blue) type, known since 1710, are undoubtedly the longest known organometallic compounds of iron.

Historically iron featured in many 'firsts' in the organometallic field. Freund (1888) alkylated ferrocyanide to obtain the first metal isocyanide complex $Fe(CN)_2(CNEt)_4$. Mond, who had described his first metal carbonyl, $Ni(CO)_4$, in the same year, prepared his second, $Fe(CO)_5$ in 1891. Along with the di- (**1**) and trinuclear carbonyls (**2**), it was to play a key role in later work. *Inter al.* it yielded the first hydrocarbon-iron complex (3) when treated with butadiene by Reihlen and coworkers in 1930, but the structure of this product

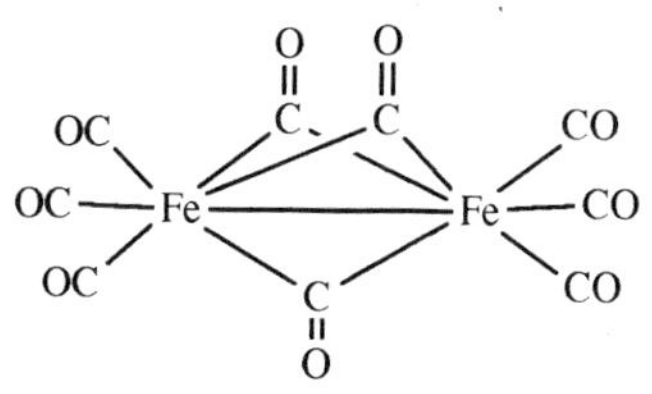

(1)

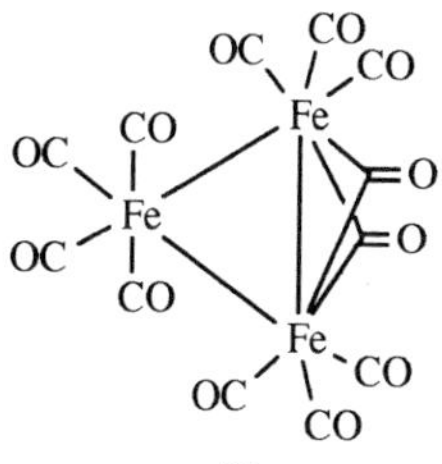

(2)

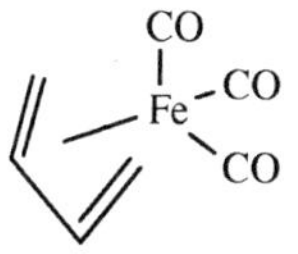

(3)

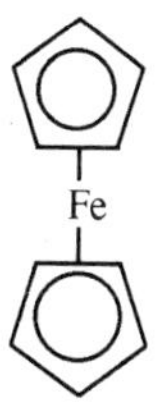

(4)

remained unknown until re-investigated in the late 1950s. The nonacarbonyl (**1**), obtained by photolysis of $Fe(CO)_5$, was the first metal carbonyl to have its structure determined by X-ray crystallography (Powell and Ewens, 1939). It was the improvement in methods of structure determination and the much enhanced understanding of chemical bonding, notably Dewar's concept of π-bonding, which permitted the explosive growth of the subject which followed the discovery of ferrocene (dicyclopentadienyliron) (**4**) in 1951. Moreover, ferrocene was not only the first of the now vast range of cyclopentadienylmetal complexes, but also the first organometallic compound with typical aromatic character. It owes much of its rich chemistry to the ease of electrophilic substitution.

4.2 Iron carbonyls

Ultimately all organo-iron compounds must derive from the metal or its simple salts, but the great majority of preparations involve transformations of other organo-iron compounds. The carbonyls serve as the most readily available and also conveniently reactive intermediates for the preparation of hydrocarbon complexes with the result that mixed hydrocarbon-iron carbonyls are known in much greater variety than 'pure' hydrocarbon complexes or complexes with other counterligands. It is convenient therefore to outline key features of iron carbonyl chemistry before dealing with other types of compounds.

The pentacarbonyl (**5**), an orange liquid, b.p. 103°C, m.p. −20°C, is readily obtained from the metal and carbon monoxide. Although its structure is that of a trigonal bipyramid, its room temperature ^{13}C NMR spectrum shows

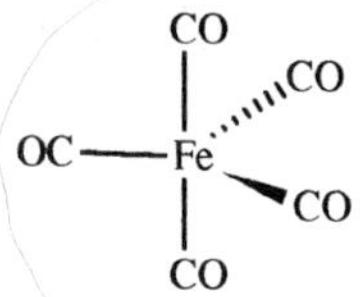

(**5**)

only a single signal, implying rapid interchange between the axial and equatorial carbonyl groups, apparently by a Berry pseudorotation mechanism. Photolysis cleaves one of the carbonyl groups and the remaining, coordinatively unsaturated fragment, $Fe(CO)_4$, combines with a molecule of the pentacarbonyl (equation 4.1) to give the binuclear nonacarbonyl (**1**). The latter, obtained in the form of beautiful golden platelets, differs from all the other simple hydrocarbon-soluble metal carbonyls in being almost completely insoluble in all common solvents. (Its molecular weight was determined using a solution in $Fe(CO)_5$.) Nevertheless, it is the most reactive of the three iron carbonyls and the behaviour of its suspensions, even in hydrocarbon solvents, suggests that it readily releases the $Fe(CO)_4$ fragment by reversal of equation 4.1.

$$Fe(CO)_4 + Fe(CO)_5 \rightleftharpoons Fe_2(CO)_9 \quad (4.1)$$

The colourless anionic species, $Fe(CO)_4^{2-}$ and $HFe(CO)_4^-$ are obtained from $Fe(CO)_5$ either with, e.g., sodium amalgam by reduction with loss of carbon monoxide (equation 4.2), or with bases by disproportionation generating carbon dioxide (e.g. equation 4.3) and/or Fe^{2+} (e.g. equation 4.4). There is evidence for initial disproportionation of $Fe(CO)_5$ in amines to give $[Fe(CO)_6]^{2+}[Fe(CO)_4]^{2-}$. Protonation of these salts yields the acidic hydrocarbonyl, $H_2Fe(CO)_4$ ($K_1 = 4 \times 10^{-5}$; $K_2 = 4 \times 10^{-14}$) and oxidation of any of these species (e.g. with MnO_2) gives the green-black trinuclear carbonyl $Fe_3(CO)_{12}$ (**2**). Like $Fe(CO)_5$ this compound is fluxional in solution; its sharp singlet ^{13}C NMR signal persists even at $-150°C$.

$$Fe(CO)_5 + 2Na(Hg) \rightarrow CO + Na_2Fe(CO)_4 \quad (4.2)$$

$$Fe(CO)_5 + 3NaOH \xrightarrow{MeOH} NaHFe(CO)_4 + Na_2CO_3 + H_2O \quad (4.3)$$

$$3\,Fe(CO)_5 + 6NH_3 + H_2O \rightarrow [Fe(NH_3)_6]^{2+}[HFe(CO)_4^-]_2 + 6CO + CO_2 \quad (4.4)$$

Equations 4.2–4.4 represent the simplest processes, but depending on the precise conditions and reactants used, reduction or base-catalysed disproportionation will also convert neutral iron carbonyls to anions of higher or lower nuclearity, for example:

$$3Fe(CO)_5 + NR_3 + 2H_2O \rightarrow [R_3NH][HFe_3(CO)_{11}] + 2CO_2 + 2CO + H_2 \quad (4.5)$$

$$Fe(CO)_5 + py \rightarrow [Fe(py)_6][Fe_4(CO)_{13}] \quad (4.6)$$

$$Fe_3(CO)_{12} + 6Na \xrightarrow{THF} 3Na_2Fe(CO)_4 \quad (4.7)$$

As with the neutral clusters, colour increases with nuclearity: the simplest anion, $Fe(CO)_4^{2-}$ is colourless, $Fe_2(CO)_8^{2-}$ is orange-red, $Fe_3(CO)_{11}^{2-}$ is dark red, and $Fe_4(CO)_{13}^{2-}$ is brownish-black.

The above routes to the polynuclear carbonyl anions undoubtedly involve reactions of neutral carbonyls with anions of lower nuclearity. An example is the formation of the tetra-iron species:

$$Fe_3(CO)_{11}^{2-} + Fe(CO)_5 \xrightarrow[THF]{25°C} Fe_4(CO)_{13}^{2-} + 3CO \quad (4.8)$$

Closely related reactions involving iron carbonyl anions and carbonyls of other metals (or iron carbonyls plus carbonyl anions of other metals) provide routes to a diversity of mixed metal carbonyls. These may involve metals in the same (e.g. equation 4.9) (Geoffroy, 1980) or in different groups (e.g. equations 4.10–4.12) and may involve metal–metal addition (e.g. equations 4.9–4.12) or substitution (equation 4.13). Equation 4.14 probably effectively involves $Co(CO)_4^-$ formed by solvent-induced disproportionation of $Co_2(CO)_8$.

$$RuOs_2(CO)_{12} + Fe(CO)_4^{2-} \rightarrow [FeRuOs_2(CO)_{13}]^{2-} \xrightarrow{H^+} H_2FeRuOs_2(CO)_{13} \quad (4.9)$$

$$Mn(CO)_5^- + Fe_2(CO)_9 \rightarrow [Fe_2Mn(CO)_{12}]^- \quad (4.10)$$

$$Mn(CO)_5^- + Fe(CO)_5 \xrightarrow{h\nu} [(OC)_4Fe{-}Mn(CO)_5]^- \quad (4.11)$$

$$Co(CO)_4^- + Fe(CO)_5 \xrightarrow{h\nu} [(OC)_4Fe(\mu\text{-}CO)Co(CO)_3]^- \quad (4.12)$$

$$Co(CO)_4^- + H_2FeRu_3(CO)_{13} \rightarrow [CoFeRu_2(CO)_{13}]^- \quad (4.13)$$

$$Co_2(CO)_8 + Fe(CO)_5 + Me_2CO \rightarrow [Co(Me_2CO)_n]^- [FeCO_3(CO)_{12}]_2 \xrightarrow{Et_4NI} [Et_4N][FeCo_3(CO)_{12}] \quad (4.14)$$

Depending on the reactivity of the carbonyl used, photochemical activation may be necessary (equations 4.11 and 4.12). Reactions between neutral carbonyls may also involve photochemical cleavage of a carbonyl-metal bond (e.g. equation 4.15). Other ligands beside CO may be displaced and this allows metals like palladium and platinum, which do not form simple carbonyls, to give mixed metal carbonyls (equations 4.16–4.19).

$$Fe(CO)_5 + Re_2(CO)_{10} \xrightarrow[\text{hexane}]{h\nu} (OC)_5Re{-}Fe(CO)_4{-}Re(CO)_5 \quad (4.15)$$

$$Fe_3(CO)_{12} + Pt(PPh_3)_4 \rightarrow (OC)Pt(PPh_3)[Fe(CO)_4]_2 \text{ (triangle: OC, } PPh_3 \text{ on Pt; } (OC)_4Fe{-}Fe(CO)_4) + Fe_2Pt(PPh_3)_2(CO)_8 \quad (4.16)$$

$$Fe(CO)_5 + Pt(C_8H_{12})_2 \rightarrow \underset{(\mathbf{6})\ (9\%)}{Pt_3Fe_3(CO)_{15}} + \underset{(\mathbf{7})\ (3\%)}{PtFe_2(CO)_8(C_8H_{12})} + \underset{(\mathbf{8})\ (40\%)}{Pt_5Fe_2(CO)_{12}(C_8H_{12})_2} \quad (4.17)$$

$$Fe_3(CO)_{11}^{2-} + 2M^{2+} \xrightarrow{MeCN/25^\circ C} [Fe_4M(CO)_{16}]^{2-} \quad (M = Pd, Pt)\ (\mathbf{9}) \quad (4.18)$$

$$[Fe_4(CO)_{13}]^{2-}[Me_3N^+CH_2Ph]_2 + K_2PdCl_4 \xrightarrow{THF} \underset{(\mathbf{10})}{[Fe_6Pd_6(CO)_{24}]^{4-}} \quad (4.19)$$

The structures of the products of equations 4.17 (Adams *et al.*, 1989) and 4.18 (Longoni *et al.*, 1980) are shown. In the palladium complex (**9**, M = Pd)

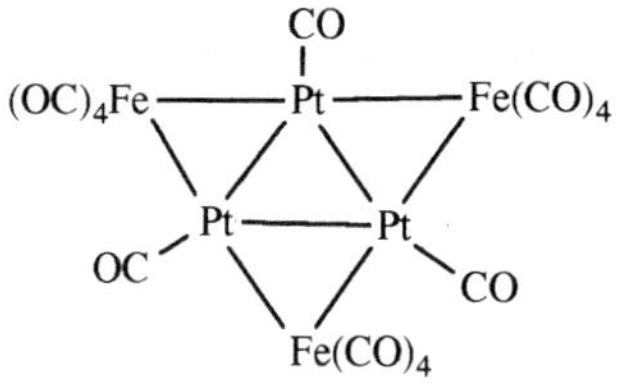

(6)

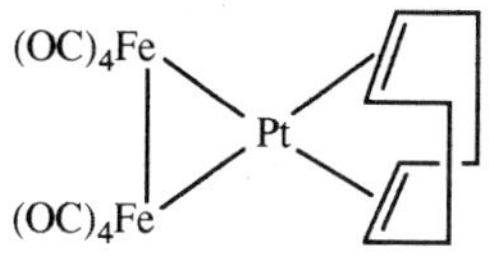

(7)

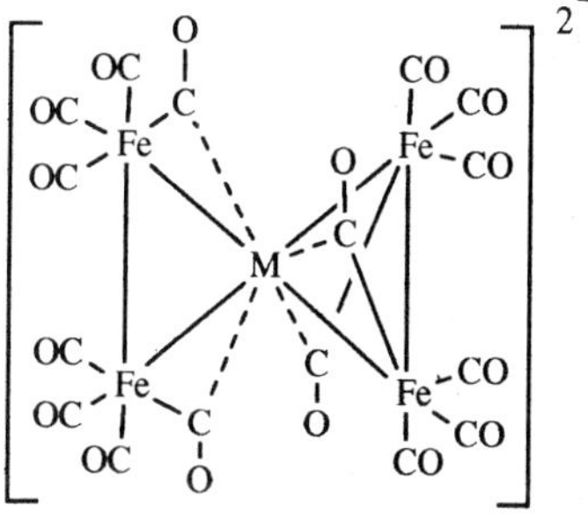

(9)

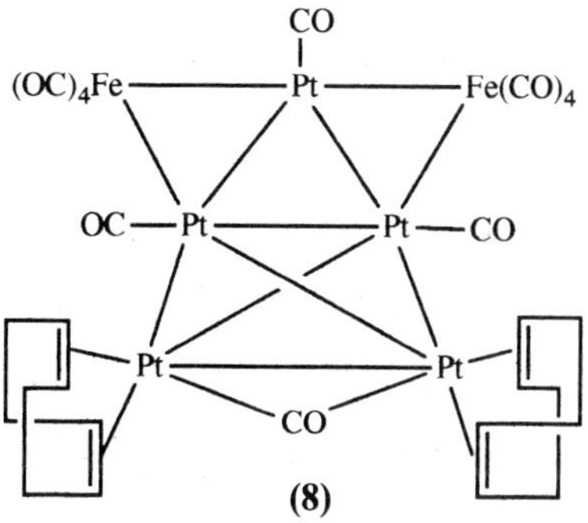

(8)

the Fe_4Pd system is effectively planar, while in the platinum analogue the two Fe_4Pt planes are twisted by 7°; in both cases four of the carbonyls are semibridging so that the metal M is surrounded by a tetrahedron of carbon atoms. The structure of the anion (**10**) is based on an octahedron of palladium atoms with each of the six lateral faces capped by an $Fe(CO)_2$ group; in addition each iron and each palladium has one doubly bridging CO group (connecting the two near-planar Fe_3Pd_3 groups) and six triply bridging CO groups cap the $FePd_2$ triangles.

In contrast to the photochemical reaction, equation 4.11, thermal reaction of $Mn(CO)_5^-$ with pentacarbonyliron, like that with nonacarbonyldiiron (equation 4.10) leads rapidly (refluxing diglyme, 5 min) to the anion $Fe_2Mn(CO)_{12}^-$, but prolonged reaction (1 h) gives, instead, the interesting carbonyl carbide cluster (**11**) (Churchill and Wormald, 1974), conveniently

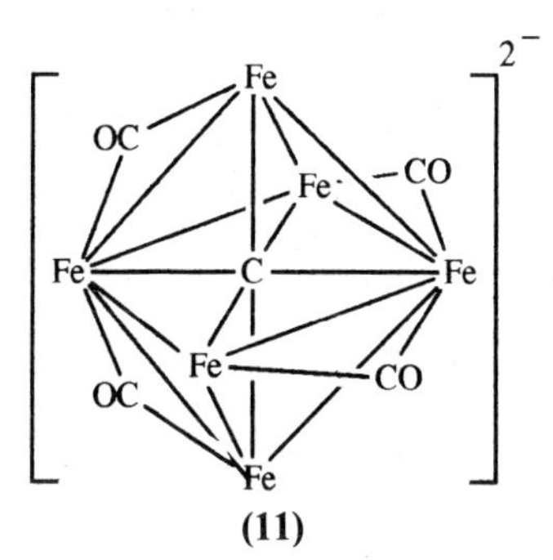

(11)

isolated as the tetraethylammonium salt. Other carbonyl anions [$Co(CO)_4^-$, $Fe(CO)_4^{2-}$, $V(CO)_6^-$ also yield this octahedral cluster (**11**), but surprisingly $(C_5H_5)Mo(CO)_3^-$ converts $Fe(CO)_5$ into a different cluster ion, $[Fe_5(CO)_{14}C]^{2-}$ (Hsieh and Mays, 1972). An Fe_4 carbonyl carbide cluster anion (**13**) and the neutral carbide (**12**) result via oxidative degradation from the Fe_6 species by the reactions in equations 4.20 and 4.21.

$$\underset{(\mathbf{11})}{[Fe_6(CO)_{16}C]^{2-}} \xrightarrow{Fe^{3+}/MeOH} [Fe_4(CO)_{12}C(COOMe)]^- \underset{MeOH}{\overset{CE_3SO_3H}{\rightleftharpoons}} \underset{(\mathbf{12})}{Fe_4(CO)_{13}C} \qquad (4.20)$$

$$[Fe_4(CO)_{12}C(COOMe)]^- + BH_3 \rightarrow [HFe_4(CO)_{12}C] \xrightarrow{KOH} \underset{(\mathbf{13})}{[Fe_4(CO)_{12}C]^{2-}} \qquad (4.21)$$

The tetranuclear carbide anion (**13**) is also obtained by the reaction sequence:

$$[Fe_4(CO)_{13}]^{2-} \xrightarrow{AcCl} [Fe_4(CO)_{12}COCOMe]^- \xrightarrow{Na} [Fe_4C(CO)_{12}]^{2-} \qquad (4.22)$$

Treatment of the tricarbonylnitrosyliron anion with dodecacarbonyltriiron yields the nitrene-bridged tetranuclear anion (**14**) according to equation 4.23 (Fjare and Gladfelter, 1981).

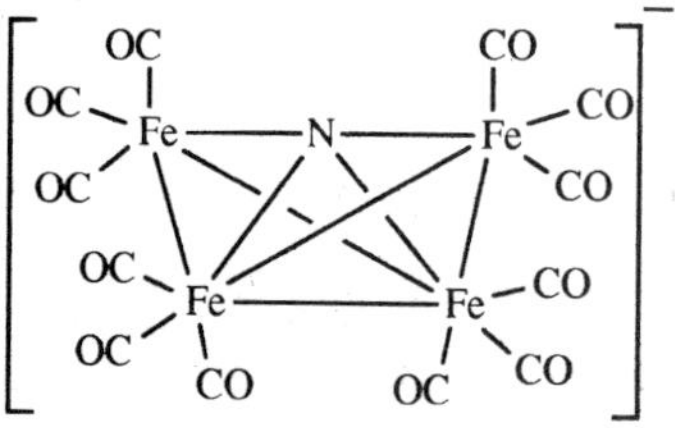

(**14**)

$$[Fe_3(CO)_{12} + [PPN][Fe(CO)_3NO] \underset{50\%}{\xrightarrow{THF}} [PPN][Fe_4N(CO)_{12}] + CO_2$$

$$(PPN = Ph_3PNPPh_3) \qquad (4.23)$$

Several related carbene-bridged mixed-metal carbonyl carbide anions, e.g. the manganese complex (**17**), have been obtained from $Fe_3(CO)_{11}^{2-}$ (**15**) via the ketenylidene complex (**16**) as shown in Scheme 4.1 (Hriljac *et al.*, 1985), while protonation of the same intermediate (**16**) yields the methylidene-bridged anion (**18**) (Kolis *et al.*, 1982).

Replacement of carbonyl groups by other donor ligands leads to a vast array of substituted carbonyls. The ligands most closely similar to carbon

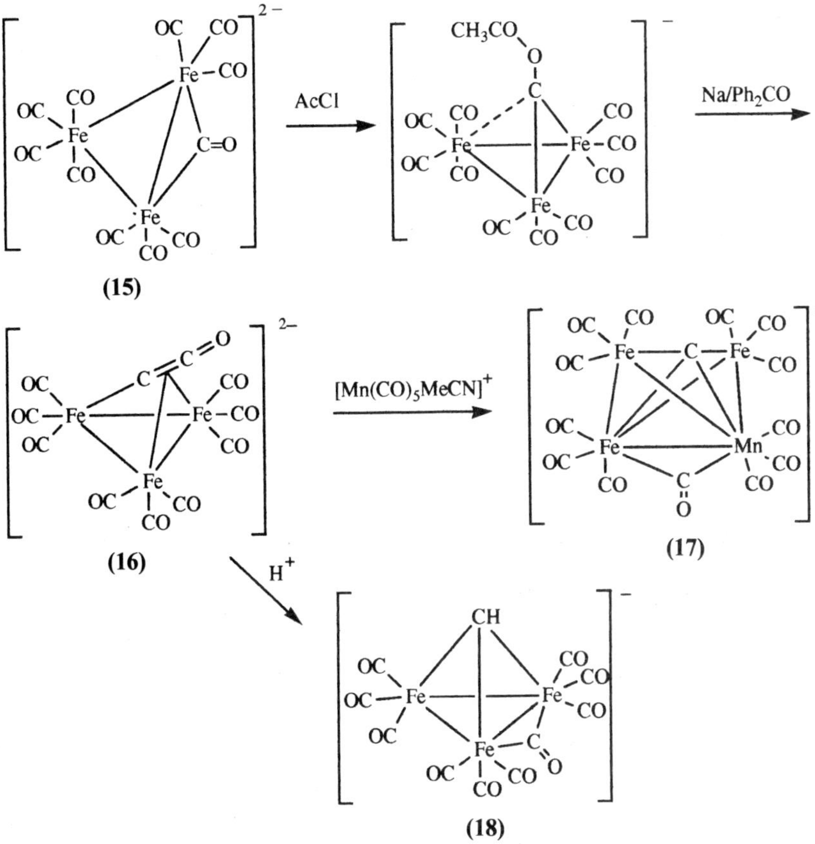

AcCl

Na/Ph$_2$CO

(15)

[Mn(CO)$_5$MeCN]$^+$

(16)

H$^+$

(17)

(18)

Scheme 4.1

monoxide are carbonyl sulphide, CS, and the isocyanides, CNR. Since the former is unstable in the free state its complexes must be prepared by indirect methods and relatively few examples are known. The simplest, $Fe(CO)_4CS$, a yellow liquid like $Fe(CO)_5$, is obtained in poor yield from the salt $Na_2Fe(CO)_4$ by treatment with thiophosgene, $CSCl_2$. Depending somewhat on R (methyl, tert-butyl and various aryl isocyanides are most commonly used), each CO group of $Fe(CO)_5$ can be replaced successively by CNR to yield all the compounds $Fe(CO)_{5-n}(CNR)_n$ where $n = 1$–5 (e.g. for $R = 2,6\text{-}Me_2C_6H_3$: *Inorganic Syntheses*, **26**, 53). That successive steps in this sequence become more difficult is attributed to the fact that isocyanides are better donors but poorer acceptors of electrons than carbon monoxide, so that after each replacement the remaining carbonyl groups are more firmly held. For most other ligands this difference is greater so that replacement of more than two carbonyl groups is rarely observed. Thus for most phosphines ($L = PR_3$) only $Fe(CO)_4L$ and $Fe(CO)_3L_2$ are formed directly from $Fe(CO)_5$

and other members of the series $Fe(CO)_{5-n}L_n$ must be obtained by alternative routes. Phosphorus trihalides and to a lesser extent phosphites, $P(OR)_3$, are relatively better acceptors and replace CO groups more easily than phosphines.

While $Fe(CO)_5$ reacts with these and many other ligands on heating, the mildest and frequently most convenient route uses $Fe_2(CO)_9$ at room temperature:

$$Fe_2(CO)_9 + L \rightarrow Fe(CO)_4L + Fe(CO)_5 \qquad (4.24)$$

Detailed procedures using these carbonyls or $Fe_3(CO)_{12}$ and also making use of catalysis by cobalt(II) chloride are available in *Inorganic Syntheses* (**8**, 185; **16**, 64; **26**, 52, 59, 353). Pyridine and related bases also yield stable, but air-sensitive complexes ($Fe(CO)_4L$, e.g. $L = C_5H_5N$) and the trimethylamine complex of this type results from the redox reaction (Elzinga and Hogeveen, 1977):

$$Fe(CO)_5 + Me_3NO \rightarrow CO_2 + Fe(CO)_4NMe_3 \qquad (4.25)$$

Reaction of pentacarbonyliron with alkaline solutions of sodium nitrite yields the salt (**19**) of the red, tetrahedral tricarbonylnitrosyliron anion, isoelectronic with $Co(CO)_4^-$ and $Ni(CO)_4$ (equation 4.26). Its reaction

$$Fc(CO)_5 + NaNO_2 + NaOH \rightarrow \underset{(\mathbf{19})}{Na[Fe(CO)_3NO]} + CO + NaHCO_3 \qquad (4.26)$$

with acid and a second molecule of nitrite (equation 4.27) or reaction of the higher carbonyls with nitric oxide yields the likewise isoelectronic dicarbonyldinitrosyliron (**20**), a deep red low-melting solid.

$$Na[Fe(CO)_3NO] + NaNO_2 + 2HOAc \rightarrow \underset{(\mathbf{20})}{Fe(CO)_2(NO)_2} + CO + 2NaOAc + H_2O \qquad (4.27)$$

Iodine or bromine oxidise pentacarbonyliron to the *cis*-tetracarbonyldihalo compounds (**21**; X = I or Br), and mercury(II) halides similarly yield the *bis*(halomercuri) derivatives (**21**; X = HgCl or HgBr).

Reaction of the pentacarbonyl with potassium cyanide in methanol converts it to the yellow tetracarbonylcyanoferrate ion (**22**) in which the cyano group has replaced one of the axial carbonyl groups. The octahedral

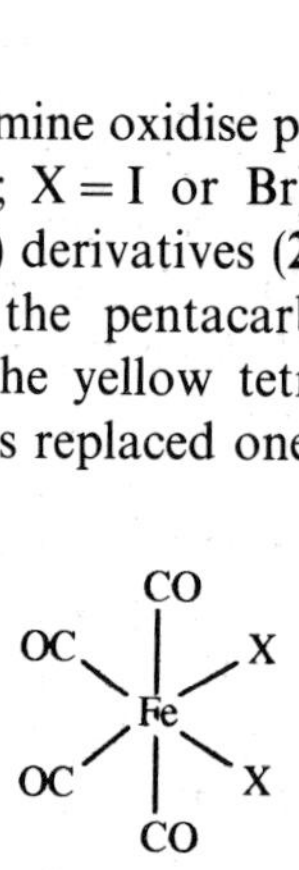

(**21**) 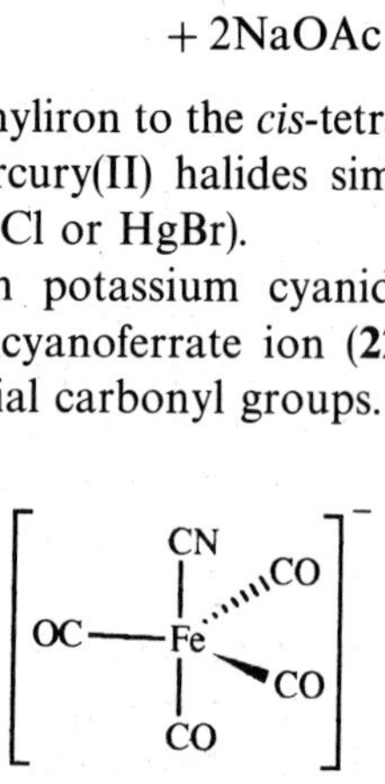 (**22**)

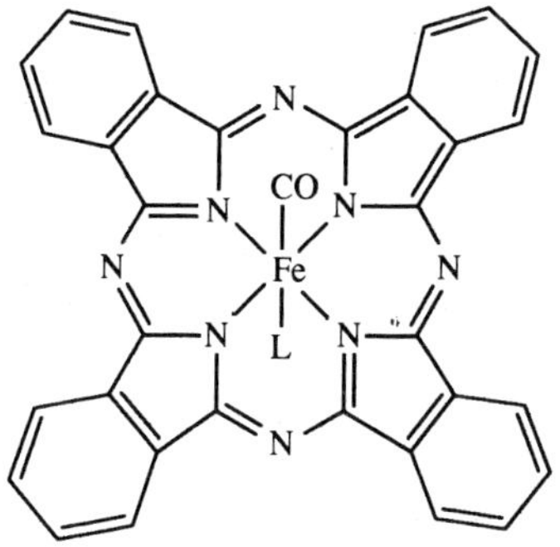

L = O=CHNMe$_2$

(23)

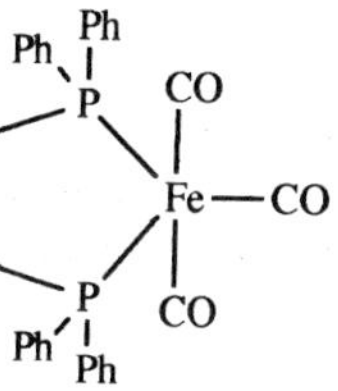

(24)

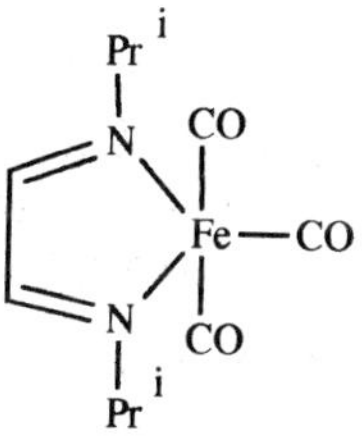

(25)

$[Fe(CN)_5CO]^{3-}$ ion is formed when carbon monoxide reacts with aquopentacyanoferrate, $[Fe(CN)_5OH_2]^{3-}$

When pentacarbonyliron is heated (110–140°C) with *o*-dicyanobenzene in dimethylformamide it gives in high yield (70%) the carbonylphthalocyanine-iron complex (**23**) (Calderazzo *et al.*, 1980) and of course the affinity of haemoglobin for carbon monoxide to give a similarly coordinated carbonyl complex is well known to be responsible for the toxicity of CO (see chapter 6).

Simpler chelating ligands commonly replace two carbonyl groups of $Fe(CO)_5$ as in the 'diphos' complex (**24**) and the diazadiene complex (**25**) (tom Dieck and Orlopp, 1975), both obtainable from the respective ligand with $Fe_2(CO)_9$ at room temperature or with $Fe(CO)_5$ on irradiation.

Two or more iron atoms or iron and other metal atoms may be held together by various atoms or groups including carbon, nitrogen, oxygen, phosphorus and sulphur ligands yielding di- and polynuclear carbonyl complexes. Thus in contrast to the formation of the chelated bis(diphenyl-phosphino) ethane complex (**24**), the lower homologue, $Ph_2PCH_2PPh_2$, reacts with $Fe_2(CO)_9$ to give the *bis*(diphenylphosphino)methane bridged complex (**26**) (Cotton and Troup, 1974). The related phosphido-bridged complex (**27**) forms via $Fe(CO)_4PPh_2H$ on treating $Fe(CO)_5$ or $Fe_2(CO)_9$ with diphenyl-

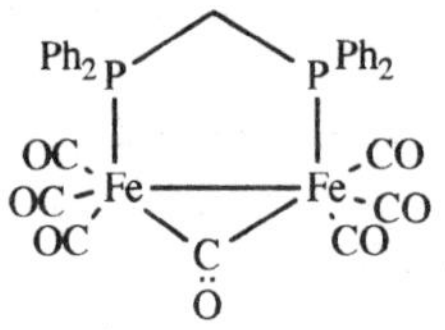

(26) 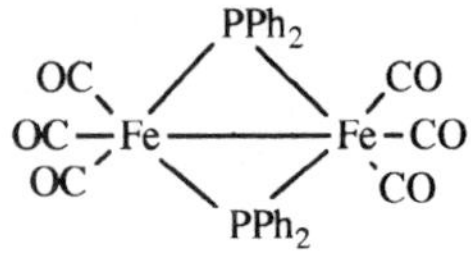 (27) 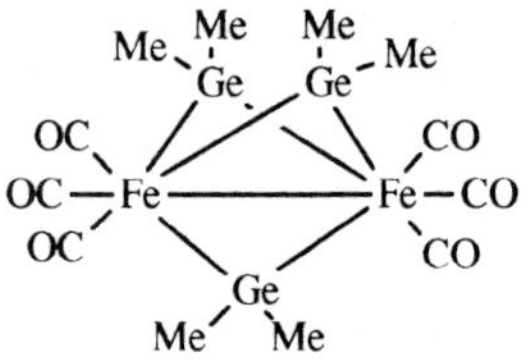 (28)

phosphine (Treichel *et al.*, 1972) or, more efficiently, from $Na_2Fe_2(CO)_8$ with chlorodiphenylphosphine, PPh_2Cl (Collman *et al.*, 1977), while the germanium complex (**28**) was prepared from $Fe_3(CO)_{12}$ with excess of dimethylgermane (Elder and Hall, 1969). The effect of the size of the bridging group on the length of the iron–iron bond is evident from a comparison of the last three complexes (2.71 Å in (**26**), 2.62 Å in (**27**) and 2.75 Å in (**28**)) with the value for $Fe_2(CO)_9$ (2.46 Å) and with the Fe–Fe distances in related nitrogen-bridged complexes (2.39 Å in (**29**), 2.47 Å in (**30**), 2.50 Å in (**31**) and

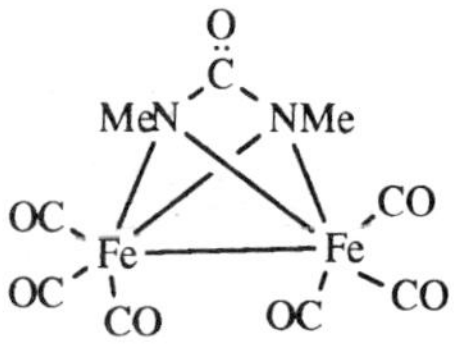

(29) 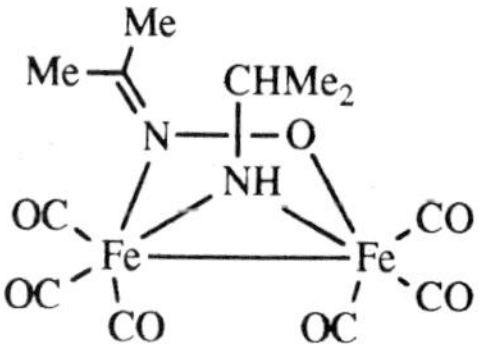 (30)

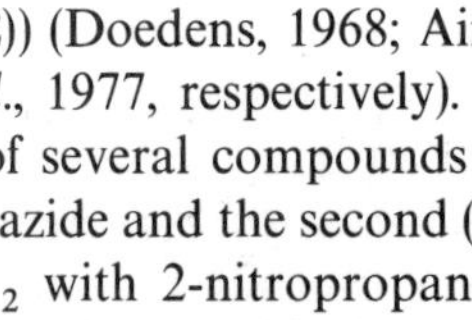

(31) 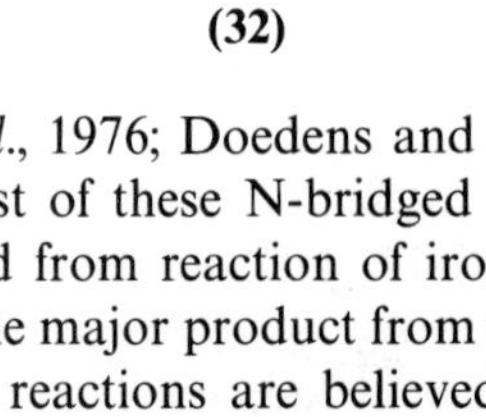 (32)

2.57 Å in (**32**)) (Doedens, 1968; Aime *et al.*, 1976; Doedens and Ibers, 1969; Cotton *et al.*, 1977, respectively). The first of these N-bridged compounds (**29**) is one of several compounds isolated from reaction of iron carbonyls with methyl azide and the second (**30**) is the major product from the reaction of $Fe_3(CO)_{12}$ with 2-nitropropane; both reactions are believed to involve nitrene intermediates with the iron carbonyl functioning also as reducing agent in the second case.

All azo-compounds (including pyrazine derivatives) apparently initially bond as in complex (**32**) but this and related complexes lose CO on warming or irradiation (Herberhold and Leonhard, 1974) and the azo-ligand rotates to the type of structure exemplified by complex (**31**) thus allowing the iron atoms to retain their complete 18 outer-electron shells.

Sulphur ligands also form closely related complexes. An example is the bridged compound (**33**), formed in 42% yield from pentacarbonyliron and 1,3-dithiane by a combination of photolysis and heating to 150°C (Lotz *et al.*, 1987).

When $Na_2Fe(CO)_4$ reacts with dichlorophenylarsine, $PhAsCl_2$, both the mononuclear arsabenzene complex (**34**) and the trinuclear bridged complex

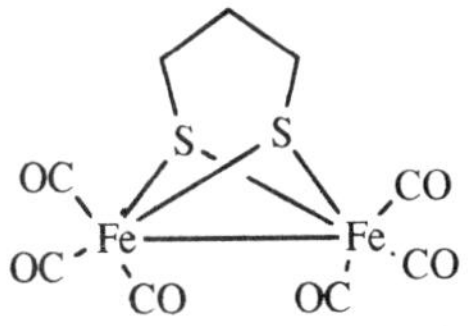

(33) 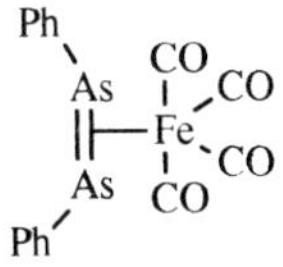 **(34)**

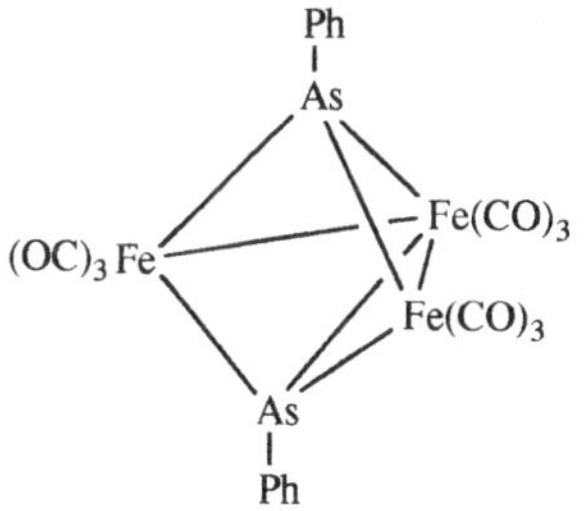

(35) 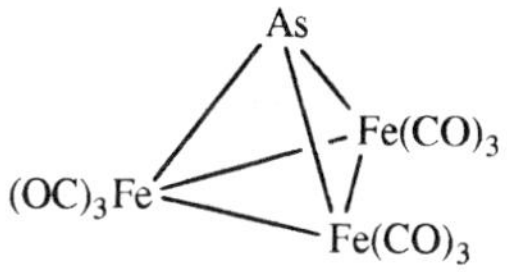 **(36)**

(**35**) are formed (Jacob and Weiss, 1977, 1978). The symmetrical arsenic-bridged trinuclear complex (**36**) results in low yield from $Fe(CO)_5 + AsF_3$ at 120°C (Delbaere *et al.*, 1973) and the tetranuclear complex (**37**) is formed from $Fe_2(CO)_9$ and methylarsine, $MeAsH_2$, on heating a benzene solution of the intermediate $Fe(CO)_4AsMeH_2$ (Röttinger and Vahrenkamp, 1978).

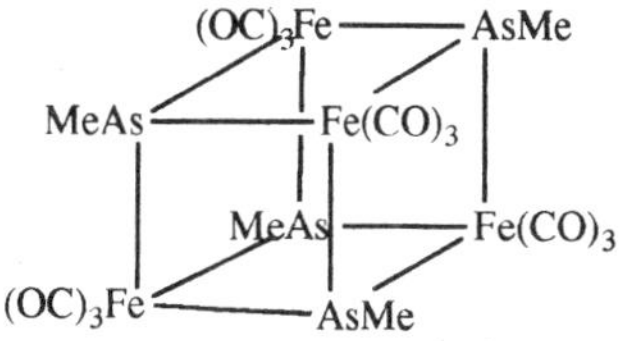

(37) 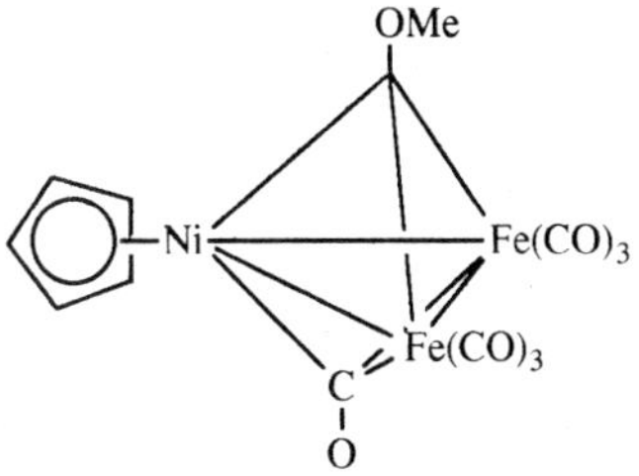 **(38)**

Both the trinuclear complexes (**35**, **36**) are examples of whole series of related compounds (Vahrenkamp, 1990) in which the bridging atoms may be *inter al.* carbon (e.g. **18**, **38**) (see also section 4.3.2), oxygen, sulphur, nitrogen or phosphorus, and in which one or two of the $Fe(CO)_3$ groups may be replaced by other organometallic groupings. Examples are the oxygen-bridged Fe_3 anion (**39**) and the mixed-metal complexes (**38**), (**40**) and (**41**).

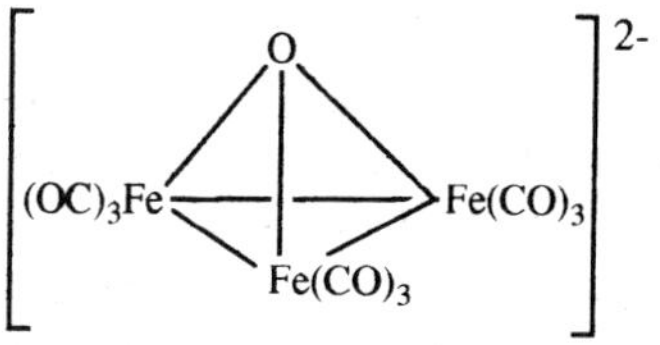

(**39**)

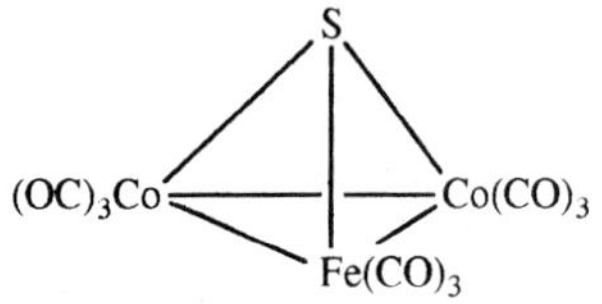

(**40**)

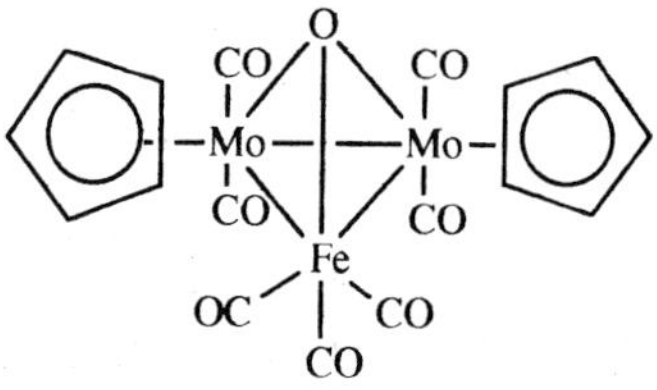

(**41**)

4.3 Cyclopentadienyliron complexes

Not only historically, but as by far the most extensively studied organo-iron complexes, the cyclopentadienyl compounds occupy a central position, making it convenient to deal with them before proceeding to other hydrocarbon ligands.

4.3.1 Ferrocene and its derivatives

4.3.1.1 Formation. Of the original two routes, the high temperature preparation from iron oxides and cyclopentadiene retains little more than historical interest. The more efficient Grignard procedure on the other hand is the basis for the formation not only of ferrocene itself, but of many derivatives, although numerous variants have been developed to maximise yield and convenience. Cyclopentadienylsodium is the most widely used reagent for preparing cyclopentadienylmetal complexes in general: it is readily formed directly from the metal and cyclopentadiene (Scheme 4.2) in suitable

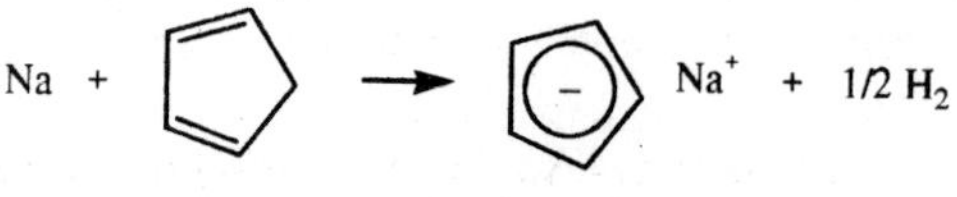

Na + C_5H_6 → $C_5H_5^-$ Na^+ + 1/2 H_2

Scheme 4.2

polar aprotic solvents, e.g. tetrahydrofuran (THF), and these solutions are sufficiently reactive to react smoothly even with metal salts which are insoluble in such media (including $FeCl_2$; *Organic Syntheses Coll.*, **IV**, 473), whereas cyclopentadienylmagnesium bromide only reacts readily with soluble metal derivatives (e.g. $FeCl_3$, $Fe(acac)_2$). However, the stability and ease of formation of ferrocene are such, that it can be obtained from the cyclopentadienyl derivatives of almost any other metal. It can also be prepared from cyclopentadiene without generating more than a small equilibrium concentration of the anion. Convenient procedures therefore include the use of relatively weak bases, e.g. amines (*Organic Syntheses Coll.*, **IV**, 476) and do not necessitate anhydrous conditions. Thus a very simple procedure (*Inorganic Syntheses* (1968), **11**, 120) uses potassium hydroxide as base and the hydrated ferrous sulphate, $FeSO_4.4H_2O$, dissolved in dimethyl sulphoxide.

For the more highly substituted cyclopentadienes, which do not react smoothly with alkali metals, the lithium salt, generated using butyllithium (or other organolithiums) is generally preferred. Lithium cyclopentadienides are also the most accessible starting materials in some other cases, e.g. tert-alkylcyclopentadienes, whose salts are formed by addition of alkyl- (or aryl-) lithiums to fulvenes (Scheme 4.3) and used directly to form 1, 1′-di-(tert-alkyl) ferrocenes (Scheme 4.4).

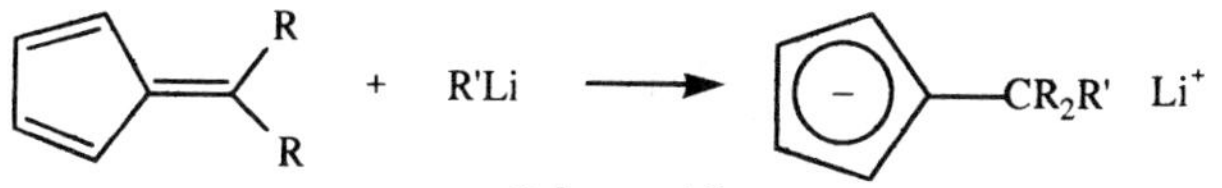

R R + R'Li → CR_2R' Li^+

Scheme 4.3

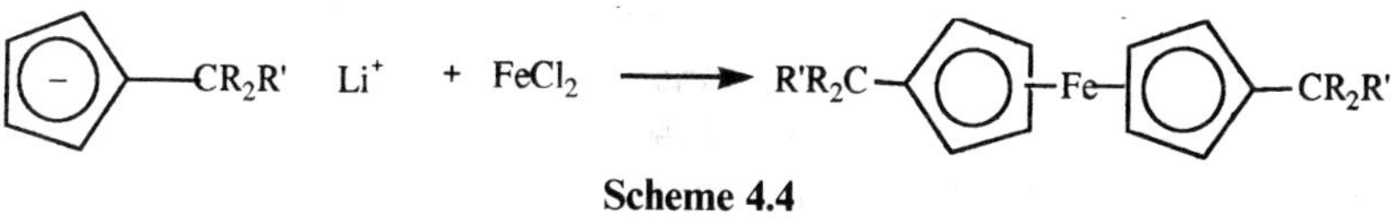

CR_2R' Li^+ + $FeCl_2$ → $R'R_2C$ Fe CR_2R'

Scheme 4.4

Many alkyl and aryl including polysubstituted cyclopentadienes as well as indenes can be converted to di- and polysubstituted ferrocenes and benzoferrocenes by suitable choice of one of the above procedures. However, mono-substituted ferrocenes are only obtained in low yield from mixtures of substituted cyclopentadienes with cyclopentadiene and few cyclopentadienes bearing functional substituents are readily available. It is therefore nearly always easier to introduce single substituents or two (or more) functional groups by making use of the high reactivity of ferrocene in electrophilic substitution reactions (see below).

4.3.1.2 Properties. Ferrocene forms readily sublimable orange crystals of m.p. 174°C. It is soluble in most organic solvents but insoluble in water and stable to 470°C; it is rather readily oxidised, especially in the presence of acids (e.g. in 90% ethanol/$HClO_4$/$NaClO_4$ its redox potential is – 0.56 V vs. the normal hydrogen electrode). The oxidised form is the ferrocenium cation, $(C_5H_5)_2Fe^+$, which owes its intense blue colour in solution and its lower stability to its paramagnetism. Its simple salts (e.g. halides, sulphate, etc.) are readily soluble in water, but it may be precipitated, e.g. as the triiodide, hexafluorophosphate, picrate, etc. (Hendrickson *et al.*, 1971). As may be expected, alkyl and other electron donating substituents facilitate and electron-withdrawing groups hinder the oxidation of the ferrocene nucleus. The salt $[(C_5Me_5)_2Fe]^{\dot{+}} - [TCNE]^{\dot{-}}$ is ferromagnetic above 60 K (Miller *et al.*, 1987).

The structure of ferrocene has been the subject of several X-ray and electron diffraction studies. They confirm that the rings lie in parallel planes with the iron atom equidistant from all carbons. There is known to be only a very small energy barrier to rotation of the two rings relative to each other and appreciable motion persists at room temperature in the crystal. Early X-ray results were interpreted as showing a staggered conformation, but the more recent low temperature (98 K) results (Seiler and Dunitz, 1982) indicate that an eclipsed structure is preferred under these conditions as also in the gas phase. In the more crowded decamethylferrocene, electron diffraction (Almenningen *et al.*, 1979) indicates a staggered conformation in the gas phase as does X-ray study in the crystal (Struchkov *et al.*, 1978).

Ferrocene does not undergo hydrogenation under conditions where benzene is readily reduced to cyclohexane, but metal/amine systems reduce it to iron metal and the cyclopentadienide ion (e.g. as in equation 4.28).

$$(C_5H_5)_2Fe + 2Li \xrightarrow{EtNH_2} Fe + 2Li[C_5H_5] \tag{4.28}$$

Excess of bromine causes oxidative cleavage yielding pentabromocyclopentane (equation 4.29)

$$2(C_5H_5)_2Fe + 13Br_2 \rightarrow 4C_5H_5Br_5 + 2FeBr_3 \tag{4.29}$$

Smaller amounts of halogen simply cause oxidation to ferrocenium salts as do nitrating agents. Since the positive charge of the ferrocenium ion prevents electrophilic attack, direct halogenation or nitration of ferrocene is prevented. But ferrocene is very readily substituted by a wide range of other electrophiles, its reactivity being comparable to that of benzene derivatives bearing strongly activating groups such as phenol or aniline. Some of the most useful substitution reactions including preferred reagents are summarised in Scheme 4.5. Included therein are two reactions, the arylation and the cyanation, which are believed to involve electron transfer followed by radical coupling, e.g. by the steps of Scheme 4.6.

Although they therefore cannot be classified as electrophilic substitution, they nevertheless probably proceed through an intermediate (here **42**) which

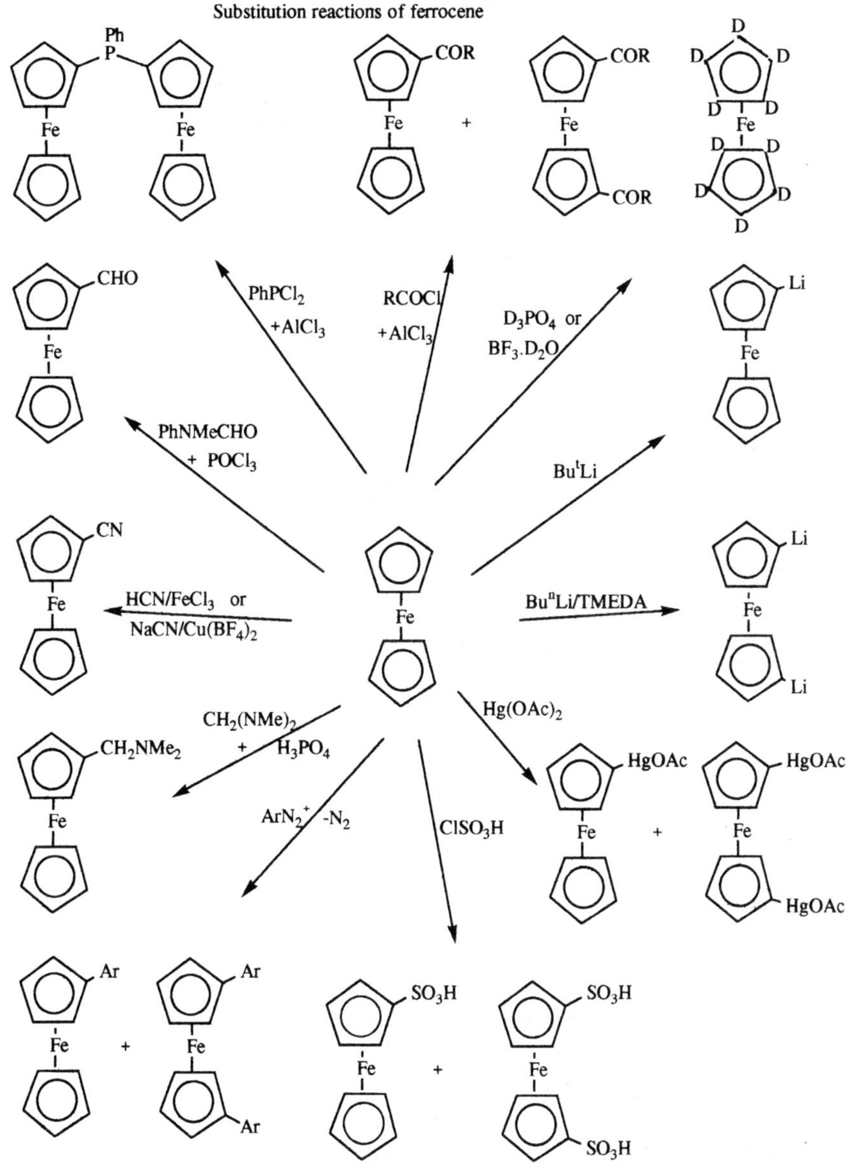

Substitution reactions of ferrocene

Scheme 4.5

is of the general Wheland type (**43**) believed to be involved in true electrophilic processes (Scheme 4.7).

Recent evidence (Cunningham, 1991) strongly suggests that the preferred mode of electrophilic attack is from the *exo* side with subsequent (or concerted) transfer of the *endo* proton (in **44**) to the iron atom. Subsequent

$C_{10}H_{10}Fe + ArN_2^+ \longrightarrow C_{10}H_{10}Fe^+ + Ar\cdot + N_2$

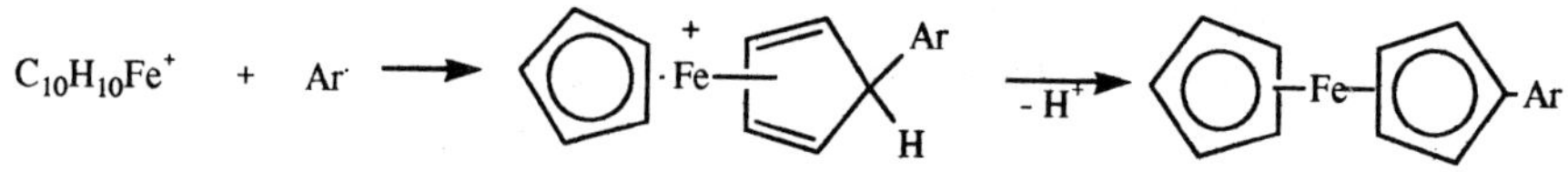

$C_{10}H_{10}Fe^+ + Ar\cdot \longrightarrow$ Fe$^+$ Ar H $\xrightarrow{-H^+}$ Fe Ar

(42)

Scheme 4.6

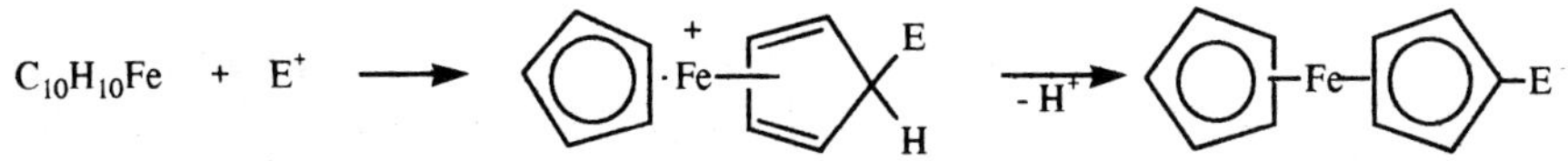

$C_{10}H_{10}Fe + E^+ \longrightarrow$ Fe$^+$ E H $\xrightarrow{-H^+}$ Fe E

(43)

Scheme 4.7

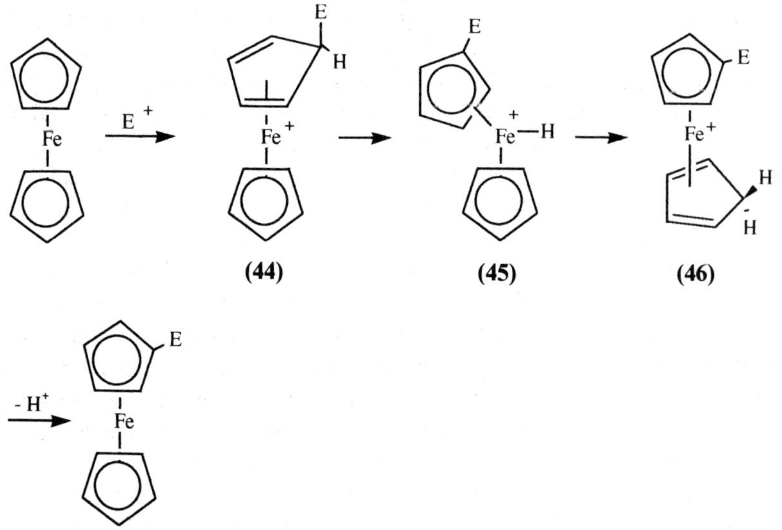

Fe $\xrightarrow{E^+}$ E H Fe$^+$ $\longrightarrow$ E Fe$^+$—H $\longrightarrow$ E Fe$^+$ H H

(44) (45) (46)

$\xrightarrow{-H^+}$ E Fe

Scheme 4.8

transfer to an unsubstituted ring carbon (as in **46**) allows the process to be completed by loss of the *exo* proton. Iron protonated species of the type (**45**; E=H) have been observed spectroscopically, but according to this mechanism, would not be formed by direct attack on the metal atom. Earlier evidence for faster *exo* than *endo* attack of electrophiles came from comparison of the rates of cyclisation of the isomers (**47A**) and (**47B**), but the fact that the latter does cyclise shows that *endo* addition of the electrophile is also possible (Rosenblum and Abbate, 1966).

CH₂CH₂COOH H Fe H CH₂CH₂COOH Fe

(**47A**) (**47B**)

The simple Friedel-Crafts acetylation was the first electrophilic substitution to be tried and it also revealed both the high reactivity of ferrocene and the transmission of substituent effects from one ring to the other. Thus, not only does ferrocene undergo acetylation under significantly milder conditions than benzene, but when it was allowed to react with a limiting amount of acetyl chloride in the presence of ten molar equivalents of anisole, no methoxyacetophenone could be detected: only acetylferrocene was formed. That the first acetyl group, by its deactivating effect, ensures that a second group enters preferentially into the unsubstituted ring [up to 2% of 1,2-diacetylferrocene is formed along with the 1,1′-isomer] seems hardly surprising, but when milder reagents (e.g. BF_3–Ac_2O) are used little or no diacetylation occurs, showing that there must be considerable deactivation even of the unsubstituted ring by the first electron-withdrawing substituent. A quantitative study suggests that ferrocene is acetylated faster than acetylferrocene by a rate factor as high as 10^4 (Rosenblum *et al.*, 1963).

The high reactivity of ferrocene is strikingly illustrated by its ability to undergo both the Vilsmeier formylation and the Mannich aminomethylation reactions. Significantly, both proceed in high yield to give only the monosubstitution product, although the aminomethylation can give moderate yields of the 1,1′-disubstituted compound under more forcing conditions. Numerous reactions of ferrocenecarboxyaldehyde have been recorded; its behaviour throughout is that of a typical aromatic aldehyde in which the formyl group is attached to an electron-rich ring. Thus it fails to undergo benzoin condensation, but will undergo a mixed benzoin condensation with benzaldehyde. A notable feature of this aldehyde is its solubility in aqueous mineral acids, apparently as the protonated cation (**48**).

$$\mathrm{FcCHO + H_3O^+ \rightleftharpoons Fc{-}\overset{+}{C}H{-}OH \leftrightarrow Fc{-}CH{=}\overset{+}{O}H} \quad (\mathrm{Fc = C_5H_5FeC_5H_4}) \quad (\mathbf{48}) \tag{4.30}$$

This is one of many examples of the high stability of α-ferrocenylcarbenium ions (Watts, 1979, 1981). Some of these carbocations have been isolated as crystalline salts and many have been studied spectroscopically in solution. X-ray crystallographic results are so far available only in two rather special cases; they reveal significant displacement of the α-carbon atom from the ring plane towards the iron atom (*c*. 20°), although the Fe–α–C distance remains too long (> 2.7 Å compared to Fe–ring–C distances of 2.0–2.1 Å) to correspond to more than a very weak bond. Nevertheless contribution from a resonance form of the type (**49B**) is considered significant.

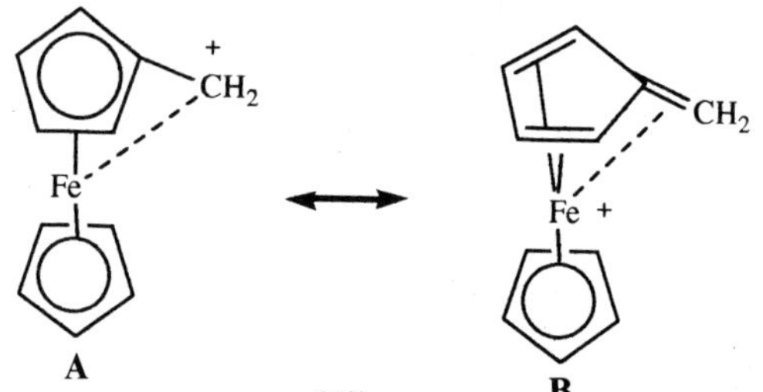

$\overset{+}{C}H_2$ Fe A ⟷ CH_2 Fe + B

(49)

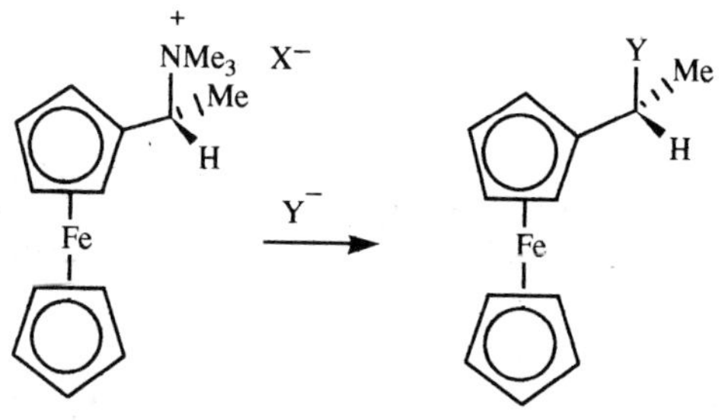

$\overset{+}{N}Me_3$ X^- Me H Fe $\xrightarrow{Y^-}$ Y Me H Fe

(50) **(51)**

It is also largely to the stability of α-ferrocenylcarbocations that the (dimethylamino)methylferrocene or rather its methiodide and analogues owe their synthetic importance. Typical transformations of these compounds are summarised in Scheme 4.9. The simple nucleophilic substitutions of the NMe_3 group have been shown to proceed by S_N1 mechanisms with anchimeric assistance from the iron atom and the Fe–C(α) bond is sufficiently strong or the C(α)–ring–C double bond character in the intermediate (**49**) sufficiently great to prevent rotation about this C(α)–ring bond before the cation is captured by the nucleophile. Hence such substitutions, e.g. of optically active ammonium salts (e.g. **50**), occurs with complete retention of configuration. The same mechanistic and stereochemical features apply to other reactions proceeding through α-ferrocenylcarbocations, e.g. the solvolysis of the ester (**51**; Y = OAc) and the anchimeric assistance referred to above was first demonstrated in the related solvolysis of the bicyclic *exo*-acetoxy compound (**56**) which reacts *c*. 2500 times faster than its *endo* isomer. However, since both reactions proceed via the same stabilised cation (**57**), both give *exo*-substituted solvolysis products: i.e. whereas the *exo*-acetate reacts with the same retention of configuration as open-chain analogues, the *endo*-acetate reacts with inversion.

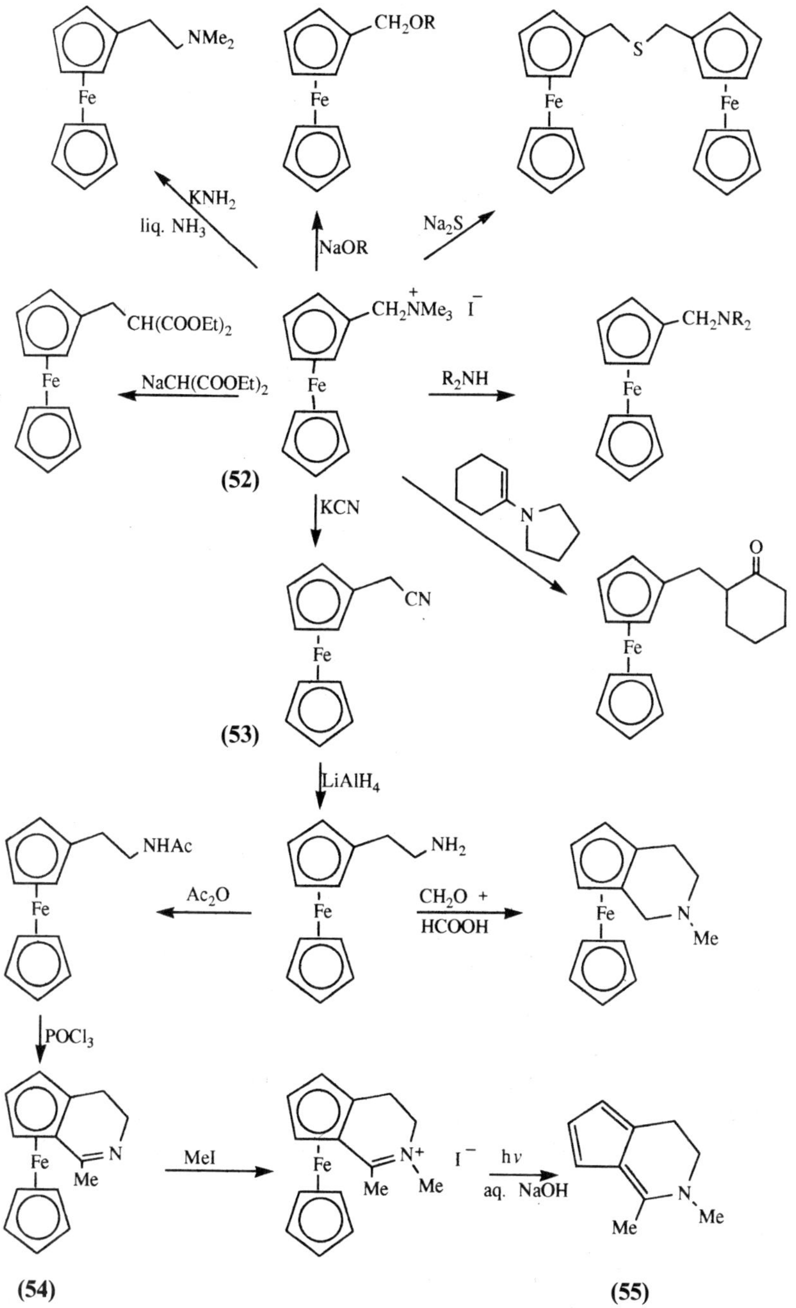

KNH$_2$ liq. NH$_3$ NMe$_2$ NaOR CH$_2$OR Na$_2$S S Fe CH$_2\overset{+}{N}$Me$_3$ I$^-$ (52) NaCH(COOEt)$_2$ CH(COOEt)$_2$ R$_2$NH CH$_2$NR$_2$ KCN CN (53) N O LiAlH$_4$ NH$_2$ Ac$_2$O NHAc CH$_2$O + HCOOH N Me POCl$_3$ (54) MeI N+ Me I$^-$ hν aq. NaOH (55)

Scheme 4.9

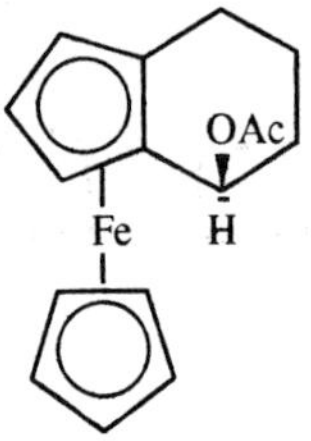

(56) 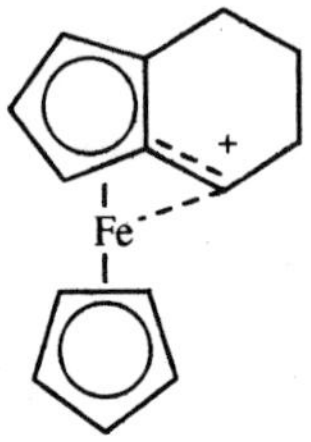 (57)

Included in Scheme 4.9 is the Stevens rearrangement which the quaternary salt (**52**) undergoes on treatment with potassium amide. This contrasts with the Hauser rearrangement of the benzyltrimethylammonium cation under such conditions, a difference which is attributed to steric factors in the transition state. Finally Scheme 4.9 includes further transformations of ferrocenylacetonitrile (**53**) showing that this intermediate allows ready access to di- and tetrahydropyridoferrocenes. The conversion of the former (**54**) *via* quaternisation and photolytic cleavage to the metal-free heterocycle (**55**) (Khand *et al.*, 1989) exemplifies a use of ferrocene derivatives in organic synthesis (cf. also Scheme 4.13). It utilises the instability to light of α-ferrocenylcarbocations or compounds having a partial positive charge on the α-carbon, in contrast to the good stability of most neutral ferrocene derivatives.

Scheme 4.10 illustrates another type of use of the ferrocenylalkylamines which depends on the directing effect of the tertiary amino group in

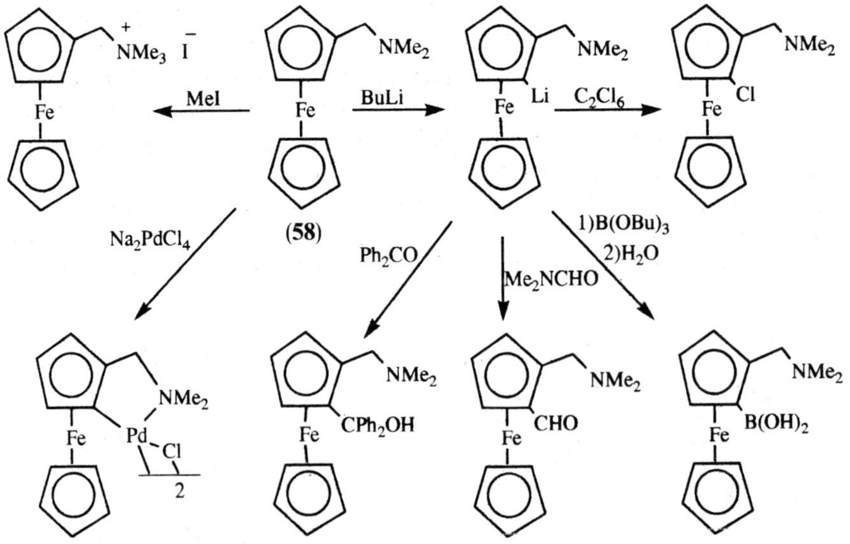

$\overset{+}{N}Me_3\ \overset{-}{I}$ MeI BuLi NMe_2 Li C_2Cl_6 Cl Fe

Na_2PdCl_4 (58) Ph_2CO Me_2NCHO 1)$B(OBu)_3$ 2)H_2O

NMe_2 Pd Cl 2 CPh_2OH CHO $B(OH)_2$

(60)

Scheme 4.10

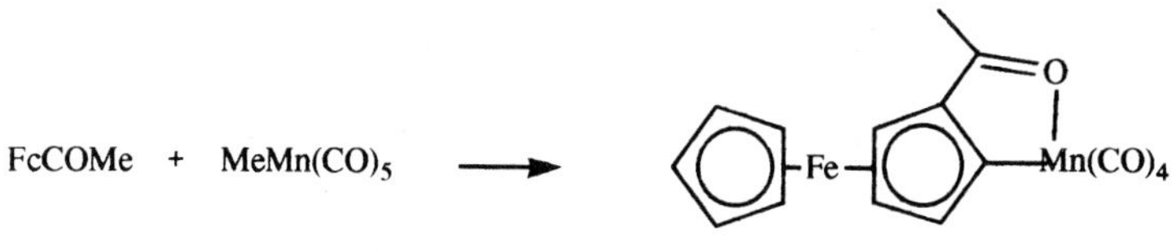

Scheme 4.11

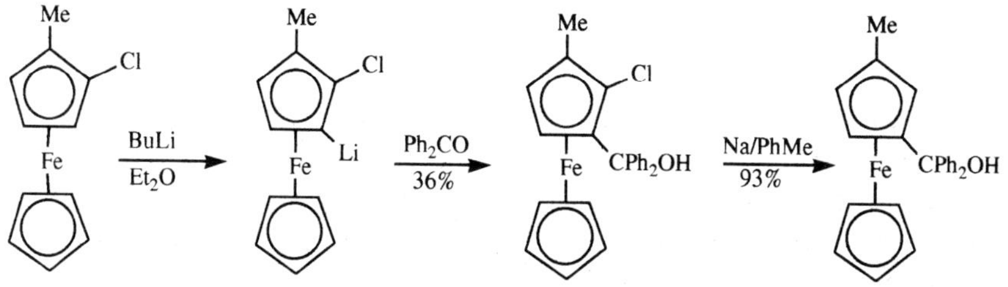

Scheme 4.12

metallation reactions. Lithiation and palladation provide a variety of attractive examples. By chelating to the metal, the nitrogen atom ensures that metallation occurs in the adjacent 2-position thus providing access to a range of homoannularly disubstituted compounds. This effect is by no means unique to nitrogen, but is shared by other donor atoms: the metallation of acetylferrocene by pentacarbonylmethylmanganese (Scheme 4.11) (Crawford *et al.*, 1975) exemplifies the effect of donor oxygen and chlorine even directs lithiation when directly attached to the cyclopentadienyl ring (e.g. Scheme 4.12) (Slocum *et al.*, 1974), providing in the example shown a selective route to 1,3- from 1,2-di-substituted ferrocenes.

A major reason for interest in homoannularly disubstituted ferrocenes (1,2,- or 1,3-) is their chirality, provided that the two substituents are not identical. The first experimental verification of such chirality was obtained by the optical resolution of the ketone (**59**) (Thomson, 1959).

One very convenient route to chiral 1,2-disubstituted ferrocenes involves the asymmetric cyclopalladation of (ferrocenylmethyl)dimethylamine (**58**), achieved in the presence of $S(+)$-N-acetylvaline to give the complex (**60**) in

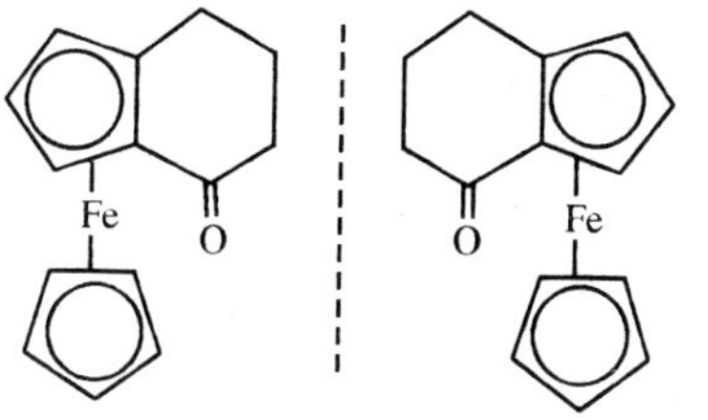

(59)

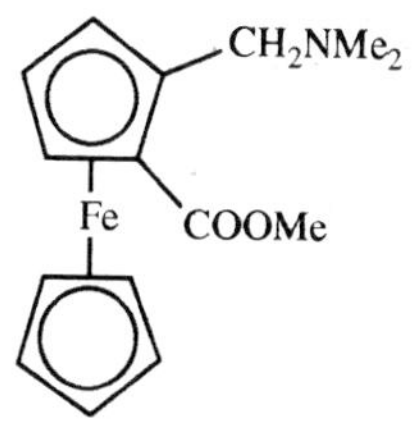

(61)

optically active form [$R(+)$- ; 85% yield; 73% ee] (Sokolov *et al.*, 1979). The metal is readily replaced, e.g. by carboxylation with CO/MeOH to give the asymmetric ester (**61**).

If the initial ring substituent is chiral, cyclometallation may occur with diastereoselection. Thus Ugi and coworkers (Marquarding *et al.*, 1970, 1977) found that *N*,*N*-dimethyl-1-ferrocenylethylamine is readily resolved as its tartrate and when the *R*-isomer (**62**) is treated with butyllithium, the (*R*,*R*)-lithio compound (**63**) is formed in 96:4 ratio with its (*R*,*S*)-diastereomer (**64**).

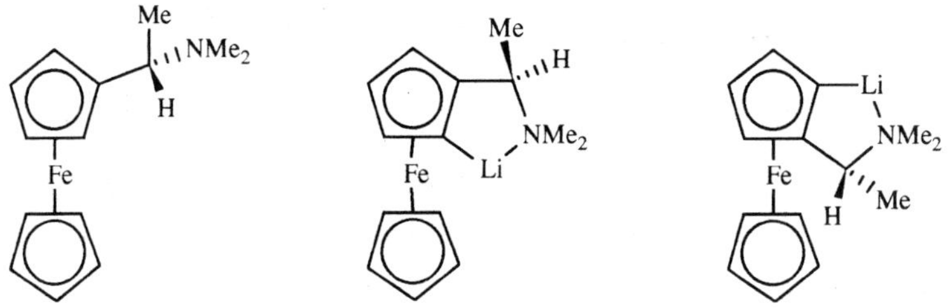

(62) **(63)** **(64)**

The simple lithio- and 1,1′-dilithioferrocenes owe their importance to the opportunities which they provide, not only to introduce a wide range of carbon and organometallic substituents, but also to obtain the halo-, nitro-, amino- and hydroxy-derivatives which are not accessible by the usual routes of benzene chemistry (Scheme 4.13). Monolithioferrocene is selectively formed (70%) using $Bu^tLi/THF/0°C$ (Rebiere *et al.*, 1990), whereas $Bu^tLi/Et_2O/20°C$ (Allcock *et al.*, 1984) or $Bu^nLi/TMEDA$ (Rausch and Ciappenelli, 1967) yield predominantly the 1,1′-dilithio compound and other conditions yield varying mixtures. As shown in Scheme 4.13, reactions with various halogenating agents, with dinitrogen tetroxide, and with *O*-alkylhydroxylamines lead respectively from the lithio- to the chloro-, bromo-, iodo-, nitro- and amino ferrocenes, although in the last two cases only the monosubstituted compounds have been obtained. Mercuration provides an alternative route to both mono- and dihalo derivatives, the mono(chloromercuri) compound being readily separated from the much less soluble 1,1′-bis-derivative.

A third route employs the reaction of the (mixed) lithioferrocenes with trialkyl borates to yield the mono- and 1,1′-di-boronic acid derivatives, which (after separation) react with copper(II) halides with displacement of the boronic acid group. Analogous reaction with copper(II) acetate converts ferrocenylboronic acid to ferrocenyl acetate from which the air-sensitive hydroxyferrocene (a 'phenol'), may be obtained. The poor stability of both this compound and of aminoferrocene is associated with the greater ease of oxidation of the ferrocene nucleus when bearing such strongly electron-donating groups. Perhaps for the same reason ferrocenylamine (aminoferrocene) cannot be diazotised, but again the lithio-derivative offers a route: its reaction with

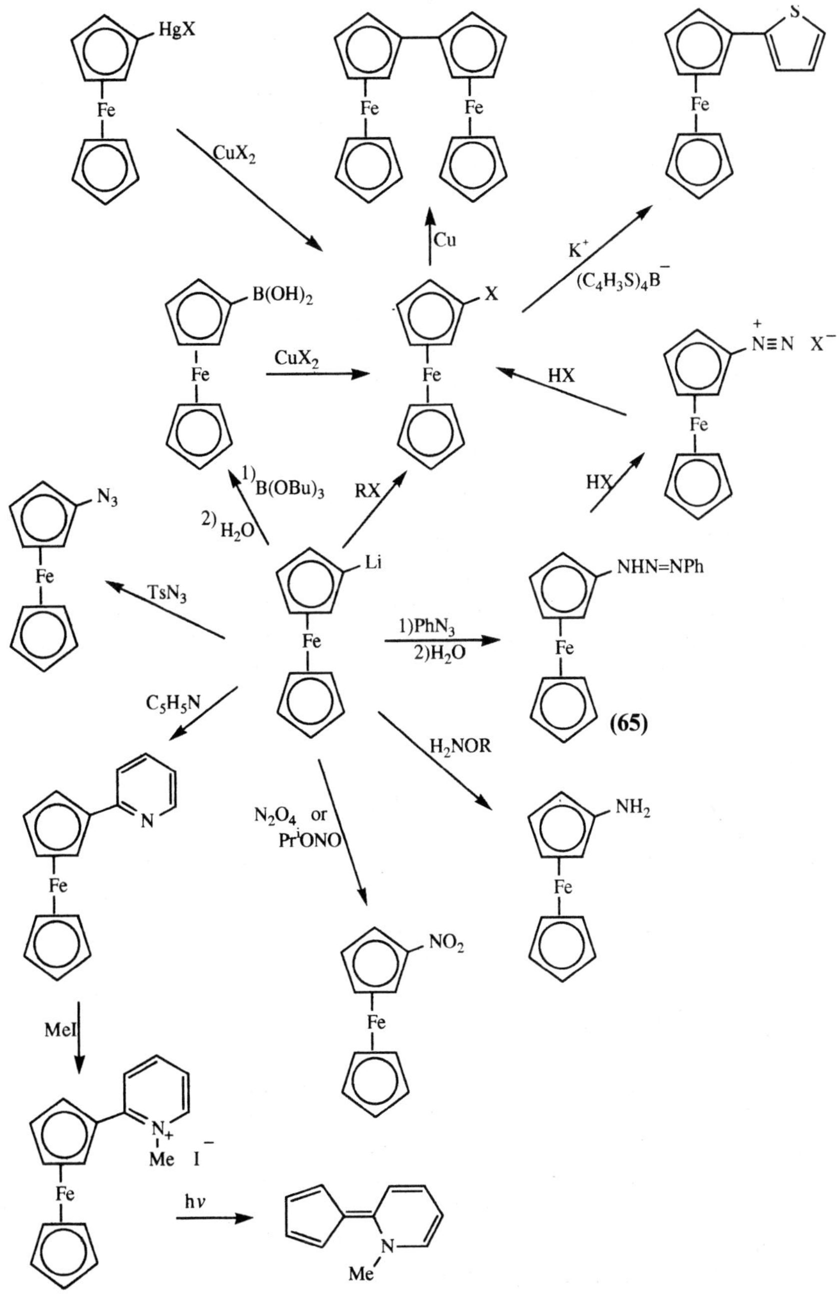
HgX
Fe
CuX2
Fe
Fe
Cu
S
Fe
K+
(C4H3S)4B−
B(OH)2
X
Fe
CuX2
Fe
+
N≡N X−
HX
Fe
1) B(OBu)3
2) H2O
RX
HX
N3
Li
Fe
TsN3
Fe
NHN=NPh
1)PhN3
2)H2O
Fe
(65)
C5H5N
H2NOR
N
N2O4 or
PriONO
NH2
Fe
Fe
NO2
Fe
MeI
N+
Me I−
Fe
hν
N
Me

Scheme 4.13

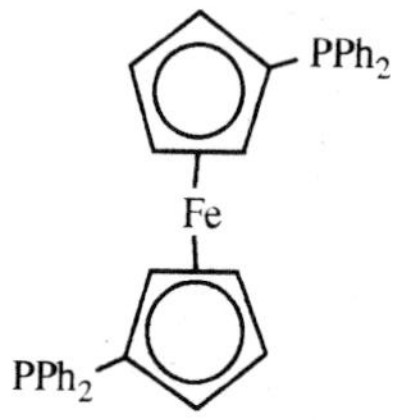

(66) 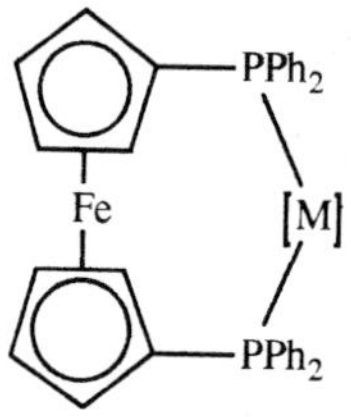 (67) 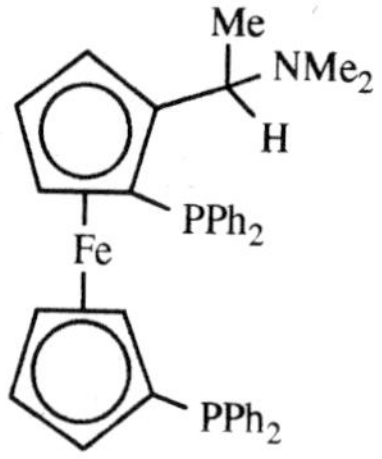 (68)

phenyl azide (azidobenzene) yields the tautomeric triazine (**65**) which is selectively cleaved by mineral acids to aniline and ferrocenediazonium cation (rather than ferrocenylamine and benzenediazonium).

Reaction of dilithioferrocene with chlorodiphenylphosphine yields the chelating diphosphine (**66**) from which a large number of metal complexes (**67**), e.g. $M = Mo(CO)_4$, $PdCl_2$, $Fe(CO)_3$, ReH_7, etc.) have been obtained. Dilithiation of the (*R*)-amine (**62**) and treatment with the chlorophosphine yields the analogous diphosphine (**68**) with efficient asymmetric induction (Appleton *et al.*, 1985): the product consists of 96% of the *R*,*R*-isomer. This and closely related mono- and diphosphines have proved to be highly effective in inducing asymmetry when employed as ligands in homogeneous transition-metal catalysed hydrogenation (Hayashi *et al.*, 1979), hydrosilylation (Hayashi *et al.*, 1980), allylic amination (Hayashi *et al.*, 1989, 1990), C-allylation (Hayashi *et al.*, 1988), Grignard coupling (de Graaf *et al.*, 1989), and aldol condensation reactions (Ito *et al.*, 1989; Togni and Pastor, 1990).

The metal complexes (**67**) are examples of ferrocenophanes, the name given to ferrocenes in which the two rings are bridged (Watts, 1967). They are classified as [*n*]-ferrocenophanes (**69**) when an *n*-atom bridge spans the two rings and [*n*,*m*]-ferrocenophanes when two nuclei are involved (**70**). The first ferrocenophane (**71**) was obtained as the only cyclisation product from the trifluoroacetic anhydride-catalysed Friedel-Crafts reaction of 3-ferrocenylpropanoic acid. Bis(ferrocenylmethyl)malonic acid, $(FcCH_2)_2C(COOH)_2$ cyclises similarly to the spiro-diketone (**72**), whereas 4-ferrocenylbutyric acid cyclises exclusively to the homo-annularly fused ketone (**59**) mentioned earlier.

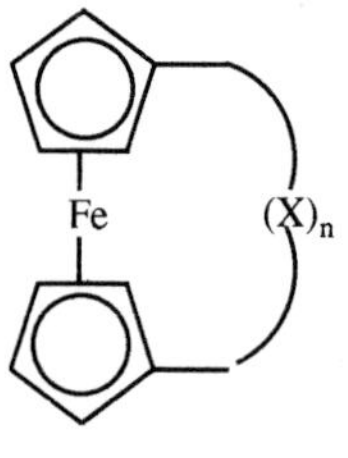

(69) 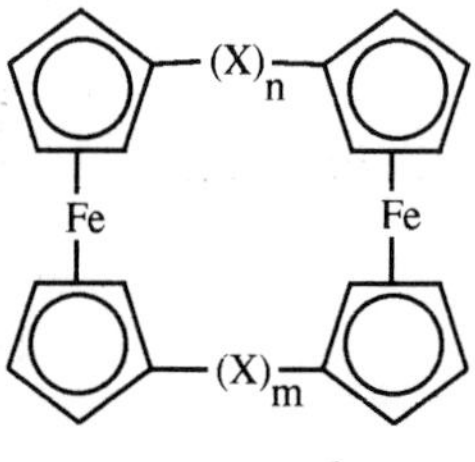 (70) 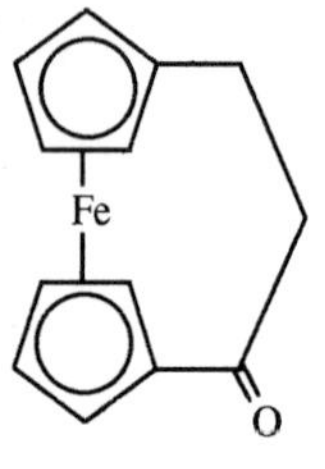 (71)

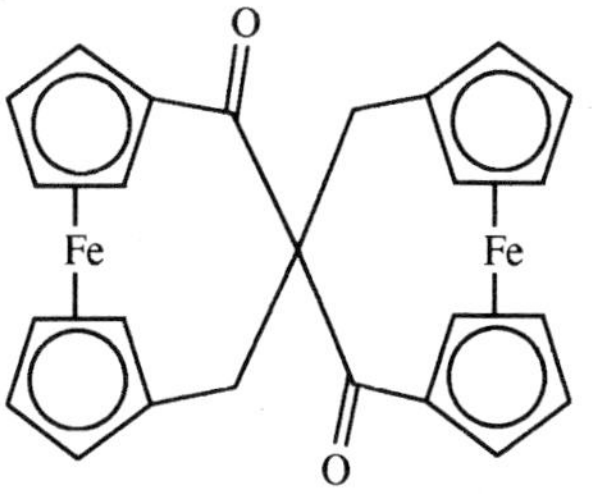

(72) 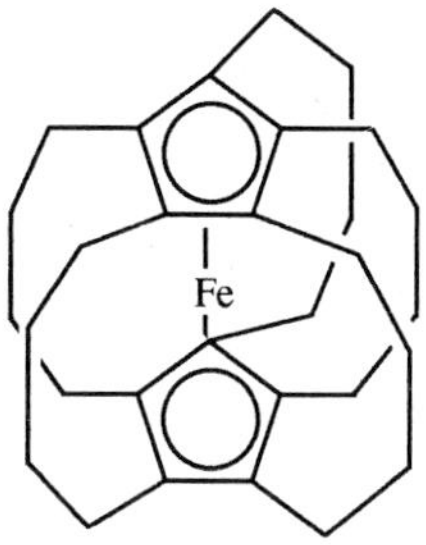 (73)

Many schemes have been devised to obtain other ferrocenophanes, including multiply bridged systems, culminating in the synthesis of the [4][4][4][4][4]-ferrocenophane (**73**) (Hisatome *et al.*, 1990). A three-carbon or larger bridge imposes little strain: e.g. the dihedral angle between the ring planes in compound (**71**) is only 9°. Smaller bridges necessitate considerable ring-tilting: e.g. 23° in 1,1,2,2-tetramethyl[2]ferrocenophane ((**74**); R = Me). The effect of this ring tilt on the chemistry of [2]ferrocenophanes is seen in the enhanced Lewis basicity, with the result that the parent compound ((**74**), R = H) forms the protonated cation (**75**) even in 0.1% ethanolic sulphuric acid (Lentzner and Watts, 1971). Carbon [1]ferrocenophanes are not known, but larger atoms readily bridge the rings to form hetero[1]ferrocenophanes. Thus 1,1′-dilithioferrocene reacts with Cp_2MCl_2 (M = Ti, Zr, Hf) to give the [1]-ferrocenophanes (**69** 28; $X_n = M(C_5H_4R)_2$) (Broussier *et al.*, 1990) with phenyl-,

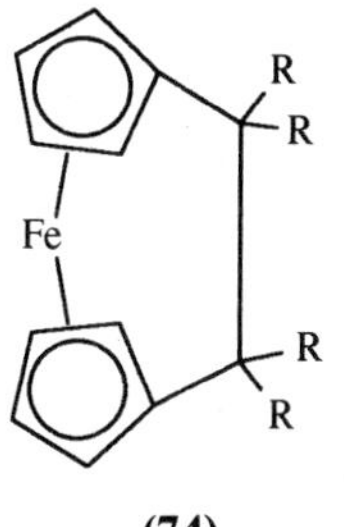

(74) 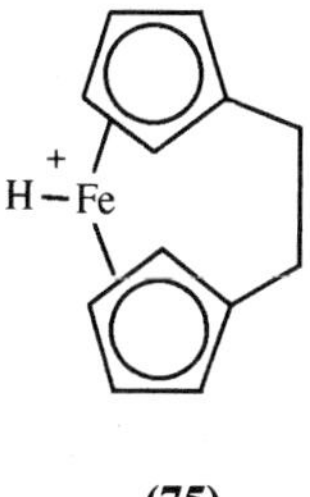 (75) 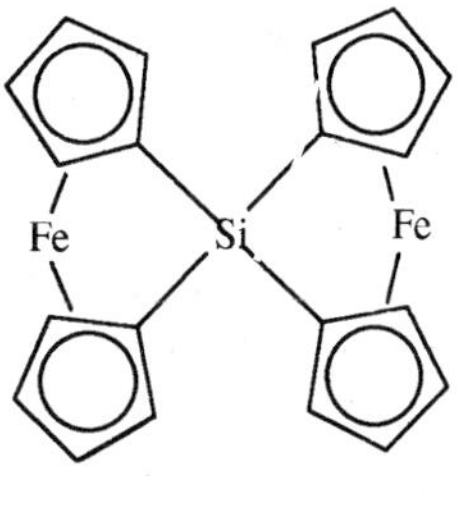 (76)

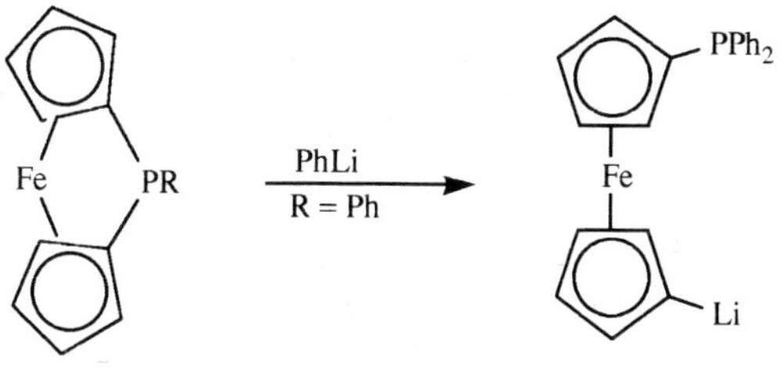

(77)

Scheme 4.14

methyl-, or tert-butyldichlorophosphine to give the [1]phosphaferroceno-phanes (**77**) (Osborne *et al.*, 1980; Seyferth and Withers, 1982; Butler *et al.*, 1983) and with silicon tetrachloride to give the spiroferrocenophane (**76**) (Osborne *et al.*, 1980). The former are ring-opened by aryllithiums (Scheme 4.14).

Dilithioferrocene is also the source of the trithia[3]ferrocenophane (**69**), $X_n = S_3$), formed on reaction with elemental sulphur, and, indirectly (via the dibromo- or di-iodo-derivative) of 1,1′-biferrocenylene ([0,0]ferrocenophane; bisfulvalenediiron; **70**, no X_n or X_m). The latter is one of the products (c. 20% yield) formed when the 1,1′-dihaloferrocenes undergo an Ullmann reaction with copper (Cowan and Le Vanda, 1972). An alternative and slightly more efficient route to this compound involves coupling of cyclopentadienylsodium to form dihydrofulvalene followed by lithiation and reaction with iron (II) chloride (Mueller-Westerhoff and Eilbracht, 1972). The oxidation of this and related [*n*,*m*]ferrocenophanes have been extensively studied because the half-oxidised, mixed-valence (Fe^{II}-Fe^{III}) monocations (**79**) allow facile electron exchange between the metal atoms; it is also notable that the dication (**78**) of biferrocenylene is diamagnetic, a fact which is attributed to metal-metal interaction through the π-system of the ligand (Mueller-Westerhoff and Eilbracht, 1973).

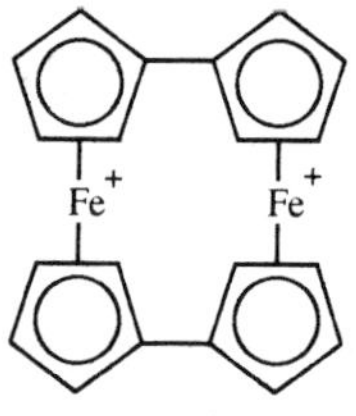

Fe$^+$ Fe$^+$ — Fe$^+$ X_n X_m Fe — Fe S CH_2 CH_2 O CH_2 n S CH_2

(**78**) 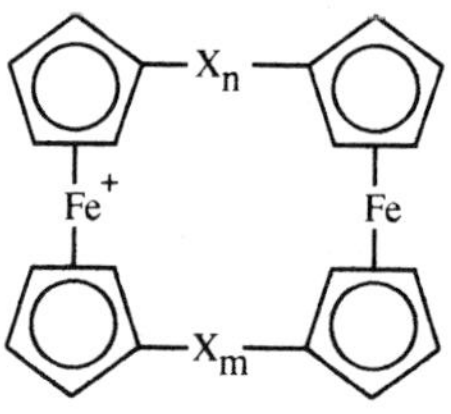 (**79**) 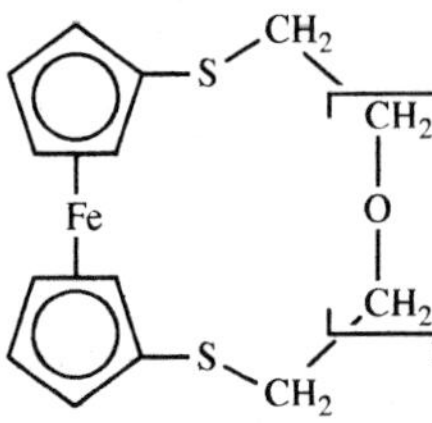 (**80**)

Finally, mention must be made of recent interest in a series of both bridged and fused macrocyclic derivatives (e.g. (**80**)) in which the ferrocene nucleus forms part of a 'crown ether' system and which display the typical polydentate ligand properties of such molecules (e.g. Sato *et al.*, 1982; Grossel *et al.*, 1991).

4.3.2 Monocyclopentadienyl iron compounds

Most monocyclopentadienyliron derivatives are formed via the dinuclear tetracarbonyl (**81**), itself obtained conveniently by heating the pentacarbonyl with dicyclopentadiene (Scheme 4.15). The latter, by its retro-Diels-Alder reaction dissociates to provide a continuous supply of monomeric cyclo-pentadiene, part of which is reduced to cyclopentene, while some of the hydrogen is set free. The steps in this process are almost certainly as shown in Schemes 4.16–4.18.

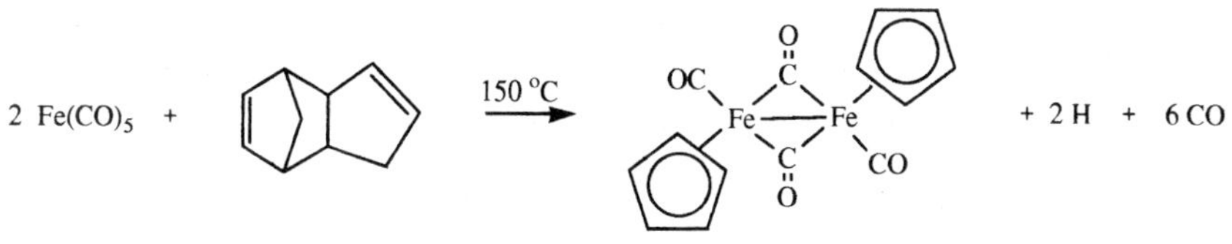

$2\ Fe(CO)_5$ + [dicyclopentadiene] $\xrightarrow{150\ ^\circ C}$ (81) + 2 H + 6 CO

(81)

Scheme 4.15

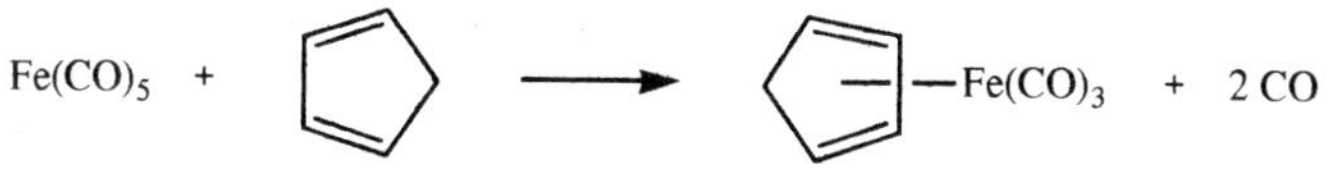

$Fe(CO)_5$ + [cyclopentadiene] ⟶ [cyclopentadiene]$-Fe(CO)_3$ + 2 CO

(82)

Scheme 4.16

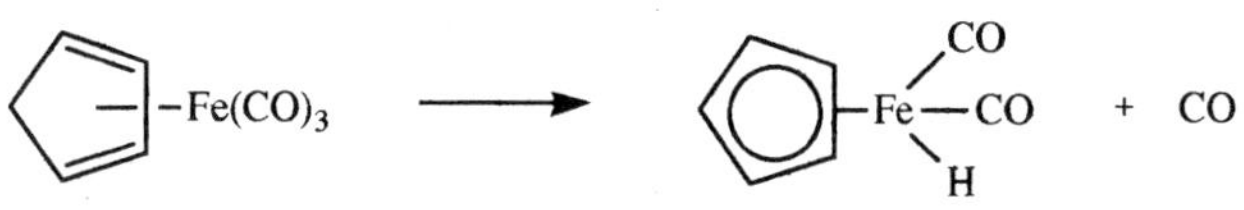

[cyclopentadiene]$-Fe(CO)_3$ ⟶ $(C_5H_5)Fe(CO)_2H$ + CO

(83)

Scheme 4.17

$(C_5H_5)Fe(CO)_2H$ ⟶ $Cp_2Fe_2(CO)_4$ + H_2

Scheme 4.18

The first intermediate (**82**) is readily isolated from low temperature (below 40°C) reaction of the monomeric diene with $Fe_2(CO)_9$ and both it and the hydride (**83**) obtainable indirectly (see below) are known to decompose to the dinuclear carbonyl (**81**) on heating. X-ray structural studies have shown that this exists in the *trans* form with the central $Fe_2(\mu\text{-}CO)_2$ system planar and a metal-metal distance of 2.534 Å. Spectroscopic studies reveal that in solution it exists in equilibrium with small amounts of the less stable *cis* isomer (**84**; X = CO) which is isolable at low temperature (Bryan *et al.*, 1970). Their interconversion involves the transient formation of the unbridged $[(C_5H_5)Fe(CO)_2\text{—}]_2$. It is noteworthy, however, that for many closely related species, including the thiocarbonyl (**84**; X = CS) (Beckman and Jacobson, 1979), the *cis* form is the more stable isomer at least in the crystal.

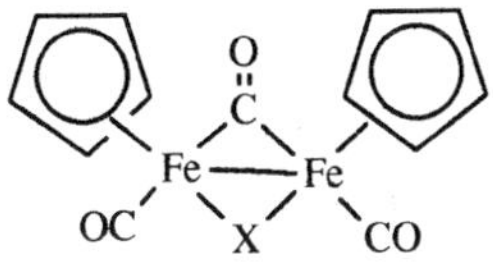

(84)

Both oxidation with halogens (equation 4.31) and reduction with sodium amalgam (equation 4.32) readily convert the tetracarbonyl (**81**) to the mononuclear dicarbonyl complexes:

$$[CpFe(CO)_2]_2 + X_2 \rightarrow \underset{(\mathbf{85})}{2CpFe(CO)_2X} \ldots \quad (4.31)$$

$$[CpFe(CO)_2]_2 + 2Na(Hg) \rightarrow \underset{(\mathbf{86})}{2Na[CpFe(CO)_2]} \ldots \quad (4.32)$$

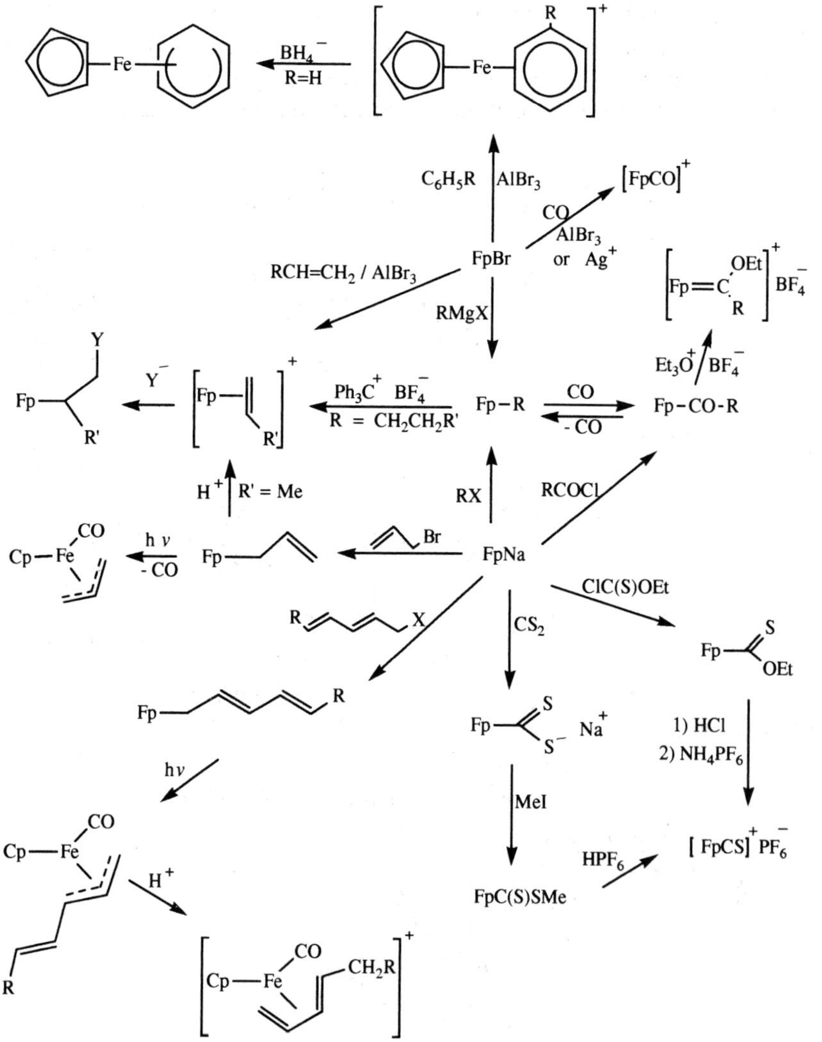

Fe
BH_4^-
R=H
R
Fe
C_6H_5R
$AlBr_3$
$[FpCO]^+$
CO
$AlBr_3$
or Ag^+
FpBr
$RCH{=}CH_2$ / $AlBr_3$
RMgX
$[Fp{=}C(OEt)R]^+$ BF_4^-
Y
Fp
R'
Y^-
Fp
R'
Ph_3C^+ BF_4^-
R = CH_2CH_2R'
Fp–R
CO
- CO
Fp–CO–R
Et_3O^+ / BF_4^-
H^+
R' = Me
RX
RCOCl
CO
Cp–Fe
h ν
- CO
Fp
Br
FpNa
ClC(S)OEt
R
X
CS_2
Fp
S
OEt
Fp
R
Fp
S
S^-
Na^+
1) HCl
2) NH_4PF_6
hν
MeI
CO
Cp–Fe
H^+
HPF_6
$[FpCS]^+ PF_6^-$
FpC(S)SMe
R
CO
Cp–Fe
CH_2R
+

Scheme 4.19

The pentamethyl derivatives $(C_5Me_5)Fe(CO)_2X$ ($X = Br_1 I$) can be obtained directly from $Fe(CO)_5$ with C_5Me_5X (Jutzi and Mix, 1990). Their further reactions have been so widely studied that a special abbreviation, 'Fp' has become accepted for the $CpFe(CO)_2$ group. Using this abbreviation, Scheme 4.19 outlines routes which lead to derivatives containing a range of different ($\eta^1 - \eta^6$) hydrocarbon ligands. Several of these classes of compounds find extensive use in organic synthetic chemistry. As these uses do not affect the cyclopentadienyl ligand, they will be discussed under the reactive ligands in later sections. The halides (**85**) and the sodium salt (**86**) have also been used to link the Fp group to a very wide range of other elements of all types. A small selection of such reactions is in Schemes 4.20 and 4.21.

Both the dimeric carbonyl (**81**) and the carbonyl halides (**85**) undergo direct replacement of one or more carbonyl groups by other donor ligands, e.g. phosphines, phosphites or isocyanides, the extent of replacement depending

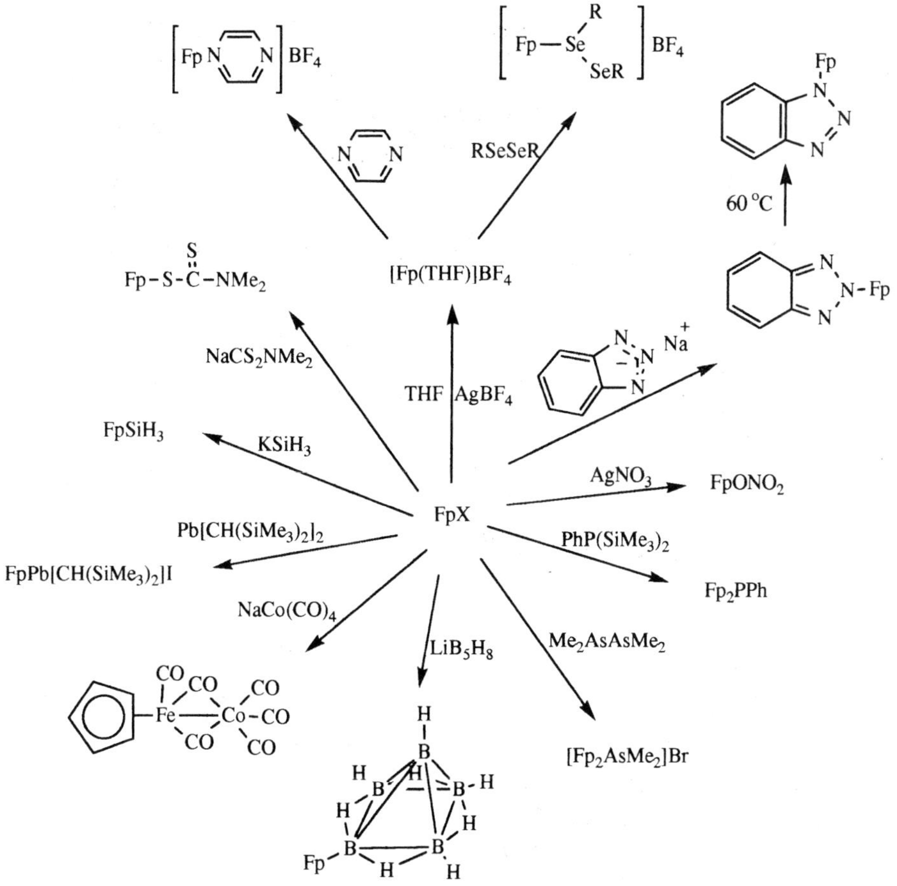

[FpN(pyrazine)N]BF_4 — [Fp—Se(R)SeR]BF_4 — RSeSeR — [Fp(THF)]BF_4 — THF, $AgBF_4$ — $Fp-S-C(=S)-NMe_2$ — $NaCS_2NMe_2$ — $FpSiH_3$ — $KSiH_3$ — FpX — Na^+ benzotriazolide — 60 °C — $AgNO_3$ — $FpONO_2$ — $Pb[CH(SiMe_3)_2]_2$ — $FpPb[CH(SiMe_3)_2]I$ — $PhP(SiMe_3)_2$ — Fp_2PPh — $NaCo(CO)_4$ — LiB_5H_8 — $Me_2AsAsMe_2$ — $[Fp_2AsMe_2]Br$ — CpFe(CO)(μ-CO)Co(CO)$_3$ — Fp-B_5H_8

Scheme 4.20

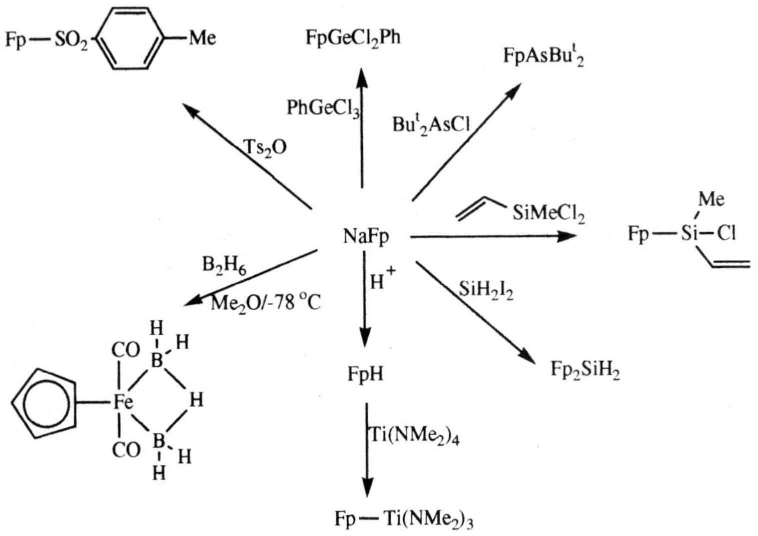

Scheme 4.21

on reaction conditions and, in the case of the halides (**85**), also on the choice of halogen, chloride being most reactive. Thus with phenylisocyanide it yields $[CpFe(CNPh)_3]Cl$ by displacement of CO and Cl, whereas the bromide yields $CpFe(CNPh)_2Br$ (Joshi *et al.*, 1963). Sodium pentacarbonylchromate, $Na_2Cr(CO)_5$, has been used to reduce the former to the neutral dinuclear complex (**87**) (Fehlhammer *et al.*, 1980).

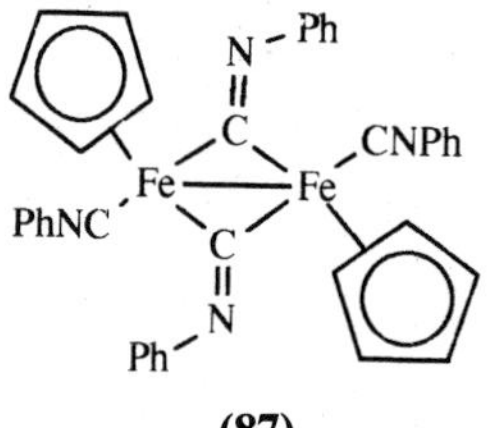

(**87**)

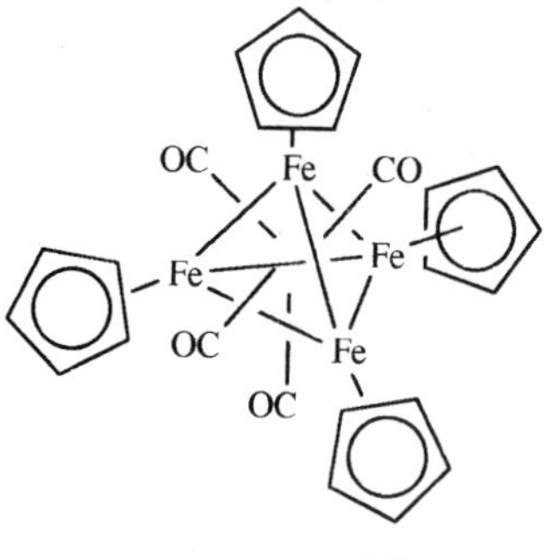

(**88**)

Loss of carbon monoxide from the dinuclear carbonyl (**81**) on heating to 500°C leads to the tetranuclear complex, $[CpFe(CO)]_4$ (**88**) which has been shown by X-ray crystallography to have face-bridging carbonyl groups. Some ferrocene is formed at the same time and this becomes the sole product (96%) if the temperature is raised to 675°C (Glidewell and McKechnie, 1987) and is also formed on heating the halides (**85**). A more controlled conversion of these

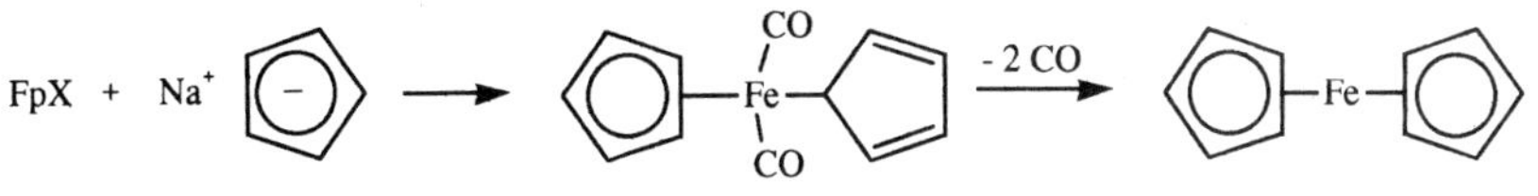

Scheme 4.22

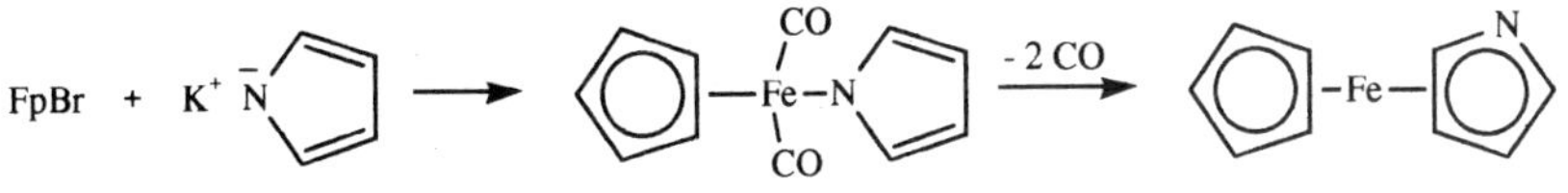

Scheme 4.23

halides to ferrocene occurs on treatment with cyclopentadienylsodium (Scheme 4.22). The moderately stable intermediate contains one η^5 and one η^1-bonded cyclopentadienyl group. The latter is highly fluxional–migration of iron around the ring occurring sufficiently rapidly at ambient temperature so that only one proton NMR signal is observed for each ring. Scheme 4.22 allows the use of differently substituted cyclopentadienyls and hence the synthesis of unsymmetrical ferrocene derivatives and has also been extended to pyrrolylpotassium [or pyrrole + diisopropylamine (Zakrewski and Giannotti, 1990)] to yield azaferrocene (**89**), again via an isolatable intermediate (Scheme 4.23).

Reaction of the dinuclear carbonyl (**81**) or its alkyl derivatives with phosphoranes (Altbach *et al.*, 1986; Caballero *et al.*, 1989) provides access to methylene-bridged analogues (**90**). An alternative route involves reduction

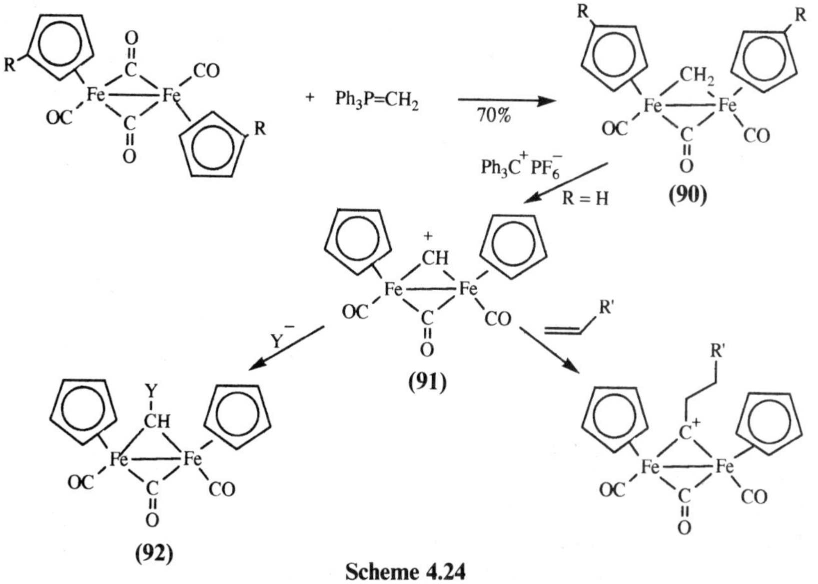

Scheme 4.24

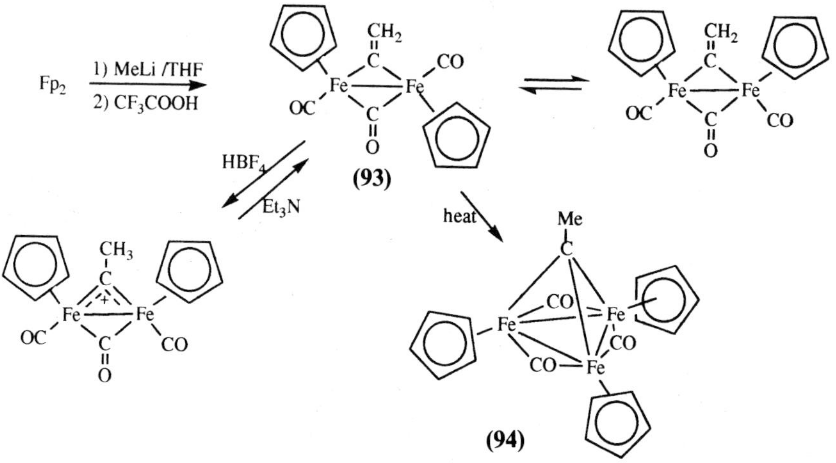

Scheme 4.25

of the carbonyl (**81**) with $[Al(OPr^i)_4]^-$ (Ortaggi and Paolesse, 1988). Chromatography below 0°C allows separation of *cis* and *trans* forms; at room temperature these equilibrate too rapidly, the *cis*-form predominating slightly. X-ray data are available for its dimethyl derivative (**90**; R = Me) (Caballero *et al.*, 1989). These methylene complexes suffer hydride abstraction on treatment with trityl salts giving the stable cations (e.g. **91**) which add alkenes, alkynes and various other nucleophiles to give substituted methylene-bridged complexes (**92**) (Casey *et al.*, 1986, 1988, 1990). The closely related vinylidene complexes (e.g. **93**) result from reaction of the tetracarbonyl (**81**) with alkyllithium and serve as an alternative source of cationic substituted methylidene complexes (Scheme 4.25) (Dawkins *et al.*, 1983). Pyrolysis converts complex (**93**) to the trinuclear 'carbyne-bridged' complex (**94**) (Brun *et al.*, 1983). A related bismuth-bridged complex (**95**) is the product of irradiating Fp_3Bi, itself obtained from $BiCl_3$ + 3FpNa.

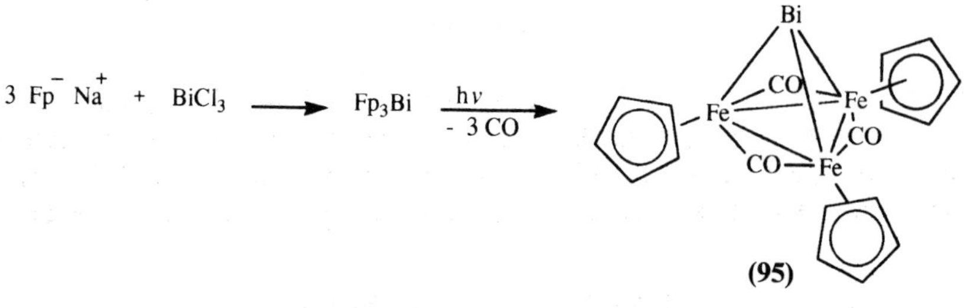

Miscellaneous other reactions whereby the tetracarbonyl (**81**) is modified by replacement of some or all of the carbonyl groups are collected in Scheme 4.26. The bridging dinitrosyl (**96**) included there is one of relatively few complexes containing iron–iron multiple bonds, confirmed in this case by a

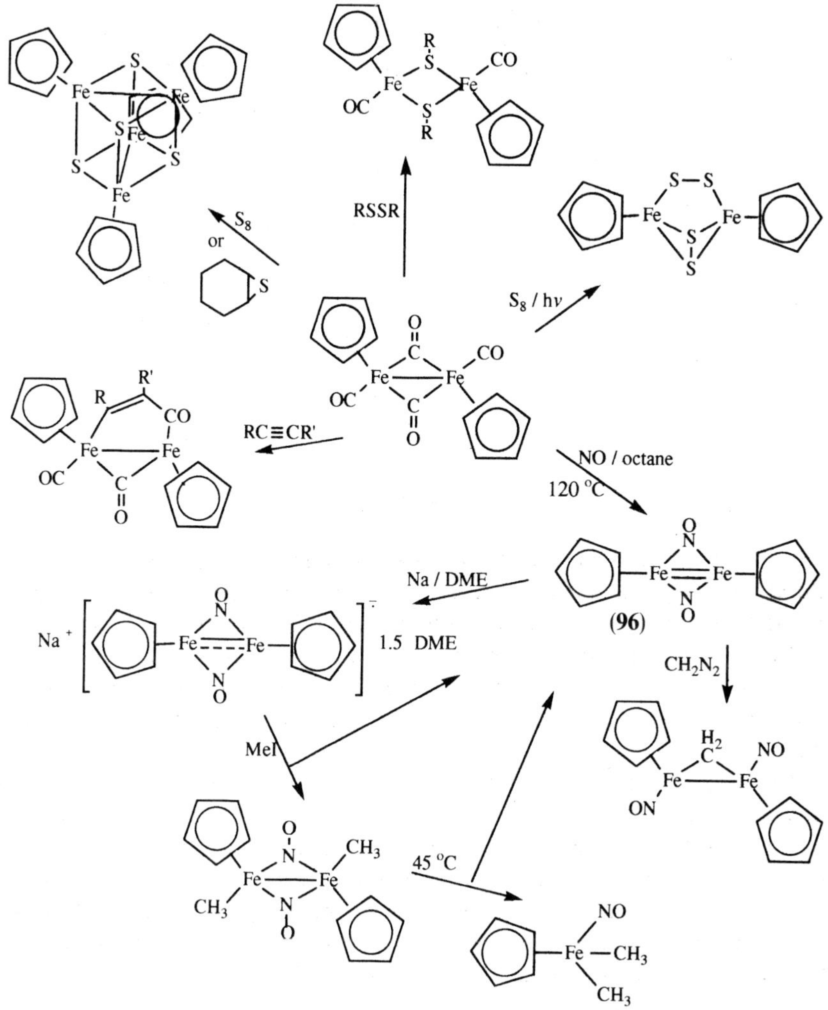

Fe S Fe S Fe S S Fe

R S CO Fe Fe OC S R

S—S Fe Fe S S

S_8 or S

RSSR

S_8 / hν

O C CO Fe Fe OC C O

R' R CO Fe Fe OC C O

RC≡CR'

NO / octane 120 °C

O N Fe Fe N O (**96**)

Na / DME

Na$^+$ [O N Fe Fe N O]$^-$ 1.5 DME

CH_2N_2

H_2 C NO Fe Fe ON

MeI

O N CH_3 Fe Fe CH_3 N O

45 °C

NO Fe —CH_3 CH_3

Scheme 4.26

metal–metal distance of 2.362 Å (Calderon *et al.*, 1974). The corresponding tricarbonyl (**97**; R = H) is only known from its spectra in rigid matrices at low temperature (Hooker *et al.*, 1983), but its decamethyl derivative (**97**, R = Me) also formed by photolysis of the corresponding tetracarbonyl has been

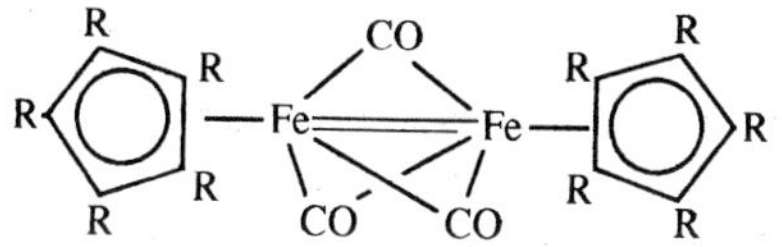

R R R R R CO Fe Fe CO CO R R R R R

(**97**)

isolated and characterised by an X-ray structure (Fe–Fe = 2.265 Å). Unlike the dinitrosyl (**96**) it is paramagnetic (Fe ≡ Fe?) (Blaha *et al.*, 1985). Examples of iron–iron triple bonds are known for cyclobutadiene complexes (q.v.).

4.4 Other hydrocarbon complexes

4.4.1 η^1-Hydrocarbon-iron complexes

As pointed out in chapter 1, the simplest, i.e. homoleptic, σ-alkyl or -aryl compounds of iron are unstable and attempts to prepare them generally lead to dimers of the hydrocarbon fragment together with metallic iron (e.g. equation 4.33):

$$2RMgCl + FeCl_2 \rightarrow Fe + R\text{-}R + 2\,MgCl_2 \qquad (4.33)$$

However, when bulky groups which cannot undergo β-elimination are used it becomes possible to prepare stable iron(IV) alkyls e.g. compound (**98**). The anionic iron(II) tetranaphthyl complex, $Li[Fe(C_{10}H_7)_4]$ is also reported to be stable at room temperature (Bagenova *et al.*, 1981).

(**98**)

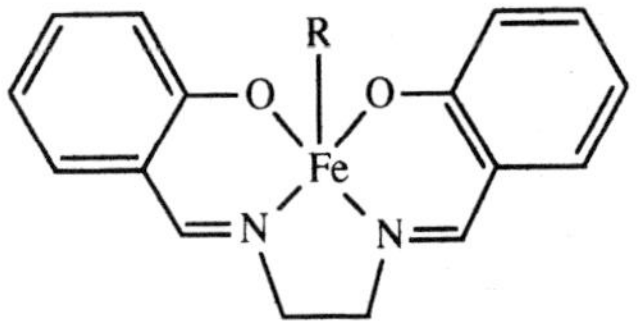

(**100**)

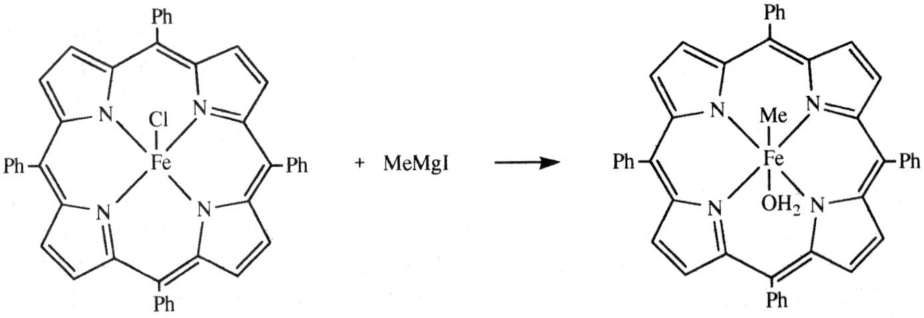

(**99**)

Attachment of quite simple alkyl, acyl and aryl groups to iron yields robust complexes when other stabilising features are present. Many such complexes containing (preferably chelating) nitrogen and phosphorus donor ligands can be prepared, simply by treating iron(ligand)$_n$-halides with Grignard, organolithium or similar organometallic reagents, e.g. complex (**99**) (Cocolios *et al.*, 1982) and the similarly formed complexes (**100** and **101**). The deep blue *cis*-diethylbis(bipyridyl)iron(II) (**101**) undergoes reversible oxidation to the dark green cation (**102**) (isolated as perchlorate salt) (Lau *et al.*, 1982).

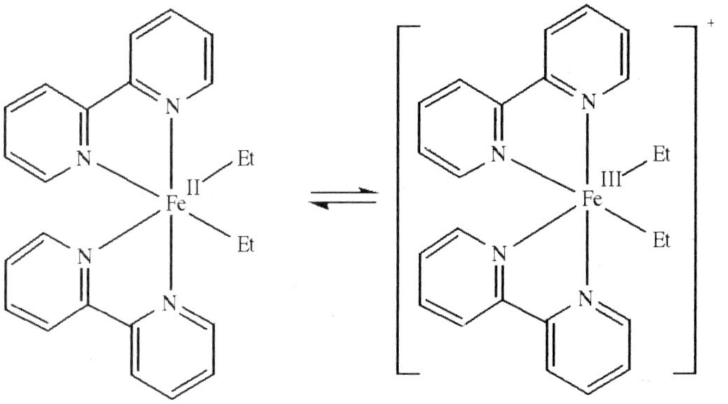

(101) **(102)**

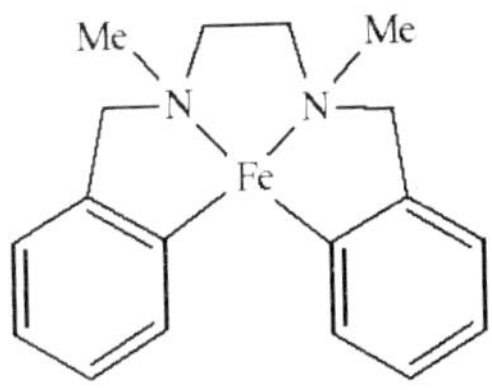

(103)

Donor ligand and alkyl group may be introduced in the same operation as in the preparation (Küpper, 1968) of the thermally stable, but air and moisture-sensitive cyclometallated complex (**103**), or as exemplified in the synthesis (Scheme 4.27) of complex (**104**); here the organoaluminium reagent serves as reducing agent as well as source of the methyl groups (Ikariya and Yamamoto, 1976). Although the dialkyl (**104**) is stable in the pure form, it loses methane in solution to yield the ortho-metallated compound (**105**) and it reacts with phenylacetylene at room temperature to give the alkynyl-hydride (**106**). The anionic iron(II) complex (**107**) is obtained by reducing neutral phthalocyanineiron(II) to the iron(0) dianion with lithium-benzophenone-ketyl, followed by addition of iodomethane (Taube and Drevs, 1977)–an example of oxidative addition, another fairly general route to metal alkyls. Thus, the phosphine complex, $Fe^0(PMe_3)_4$, adds bromomethane to give $MeFe(PMe_3)_4Br$ (Karsch, 1977), a compound also formed, together with the *cis*-dimethyl complex, $Me_2Fe(PMe_3)_4$ from $Fe(PMe_3)_2Br_2 + PMe_3 + MeLi$. The stabilising effect of bulky hydrocarbon ligands is again seen in the

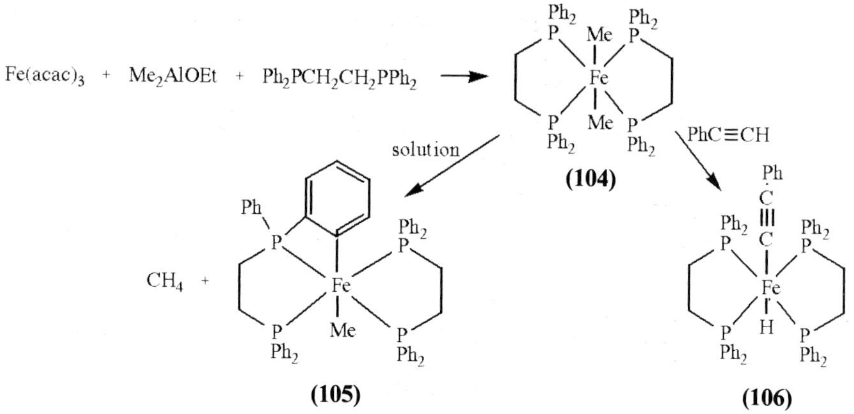

$Fe(acac)_3 + Me_2AlOEt + Ph_2PCH_2CH_2PPh_2 \rightarrow$ (104)

solution → CH_4 + (105)

$PhC\equiv CH$ → (106)

Scheme 4.27

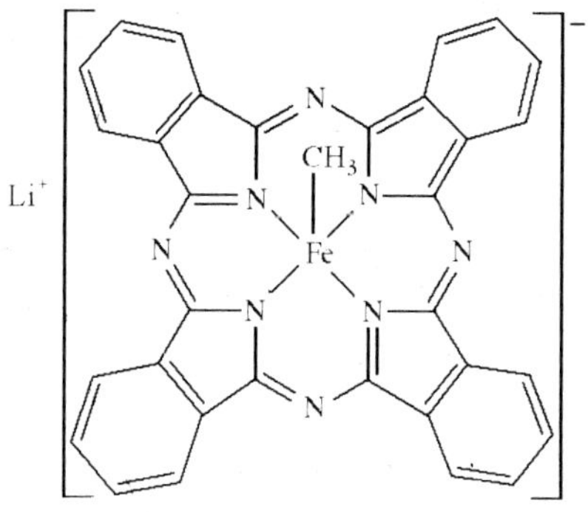

Li^+ (107)

bis-mesityl complexes, $(1,3,5\text{-}Me_3C_6H_2)_2Fe(PR_3)_2$, with only two P-donor ligands (Seidel and Lattermann, 1982).

Perfluoroalkyl and -aryl groups form the most stable η^1-iron complexes and provided some of the first examples of carbonyl-stabilised alkyls by the oxidative addition reactions of perfluoroalkyl iodides (Scheme 4.28) and of perfluoroalkenes to iron carbonyls (Schemes 4.29 and 4.30). The iodides, $R_fFe(CO)_4I$ are generally isolated in *cis* forms but probably exist in equilibrium with the *trans* isomers in solution. Both forms have been isolated for the perfluoropropenyl complex, $CF_3CF=CFFe(CO)_4I$, obtained by reaction of $Fe_2(CO)_9$ with perfluoroallyl iodide (Stanley and McBride, 1976); the *trans* isomer was shown to isomerise to the *cis* form in solution at room temperature. Scheme 4.28 includes some further transformations of the initially formed perfluoropropyltetracarbonyliodoiron (**108**). Addition of tetrafluoroethylene to the hydrocarbonyl, $H_2Fe(CO)_4$, yields *cis*-$(OC)_4Fe(CF_2CF_2H)_2$ for which X-ray crystal structure determination shows an Fe–C

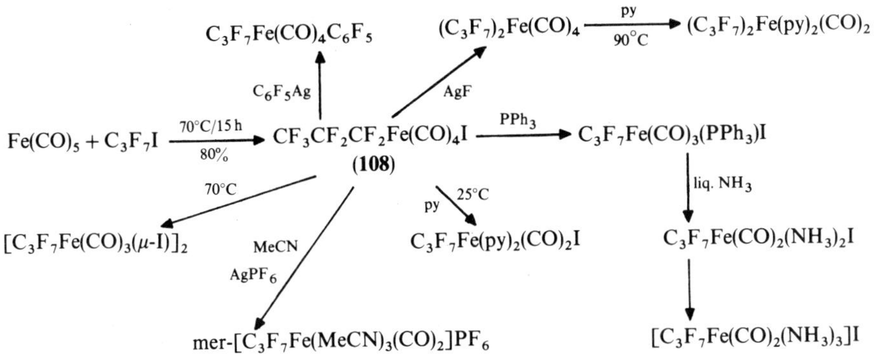

Scheme 4.28

Scheme 4.29

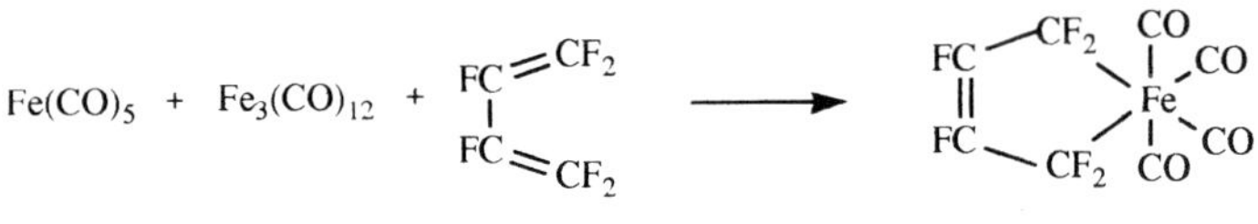

(109)

Scheme 4.30

bond length of 2.068 Å (Churchill, 1967), slightly longer than that (2.00 Å) found in the ferracyclopentene (**109**) (Hitchcock and Mason, 1967). Another route to such complexes is the reaction of 'Collman's reagent' $Na_2Fe(CO)_4$, with perfluoroacyl chlorides, R_fCOCl, to yield the diacyls, $(R_fCO)_2Fe(CO)_4$, which lose carbon monoxide in refluxing heptane or, more slowly *in vacuo* at 40–50°C yielding the dialkyls, $(R_f)_2Fe(CO)_4$.

Of greater synthetic value are the anionic alkyl and acyl complexes formed when Collman's reagent is treated with alkyl or acyl halides (Scheme 4.31) (Collman, 1975). Insertion of carbon monoxide or phosphine converts the alkyl to acyltetracarbonyl – viz. acyltricarbonylphosphine-iron(0) anions and the former are also produced (as lithium salts) when pentacarbonyliron is treated with alkyllithium:

$$Fe(CO)_5 + RLi \rightarrow Li[RCOFe(CO)_4] \quad (4.34)$$

These acyl complexes provide efficient sources of aldehydes, ketones, and carboxylic acids or their derivatives as outlined in Scheme 4.31. They are

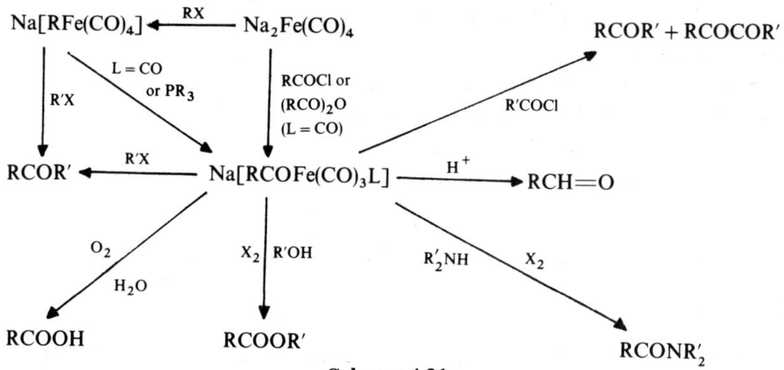

$Na[RFe(CO)_4]$ ← RX — $Na_2Fe(CO)_4$

$RCOR' + RCOCOR'$

L = CO or PR_3

R'X

RCOCl or $(RCO)_2O$ (L = CO)

R'COCl

RCOR' ← R'X — $Na[RCOFe(CO)_3L]$ — H^+ → RCH=O

O_2 / H_2O → RCOOH

X_2 / R'OH → RCOOR'

R'_2NH / X_2 → $RCONR'_2$

Scheme 4.31

Fp—CH₂CH=CHBr ← Et_3N — $[Fp—(CH_2{=}CHCH_2Br)]^+ Br^-$

Br_2, $MeCO^+$, $Me_3O^+ BF_4^-$

$[Fp—(CH_2{=}CHCH_2COMe)]^+$

$[Fp—(CH_2{=}CHCH_2Me)]^+ BF_4^-$

TCNE

Fp—CH₂CH=CH₂ **(110)**

Fp^+—CH₂=CH₂

+ FpI

NaI

H^+ / solv.

+ Fp^+(solv.)

RNCO

$(NC)_2C{=}CHCN$

+ $AlBr_3$

Ce^{IV}/ CO | ROH

ROCO

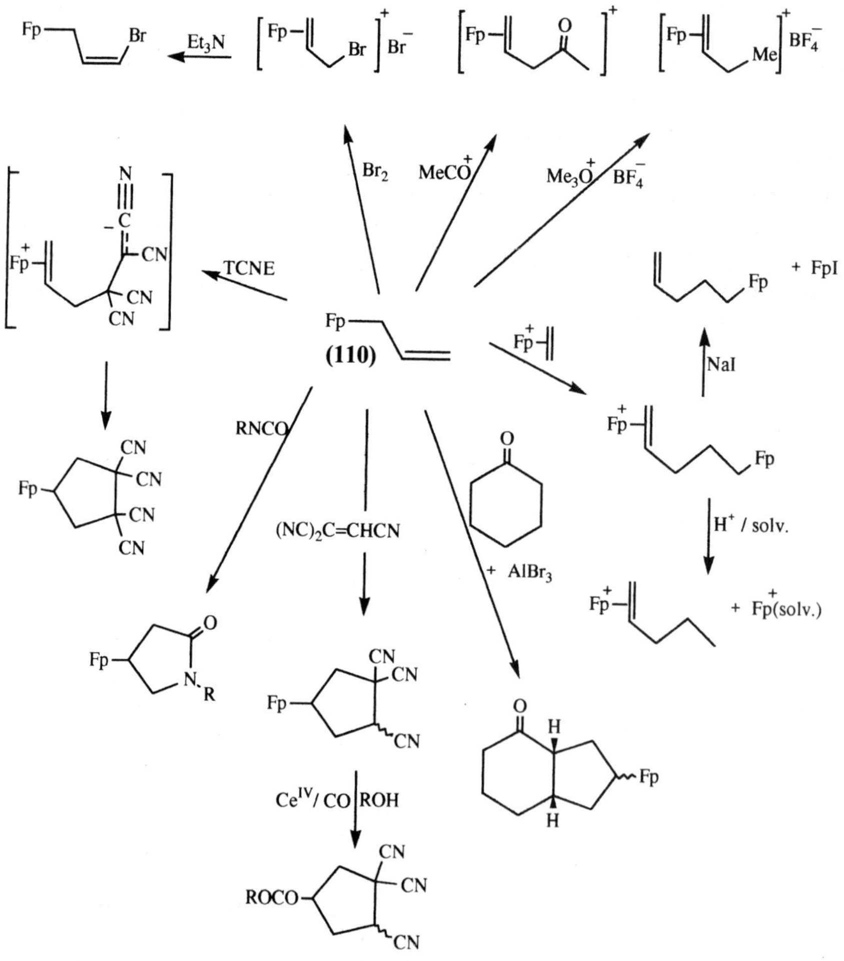

Scheme 4.32

usually prepared *in situ*, but both the alkyl- and acyltetracarbonyliron anions can be isolated, e.g. as $(Ph_3P)_2N^+$ or tetraalkylammonium salts.

Isoelectronic neutral alkyl and acyl complexes result from the sodium tricarbonylnitrosylferrate and alkyl halides:

$$NaFe(NO)(CO)_3 + RX \rightarrow RFe(NO)(CO)_3 \xrightarrow{PPh_3} RCOFe(NO)(CO)_2PPh_3 \quad (4.35)$$

Complexes in which η^1-carbon ligands are stabilised *inter al.* by the presence of cyclopentadienyl ligands have already found repeated mention in section 4.3.2, notably in Scheme 4.19. An extensive chemistry of the dicarbonylcyclopentadienyliron (Fp) linked alkyls has been developed, chiefly by Rosenblum and his school, in which the ability of the η^1-allyl complex (**110**) to add not only protons (Scheme 4.19) but a wide variety of electrophiles (Cutler *et al.*, 1976) and that of the resultant η^2-alkene complexes to add many nucleophiles play a major role. Examples are shown in Scheme 4.32 which includes both cationic electrophiles and neutral electrophiles (e.g. TCNE); the latter yield transient dipolar intermediates which cyclise to the isolated products. Moreover, since the alkyls are readily cleaved by acids and the η^2-alkene complexes by iodide ion, all such reactions have considerable potential value in organic synthesis. As a very simple example, Scheme 4.33 shows how cyclopentene may be converted to either cyclopentyl- or cyclopentenylmalonic ester. Alternatively the alkyls (or aryls) may be transformed into acyls by carbon monoxide insertion, either as a separate process or spontaneously during oxidative degradation to carboxylic acid derivatives. This is ex-

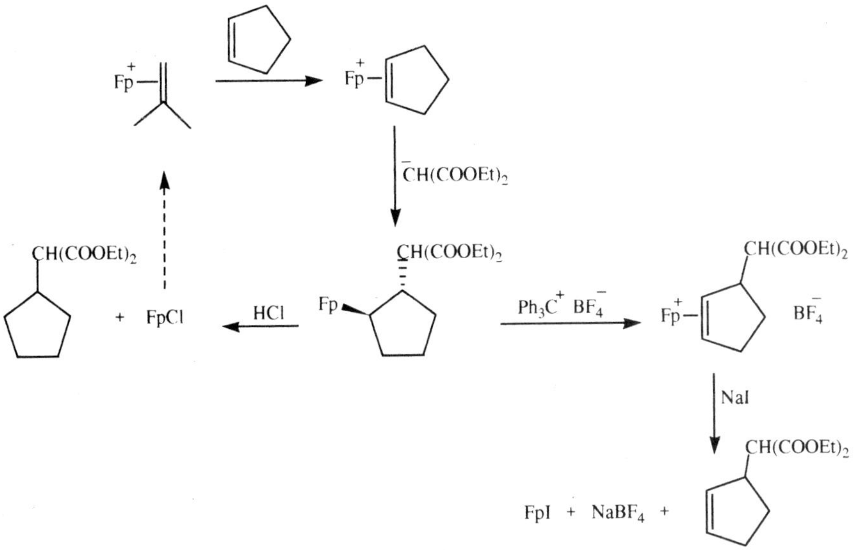

$\overset{+}{Fp}$ $CH(COOEt)_2$ $CH(COOEt)_2$ + FpCl HCl Fp $\overset{+}{Ph_3C}$ $\overset{-}{BF_4}$ $\overset{+}{Fp}$ $\overset{-}{BF_4}$ NaI FpI + $NaBF_4$ +

Scheme 4.33

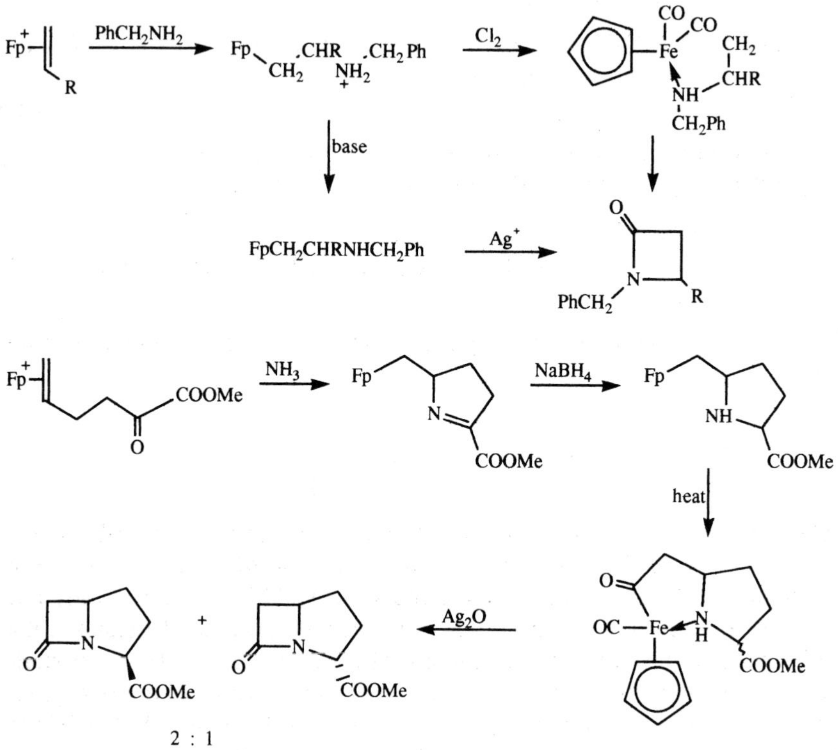

Scheme 4.34

emplified in the efficient carboxylation of the glucose derivative (**111**) (Baer and Hanna, 1982) and in the two routes for the production of β-lactams (Scheme 4.34).

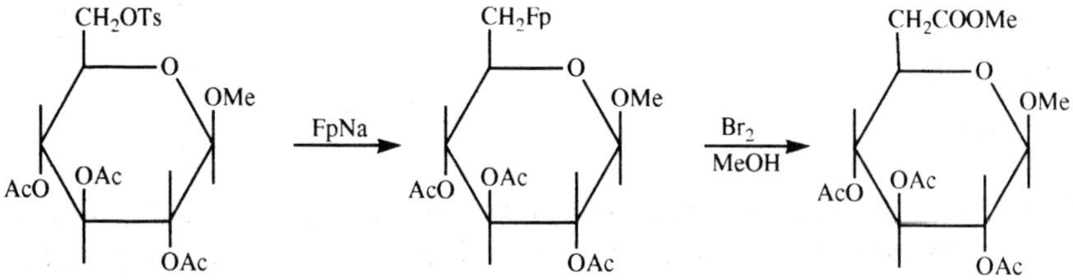

(**111**)

For the ligand (L) promoted carbonyl insertion, the clean intramolecular migration mechanism is emphasised by the retention of configuration observed when an asymmetric carbon of the alkyl group is initially attached to iron:

$$\text{Fp–*R} + \text{L} \rightarrow \text{C}_5\text{H}_5\text{Fe}\begin{matrix}\diagup\text{CO}\\ \text{—L}\\ \diagdown\text{CO*R}\end{matrix} \xrightarrow[\text{MeOH}]{\text{Br}_2} \text{C}_5\text{H}_5\text{Fe(CO)(L) Br} + \text{*RCOOMe} \quad (4.36)$$

The phosphine-promoted carbonyl insertion can be reversed using copper(I) as a phosphine-abstracting agent (Levitre *et al.*, 1986):

$$\text{CpFe}\begin{matrix}\diagup\text{CO}\\ \text{—PPh}_3\\ \diagdown\text{COMe}\end{matrix} + [\text{Cu(MeCN)}_2]\text{PF}_6 \rightarrow [\text{Cu(PPh}_3)_n]\text{PF}_6 + \text{FpMe} \quad (4.37)$$

The η^1-acyl complexes of the type Fp-$COCH_2R$ yield carbanions (enolate ions)

$$\text{Fp–CO}\bar{\text{C}}\text{HR} \leftrightarrow \text{Fp–C}\begin{matrix}\diagup\text{O}^-\\ \diagdown\!\!\diagdown\text{CHR}\end{matrix}$$

when treated with strong bases, e.g. lithium diisopropylamide (LDA), and hence can be alkylated or undergo aldol type or Michael condensations (Aktogu *et al.*, 1982, 1984; Liebeskind and Welker, 1983). Moreover, when one carbonyl group on iron is replaced by triphenylphosphine, the result is not only to make the iron atom chiral, but also to provide a substantial steric effect. As illustrated in structure (**112**), one phenyl group of the phosphine ligand will always lie under the enolate group and force any electrophile to approach from the opposite side. If additionally, the enolate group adopts a strongly preferred conformation, this leads to a high level of diastereoselectivity as e.g. in the alkylation reaction (Scheme 4.35) (Baird and Davies, 1983).

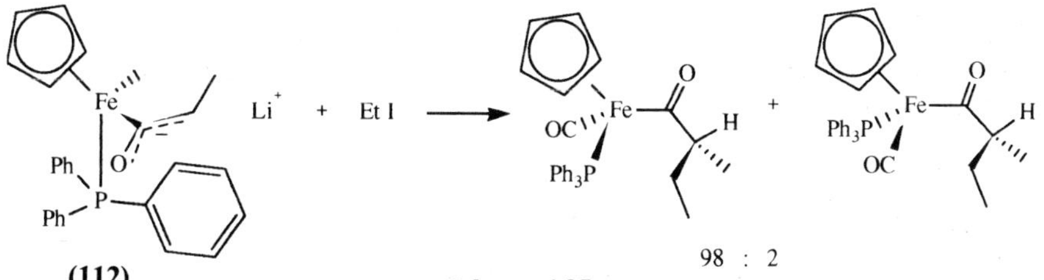

Li$^+$ + Et I ⟶ ... + ...

98 : 2

(112)

Scheme 4.35

In practice, the preferred orientation of the enolate group depends strongly on the counterion and replacement of e.g. lithium by other metals (e.g. Al, Sn, Cu) may enhance or reverse the diastereoselectivity (see example in Scheme 4.36). Further, in the example of Scheme 4.35, the initial enolate is obtained from the propanoyl complex, $CpFe(CO)(PPh_3)COCH_2CH_3$, itself obtainable from the enolate of the acetyl complex $CpFe(CO)(PPh_3)COCH_3$ with iodomethane. Clearly the diastereoselectivity will also be reversed in this case by alkylating in the reverse sequence (first with EtI and then MeI).

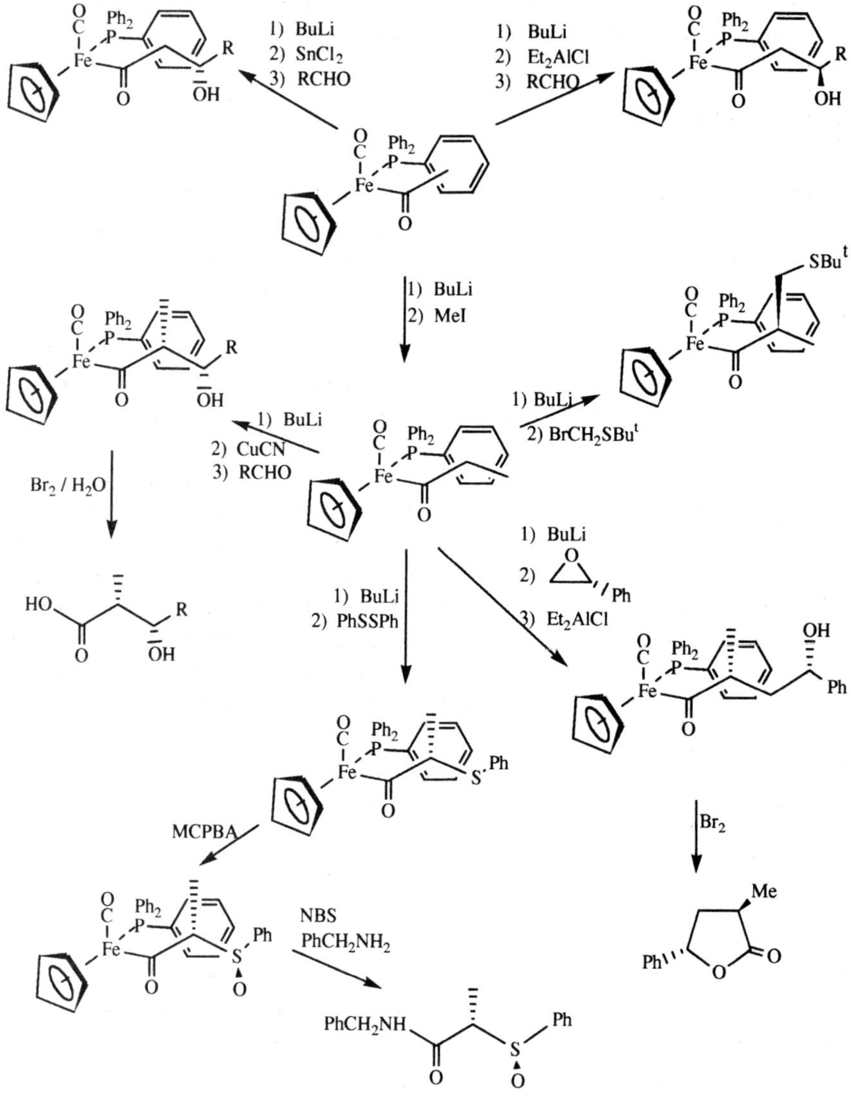

Scheme 4.36

While the above examples emphasise the diastereoselectivity achieved when racemic metal complexes are used, it also follows that use of resolved acyl complexes, $CpFe(CO)(PPh_3)COR$, will result in such reactions leading to high levels of enantioselectivity. The success, chiefly of S.G. Davies and his coworkers (Davies *et al.*, 1988, 1990) in developing a whole range of procedures based on this principle has led to the optically resolved complex $C_5H_5Fe(CO)(PPh_3)COMe$ becoming a commercially available starting material (from Oxford Chirality). A small selection of the stereoselective syntheses developed by Davies utilising these acyliron compounds is shown in Scheme 4.36.

4.4.2 η^2-Carbon–iron complexes

Apart from the carbonyls and closely related complexes (thiocarbonyls and isocyanide complexes) already covered in section 4.2, the η^2-complexes comprise the η^2-alkene, η^2-alkyne and η^2-carbene complexes and will be treated in that order.

4.4.2.1 η^2-Alkene complexes. Simple alkenes on photolysis with pentacarbonyliron, or, more efficiently, thermal reaction with enneacarbonyldiiron, afford (η^2-alkene)tetracarbonyliron complexes, e.g. the ethylene complex (Scheme 4.37). Alkenes with electron-withdrawing groups, – e.g. maleic anhydride, yield more stable complexes of this type than simple alkenes. Dienes react similarly, e.g. butadiene giving both mono- and bis-tetracarbonyliron complexes (Scheme 4.37). The former loses carbon monoxide on warming to give the η^4-complex $C_4H_6Fe(CO)_3$ (**3**). If the diene is unconjugated but capable of chelating it will give a bis-η^2-complex as exemplified for 1,5-cyclooctadiene (Scheme 4.38). Although they may be isolated and handled at room temperature, none of these η^2-alkene–iron carbonyls is very stable and their chemistry has not been studied extensively.

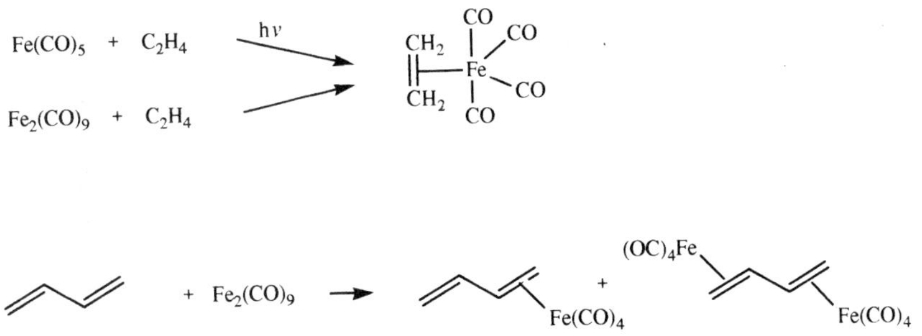

$Fe(CO)_5 + C_2H_4 \xrightarrow{h\nu}$ $(\eta^2\text{-}CH_2{=}CH_2)Fe(CO)_4$

$Fe_2(CO)_9 + C_2H_4 \rightarrow$ $(\eta^2\text{-}CH_2{=}CH_2)Fe(CO)_4$

butadiene $+ Fe_2(CO)_9 \rightarrow$ $(\eta^2\text{-}C_4H_6)Fe(CO)_4$ + $(OC)_4Fe(\eta^2,\eta^2\text{-}C_4H_6)Fe(CO)_4$

Scheme 4.37

1,5-cyclooctadiene $+ Fe_2(CO)_9 \rightarrow Fe(CO)_5 + CO +$ $(\eta^2,\eta^2\text{-}C_8H_{12})Fe(CO)_3$

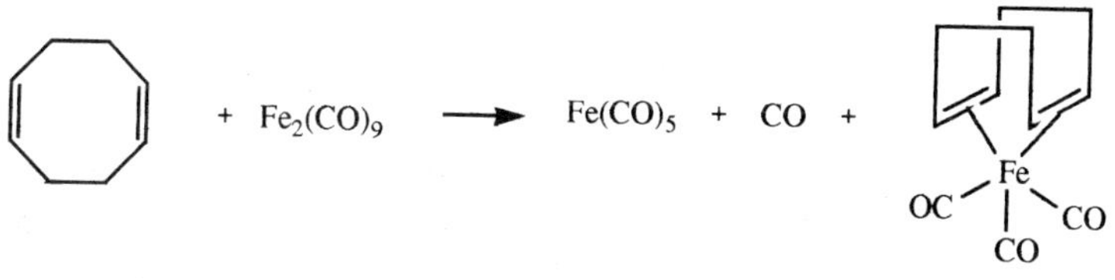

Scheme 4.38

A much more stable class of η^2-complexes of iron are the cationic cyclopentadienyls $[C_5H_5Fe(CO)_2(\eta^2\text{-alkene})]^+$ whose ready formation from and conversion back to η^1-complexes (by hydride abstraction and by addition of nucleophiles respectively) has been discussed in section 4.4.1. A valuable alternative to the hydride abstraction method of preparation is provided by protonation of β-hydroxyalkyls, themselves readily available from epoxides (Scheme 4.39).

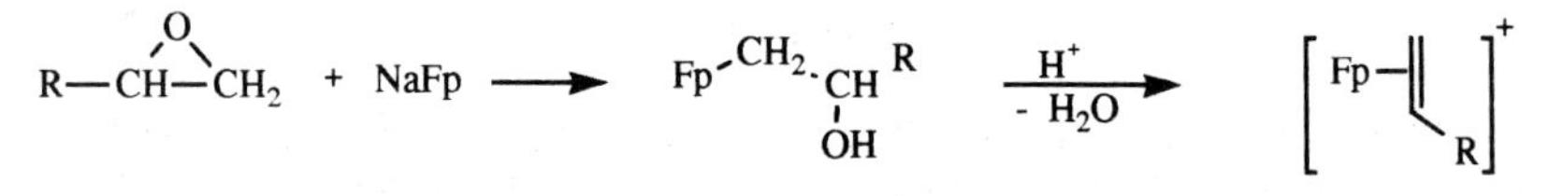

Scheme 4.39

The use of this sequence to deoxygenate epoxides can be controlled to yield either *cis* or *trans* alkenes since acid catalysed (Scheme 4.40a) and thermal eliminations (Scheme 4.40b) follow stereochemically different mechanisms.

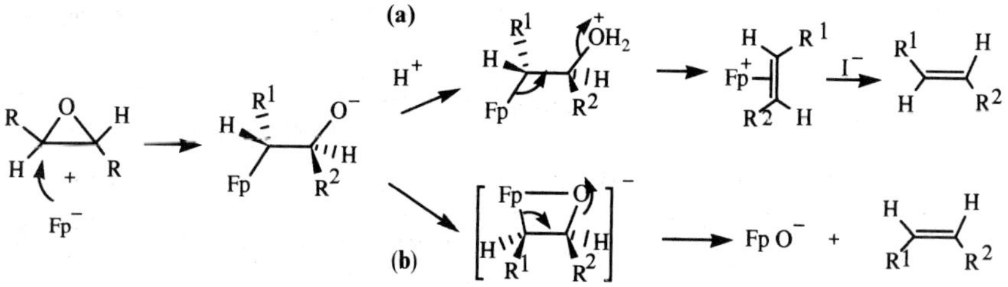

Scheme 4.40

Another valuable preparative route is the facile alkene exchange reaction, commonly starting with the isobutene complex; an example has been included in Scheme 4.33. Alternatively, cationic complexes with readily displaceable solvent molecules (e.g. Me_2CO, CH_2Cl_2, THF) as ligands are generated and used to obtain alkene complexes by exchange (e.g. Scheme 4.41) (Schmidt and Thiel, 1981).

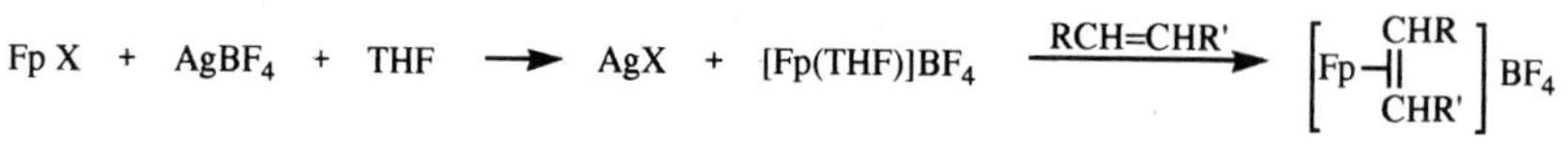

$$\text{Fp X} + \text{AgBF}_4 + \text{THF} \longrightarrow \text{AgX} + [\text{Fp(THF)}]\text{BF}_4 \xrightarrow{\text{RCH=CHR'}} \left[\text{Fp}\text{—}\|\begin{smallmatrix}\text{CHR}\\ \text{CHR'}\end{smallmatrix}\right]\text{BF}_4$$

Scheme 4.41

In the pentamethylcyclopentadienyl series the aquo complex (Scheme 4.38) has been prepared in high yield from the methyl derivative and found to be air-stable and suitable for displacement of water by alkenes (Scheme 4.42) (Tahiri *et al.*, 1990).

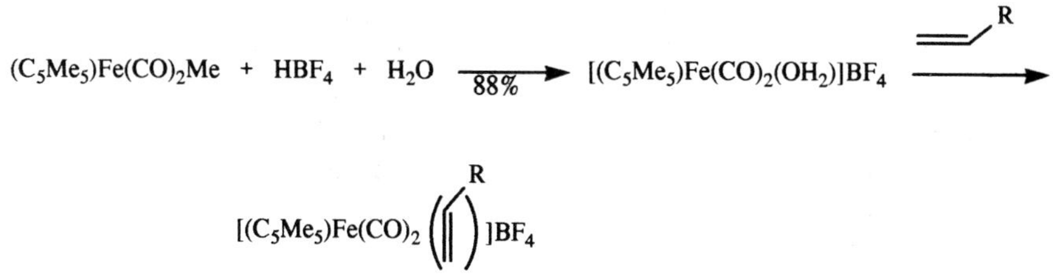

$(C_5Me_5)Fe(CO)_2Me + HBF_4 + H_2O \xrightarrow{88\%} [(C_5Me_5)Fe(CO)_2(OH_2)]BF_4 \xrightarrow{\text{=/R}} [(C_5Me_5)Fe(CO)_2(\text{alkene-R})]BF_4$

Scheme 4.42

Nucleophilic additions to such cationic alkene complexes have been adequately exemplified in Schemes 4.15 and 4.32–4.34. When a hydrogen atom alpha to the double bond is available, it is lost on treatment of the cationic complex with weakly nucleophilic bases as in Scheme 4.43; this results in the production of σ-allylic complexes, effectively reversing one of the methods of preparing the alkene complexes.

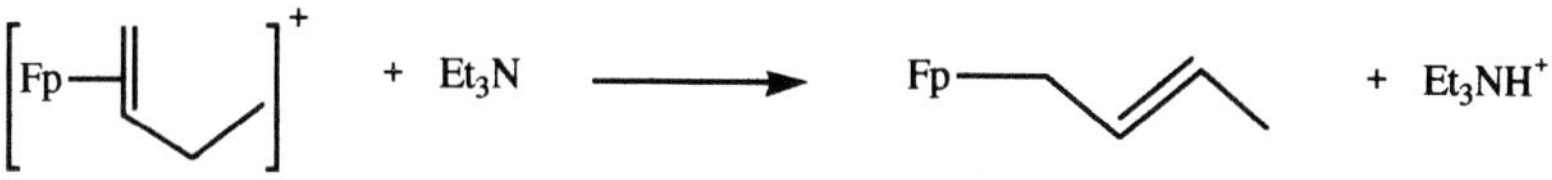

$[Fp(\text{alkene})]^+ + Et_3N \longrightarrow Fp\text{-allyl} + Et_3NH^+$

Scheme 4.43

A significant use of such alkene complexes is as a device for protecting double bonds: In the first example shown (Scheme 4.44), the more reactive double bond of the diene is selectively complexed by the alkene exchange route, allowing the less reactive one to be hydrogenated. The metal-free terminal alkene can then be liberated by reaction with iodide. In the second example (Scheme 4.45) the olefin is protected from addition during ring bromination and the free bromoaryl-olefin is then liberated in overall 80% yield (including the isobutene exchange step).

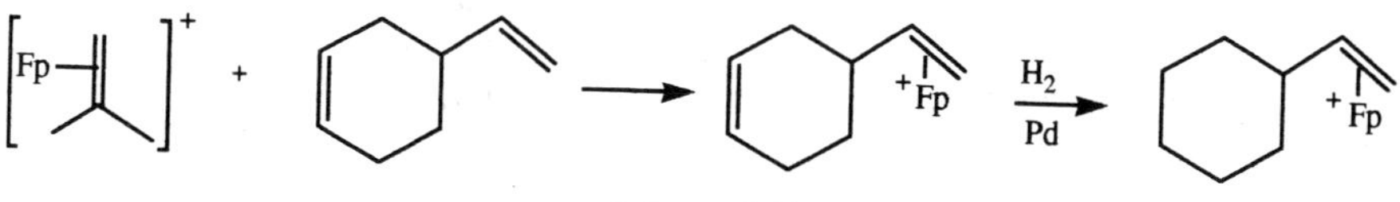

Scheme 4.44

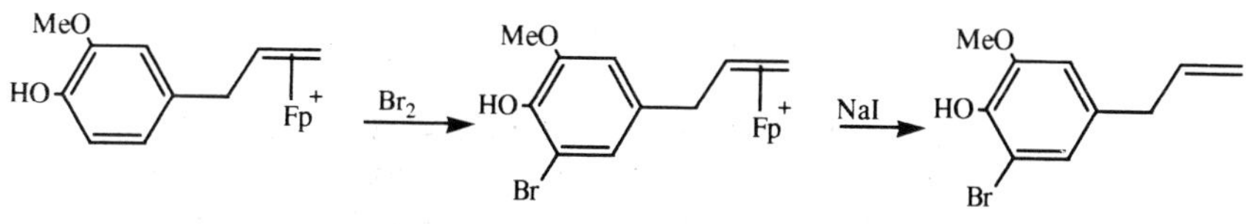

Scheme 4.45

4.4.2.2 η^2-Alkyne complexes. Like the preceding alkene complexes, the analogous cationic η^2-alkyne cyclopentadienyliron complexes are most commonly prepared by exchange reactions from the isobutene or solvate complexes. They have not normally been isolated and characterised, but have proved valuable in transformations (e.g. of alkynes to functionally substituted haloalkenes) involving nucleophilic addition. Scheme 4.46 gives examples. These show that for internal alkynes the expected alkene or alkenyl complexes

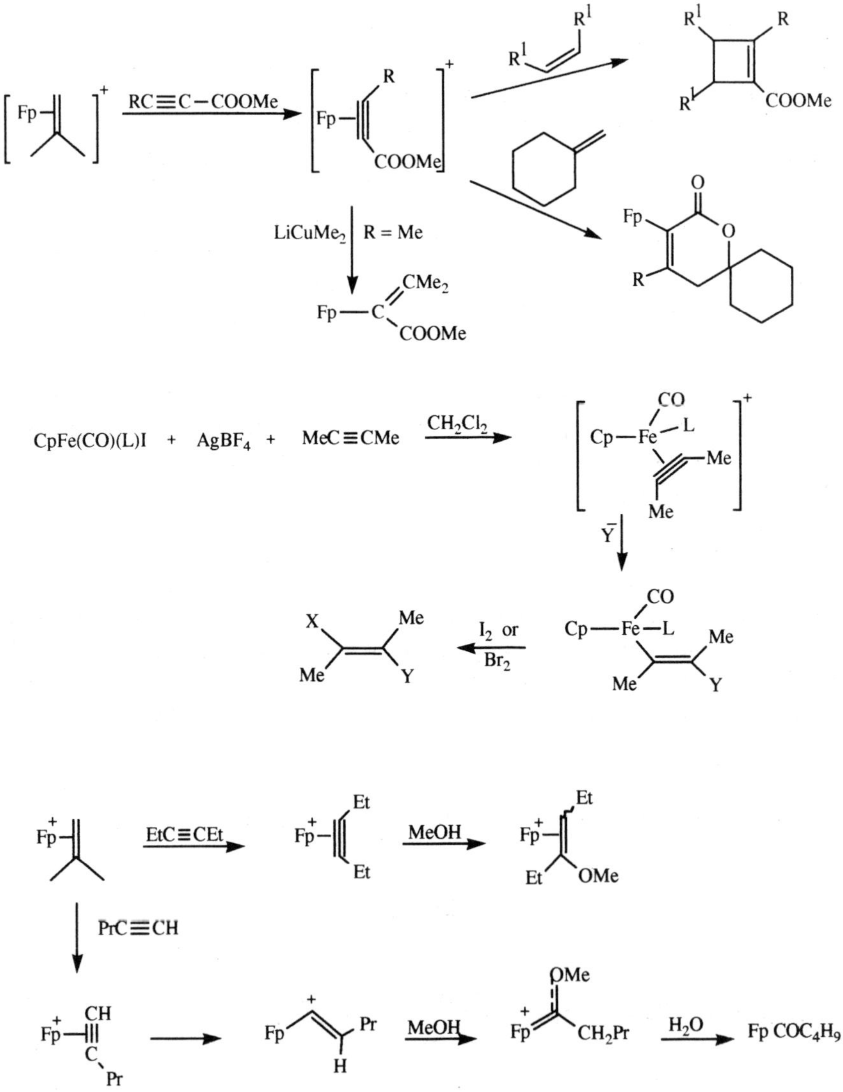

Fp +
$RC{\equiv}C{-}COOMe$
Fp R COOMe +
R^1 R^1
R^1 R R^1 COOMe
LiCuMe$_2$ R = Me
Fp—C CMe_2 COOMe
Fp O O R

$CpFe(CO)(L)I$ + $AgBF_4$ + $MeC{\equiv}CMe$ $\xrightarrow{CH_2Cl_2}$
Cp—Fe CO L Me Me +
$\bar{Y}$
Cp—Fe CO L Me Me Y
I_2 or Br_2
X Me Me Y

Fp + EtC≡CEt Fp + Et Et
MeOH
Fp + Et Et OMe
PrC≡CH
Fp + CH C Pr
Fp + Pr H
MeOH
Fp + OMe CH_2Pr
H_2O
Fp COC_4H_9

Scheme 4.46

result from such additions. However, the less stable complexes formed by terminal alkynes may rearrange before addition can occur, even when, for example, an alcohol is present during the alkene/alkyne exchange; the isolated products then result from addition to the isomerised cation (e.g. Scheme 4.46) (Bates *et al.*, 1981).

Apart from the above cyclopentadienyliron derivatives very little is known about η^2-alkyne–iron complexes. Reactions of the dinuclear nitrogen complexes, $(OC)_2L_2Fe{-}N{\equiv}N{-}Fe{-}L_2(CO)_2$ [L = PEt_3 or $P(OMe)_3$] with several alkynes result in replacement of N_2 to yield complexes $(OC)_2L_2Fe(RC{\equiv}CH)$. A pentagonal bipyramidal structure with apical phosphine ligands has been confirmed for the complex $(OC)_2L_2Fe(PhC{\equiv}CH)$. Depending on the nature of both L and R the complexes are in equilibrium with vinylidene [$(OC)_2L_2Fe{=}C{=}CHR$] and/or acetylide tautomers [$(OC)_2L_2Fe(H)C{\equiv}CR$] (Birk *et al.*, 1988; Grössman *et al.*, 1991). By addition of an acetone solution of $FeCl_2$ to the titanium complex $(\eta^5{-}Me_3SiC_5H_4)_{23}Ti(C{\equiv}CSiMe_3)_2$ a red crystalline compound is formed which has been shown to have the $FeCl_2$ group π-bonded to both triple bonds with tetrahedral coordination about iron; despite being an outer-orbital (d^6) complex with four unpaired electrons it is moderately air as well as thermally stable (Lang *et al.*, 1991).

4.4.2.3 η^2-Carbene–iron complexes. The classic Fischer synthesis of carbene complexes, so successful for metals of the chromium group, involves the steps shown in Scheme 4.47. Trimethyl- or triethyloxonium salts, the preferred alkylating agents (i.e. $X = R'_2O^+$ in step b of Scheme 4.47) when $[M] = Cr(CO)_5$ or $W(CO)_5$ proved successful in the iron series for the carbonylnitrosyliron case (Scheme 4.48) or in the preparation of the aminoalkoxycarbene complex (**113**) (Fischer *et al.*, 1972). But the acyltetracarbonyliron anions react cleanly with alkyl halides to give ketones (Scheme 4.31) presumably via initial alkylation on the metal atom and even with trialkyloxonium salts *O*-alkylation appears to be only the minor pathway. However efficient *O*-alkylation occurs when the hardest leaving groups X are

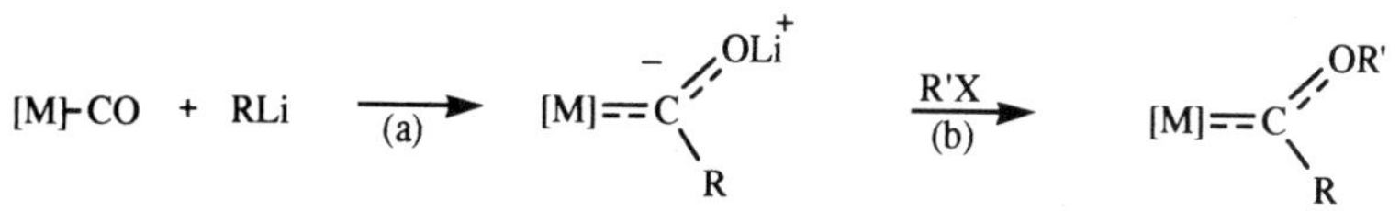

$$[M]{-}CO + RLi \xrightarrow{(a)} [M]{=}C^{-}(OLi^{+})(R) \xrightarrow[(b)]{R'X} [M]{=}C(OR')(R)$$

Scheme 4.47

$$Fe(CO)_2(NO)_2 \xrightarrow{PhLi} Fe(CO)(NO)_2{=}C^{-}(OLi^{+})(Ph) \xrightarrow{Et_3\overset{+}{O}\ \overset{-}{BF_4}} Fe(CO)(NO)_2{=}C(OEt)(Ph) + LiBF_4 + Et_2O$$

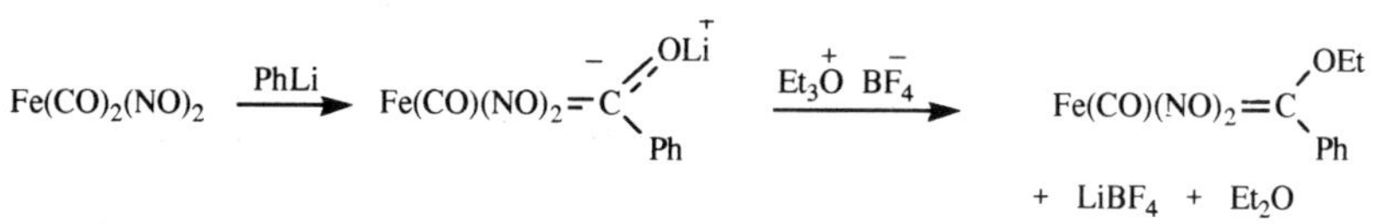

Scheme 4.48

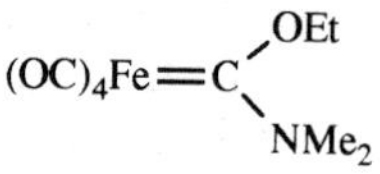

$(OC)_4Fe{=}C(OEt)(NMe_2)$

(113)

employed. The fluorosulphonates $MeOSO_2F$ (Conder and Darensbourg, 1974) and $EtOSO_2F$ (Semmelhack and Tamura, 1983) are the preferred reagents, the ethyl compound being slightly better than the lower homologue and *O*-alkylation being further favoured by replacing lithium in the salts $Li[Fe(CO)_4COR]$ by tetramethylammonium and by adding HMPA to the (ether) reaction medium.

An interesting alternative route giving this type of carbene complex, albeit in only moderate yields, involves the direct photochemical carbene transfer from molybdenum to iron as illustrated by the preparation of the dimethylamino complex (**114**) (Fischer *et al.*, 1972), more efficiently available by nucleophilic substitution from the methoxy complex (**115**) (Semmelhack

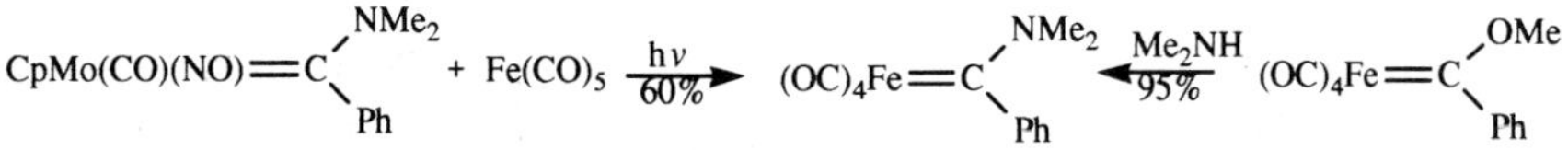

$$CpMo(CO)(NO){=}C(NMe_2)(Ph) + Fe(CO)_5 \xrightarrow[60\%]{h\nu} (OC)_4Fe{=}C(NMe_2)(Ph) \xleftarrow[95\%]{Me_2NH} (OC)_4Fe{=}C(OMe)(Ph)$$

(114) **(115)**

and Park, 1986). Other routes to such carbene complexes include the reaction of $Na_2Fe(CO)_4$ with halogenopyridinium and related salts as in Scheme 4.50 (Green *et al.*, 1975) and the sulphur extrusion involved in the formation of the trinuclear complex (**116**) (Scheme 4.51) (Benoit *et al.*, 1982).

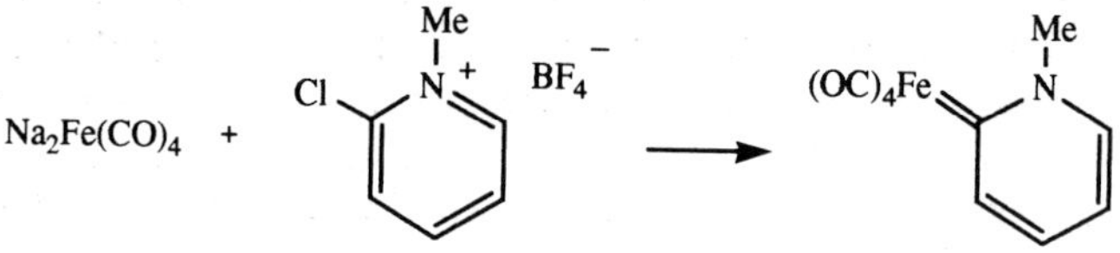

$Na_2Fe(CO)_4$ + 2-chloro-N-methylpyridinium BF_4^- → $(OC)_4Fe$=C (N-Me pyridylidene)

Scheme 4.49

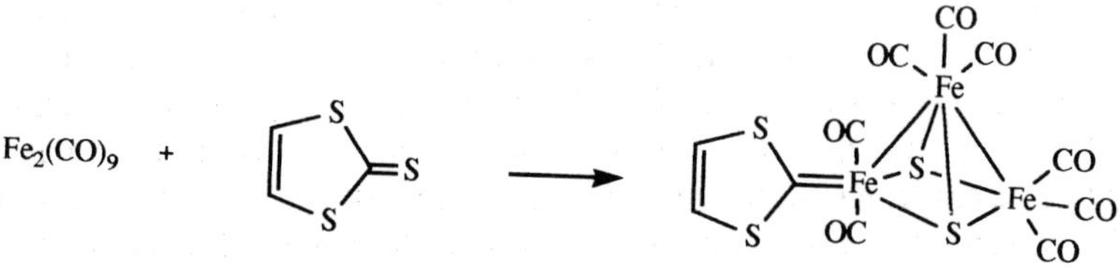

$Fe_2(CO)_9$ + 1,3-dithiole-2-thione → **(116)**

Scheme 4.50

Tetracarbonyliron-carbene complexes appear to be known only with at least one donor atom attached to the carbene-carbon and hence only partial Fe–C double bond character. Typical Fe–$C_{(carbene)}$ distances, e.g. 2.007 Å in compound (**117**) (Huttner and Gartzke, 1972) are therefore greater than Fe–CO distances.

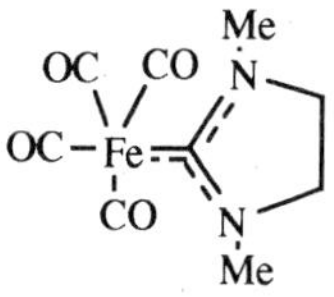

(117)

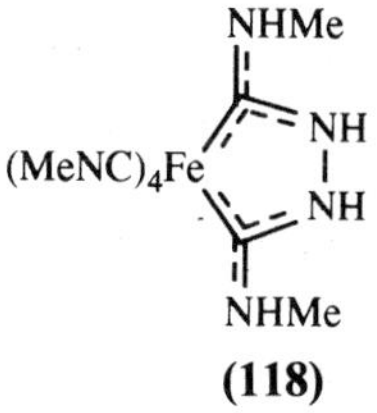

(118)

Related complexes with other counterligands are exemplified by the compound (**118**) obtained from $[Fe(CNMe)_6]^{2+}$ with hydrazine (Balch and Miller, 1972) and the vinylidene complex (**119**) resulting from rearrangement of phenylacetylene during reaction (Scheme 4.51) (Bellerby and Mays, 1976).

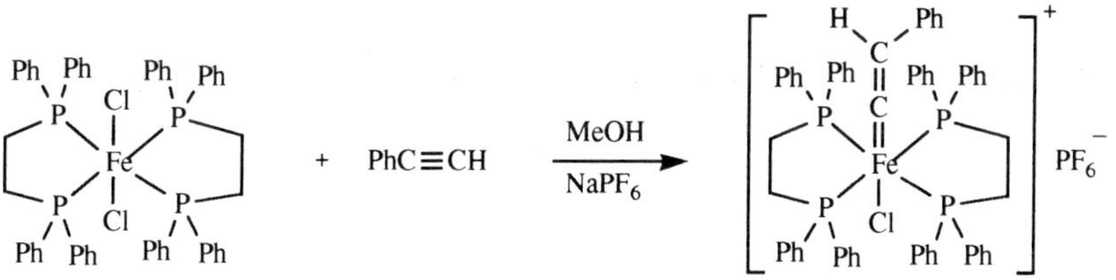

(119)

Scheme 4.51

As is the case for the η^2-alkene complexes, the chemistry of cationic cyclopentadienyl iron-η^2–carbenes is much better developed than that of neutral carbonyliron-carbenes. They are formed when a positive charge is generated on the α-carbon atom by either protonation and methanol elimination from compounds of the type FpCRR′(OMe) or by alkylation or protonation on X of the acyl-, thioacyl-, imino- and vinyl-compounds, Fp–C(R)=X (X = O, S, NR′, CHR′) or other closely similar processes. An example of alkylation of an acyl complex was included in Scheme 4.19; the two routes are used in sequence to obtain carbene complexes lacking a donor group by reducing the initial alkoxycarbene complex before alkoxide abstraction. In Scheme 4.52 (Brookhart *et al.*, 1983) this is exemplified together with the use of a homochiral phosphine ligand to obtain an asymmetric carbene complex. The reduction step is an example of the ability of the cationic carbene complexes to add nucleophiles (cf. also Reger and Swift, 1984): apart from hydride (as $NaBH_4$, $NaBH_3CN$, $LiBHEt_3$, etc.) organometallic nucleophiles, e.g. alkyllithiums and cuprates, add readily as

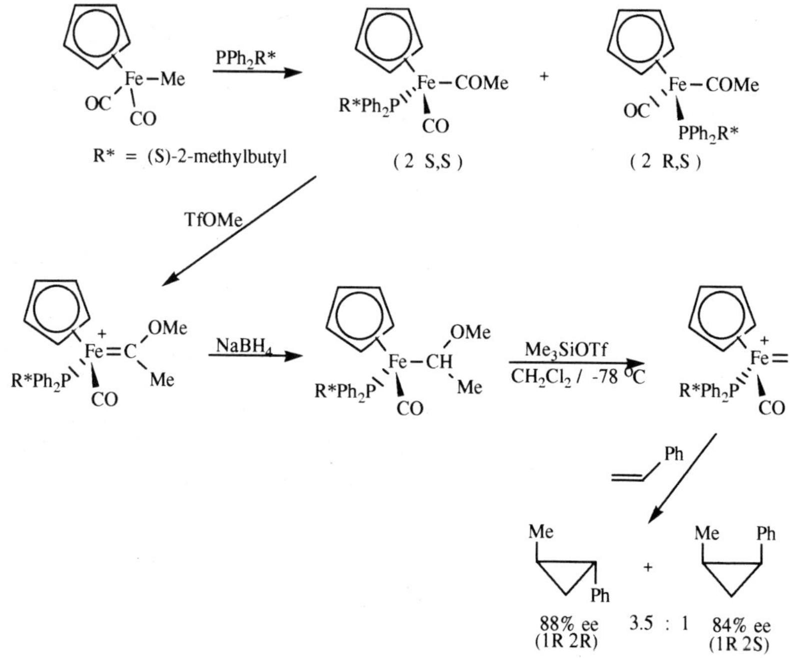

Scheme 4.52

do mercaptide ions. Those carbene complexes lacking donor groups are valuable cyclopropanating agents by transfer of the carbene moiety to alkenes and alkynes (Brookhart, 1990; Helquist, 1991). As shown in Scheme 4.52 this occurs with substantial stereoselectivity and enantioselectivity when a homochiral complex is used. In these reactions the alkene is normally present while the carbene compound is being generated to ensure reaction before rearrangement of the carbene to an η^2-alkene complex. For carbene compounds $[Fp–CRR']^+$ (R = alkyl) this occurs by hydrogen migration (e.g. Scheme 4.53) (Casey *et al.*, 1982; Kremer *et al.*, 1982) or by alkyl migration (e.g. Scheme 4.54) (Bly *et al.*, 1988) and if R = alkyl and R′ = H, so rapidly that the carbene complex cannot be observed spectroscopically even at −78 °C. Surprisingly, the cyclopropylcarbene $[Fp{=}CHC_3H_5]^+$ is relatively stable and has been so characterised (Brookhart *et al.*, 1985) and the benzyl-

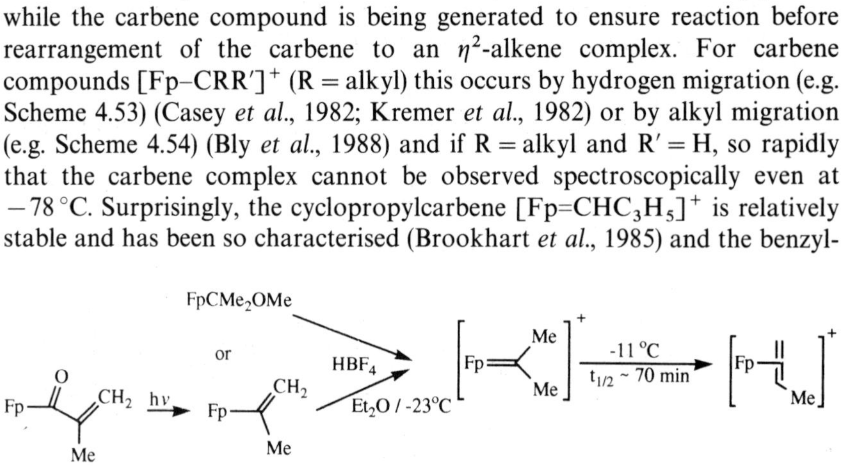

Scheme 4.53

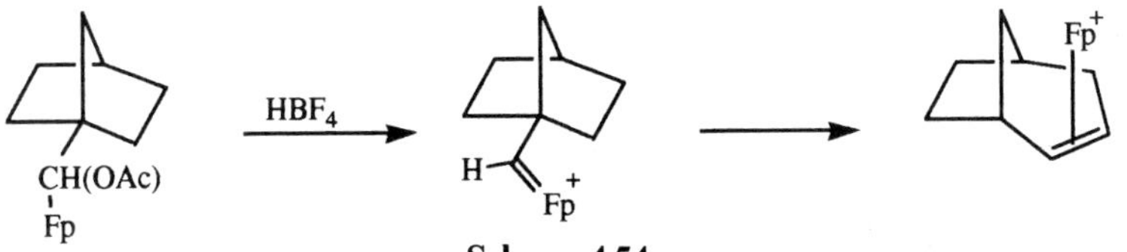

Scheme 4.54

idene compound $[Fp{=}CHPh]PF_6$ which cannot rearrange in this way has been isolated crystalline (Brookhart *et al.*, 1980). Phosphine substituted complexes $[CpFe(CO)(PR_3){=}CHR]^+$ are also relatively stable, rearranging on warming (Brookhart, Tucker and Husk, 1983). Scheme 4.53 also includes the alternative generation of the carbene cation by proton addition to an η^1-vinyl compound.

In place of methoxide abstraction use of the corresponding phenylthio compounds offers the advantage that on methylation they yield sulphonium salts (e.g. **120**) which are stable at room temperature but decompose (without

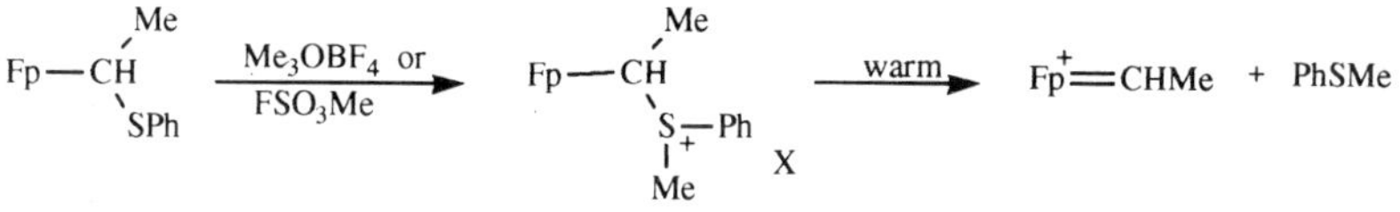

(120)

requiring acid) to the carbene complexes on warming (Kremer and Helquist, 1985).

Various other routes which have been employed to generate such cationic carbene complexes probably have less general applicability. However, the formation of vinylidenecarbene complexes by Scheme 4.55 (Boland-Lussier and Hughes, 1982) deserves mention because of the synthetically promising imine additions to the product (e.g. Scheme 4.56) which have been used *inter al.* to obtain penicillin analogues (Barrett *et al.*, 1988).

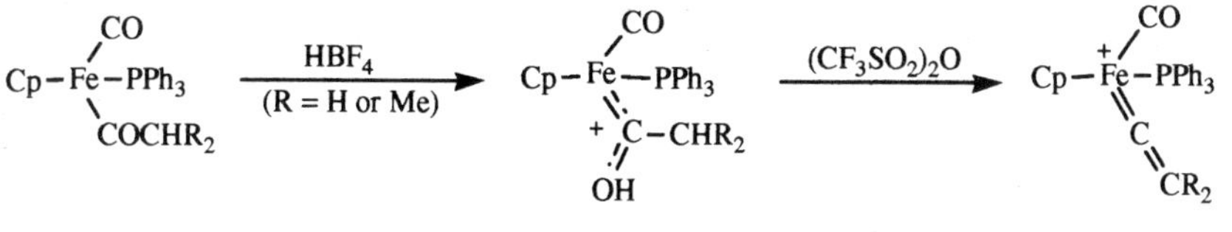

Scheme 4.55

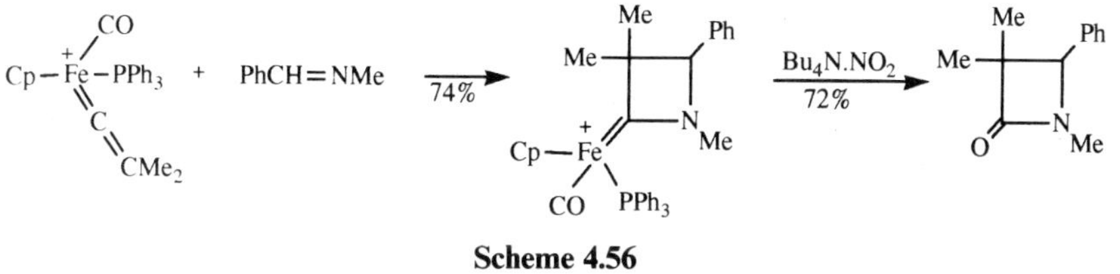

Scheme 4.56

4.4.3 η^3-Allyl iron complexes

4.4.3.1 Ligands which coordinate only through η^3-allylic groups. η^3-Allyliron complexes were first recognised as products of the addition of hydrogen chloride to η^4-diene-tricarbonyliron complexes. Thus, the butadiene complex (**3**) yields the η^3-crotyliron compound (**121**), a yellow–brown solid, stable at room temperature, but decomposing in boiling benzene according to equation 4.38 with partial regeneration of the diene complex (Impastato and Ihrman, 1961).

$$2C_4H_7Fe(CO)_3Cl \rightarrow C_4H_6Fe(CO)_3 + FeCl_2 + C_4H_8 + 3CO \qquad (4.38)$$

Analogous addition to the *anti*-pentadiene complex has been shown to be accompanied by *anti* to *syn* isomerisation (Scheme 4.57) (Gibson and Erwin,

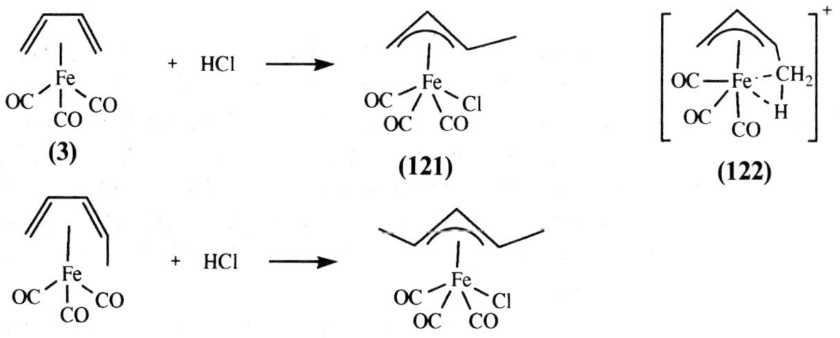

(3) + HCl ⟶ (121)

(122)

Scheme 4.57

1975). When acids with weakly coordinating anions are added to the butadiene complex (**3**) at low temperature, the initial product is the crotyltricarbonyliron cation; spectroscopic study of this formally electron-deficient species indicates that there is agostic Fe–CH bonding as shown in structure **122** (Brookhart *et al.*, 1976). This and related cations are unstable and the isolated products from such additions are the 18-electron tetracarbonyls (**123**) formed by disproportionation (Scheme 4.58) (Gibson

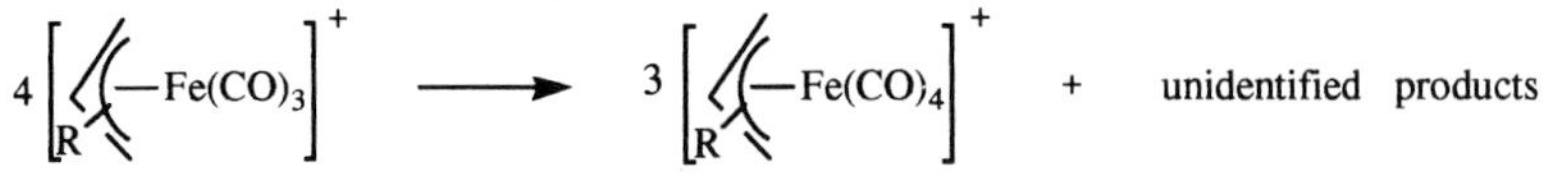

$$4\left[(\eta^3\text{-RC}_3H_4)Fe(CO)_3\right]^+ \longrightarrow 3\left[(\eta^3\text{-RC}_3H_4)Fe(CO)_4\right]^+ + \text{unidentified products}$$

(123)

Scheme 4.58

and Vonnahme, 1972) unless good donor solvents or other ligands are available to form cations of the type $[(\eta^3\text{-RC}_3H_4)Fe(CO)_3L]^+$. The tetracarbonyl cations (**123**) also result smoothly by protonation of the $(\eta^2\text{-diene})Fe(CO)_4$ complexes (Gibson and Vonnahme, 1974) or the η^2-allyl alcohol or acetate complexes $(ROCH_2CH{=}CH_2)Fe(CO)_4$ (R = H or Ac) (Dieter and Nicholas, 1981). Addition of other electrophiles to the diene complexes, notably of acyl cations to give α-acyl-η^3-allyltricarbonyliron cations is described in section 4.4.4.

The alternative route to allyltricarbonyliron compounds including the parent halides (**124**; L=CO; X=Cl, Br, or I) is by direct reaction of allyl

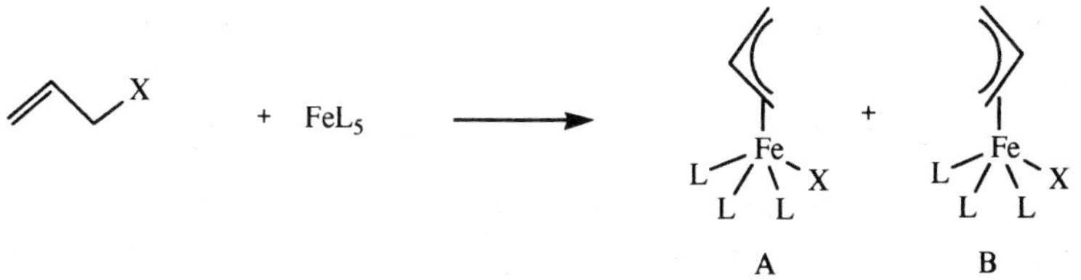

$$CH_2{=}CHCH_2X + FeL_5 \longrightarrow (\eta^3\text{-}C_3H_5)FeL_3X \ (\mathbf{A}) + (\eta^3\text{-}C_3H_5)FeL_3X \ (\mathbf{B})$$

(124)

halides with $Fe_2(CO)_9$ or an irradiation with $Fe(CO)_5$. This route has been extended to the phosphorus trifluoride analogue ($L = PF_3$) which requires irradiation (Kruck and Knoll, 1973) and to the trimethylphosphite analogue $[L = P(OMe)_3]$ which yields the cationic product $[(\eta^3\text{-}C_3H_5)FeL_4]^+$ (Muetterties and Rathke, 1974). The chloride (**124**, L = CO, X = Cl) has also been obtained by ligand transfer on reacting $[(\eta^3\text{-}C_3H_5)PdCl]_2$ with $Fe_2(CO)_9$ (Nesmeyanov *et al.*, 1969). Such molecules (**124**) exist as two rotamers (Faller and Adams, 1979).

Among derivatives of allyltricarbonylhalogeno–iron prepared by the first of these routes [from $CH_2{=}C(OSiMe_3)CH_2Br + Fe_2(CO)_9$] is the remarkably acidic (pK_a *ca.* 5.2) 2-hydroxy derivative (**125**) (Frey *et al.*, 1990).

$$HO\text{-}(\eta^3\text{-}C_3H_4)\text{—}Fe(CO)_3Br$$

(125)

The addition of iron hydride species to dienes has been used to generate the trichlorosilyl derivatives (**126**) (Connolly, 1984) and the cationic trimethylphosphine complex (**128**) as shown, the latter being in equilibrium with the

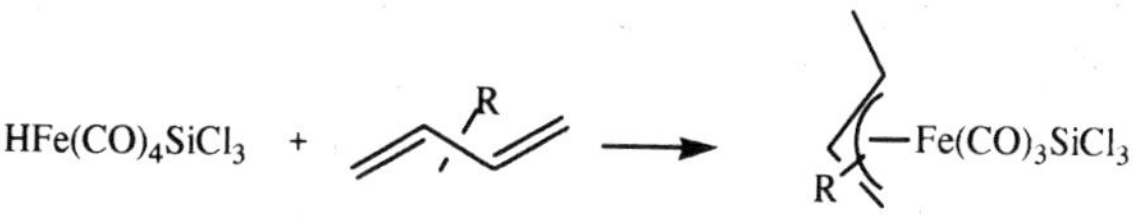

$$HFe(CO)_4SiCl_3 + \text{diene (R)} \longrightarrow (\eta^3\text{-allyl, R})\text{—}Fe(CO)_3SiCl_3$$

(126)

$$[HFe(PMe_3)_4(solv.)]^+ + \text{isoprene} \longrightarrow [(\text{diene})\text{—}Fe(PMe_3)_3H]^+ \ \underset{}{\overset{PMe_3}{\rightleftharpoons}} \ [(\eta^3\text{-allyl})\text{—}Fe(PMe_3)_4]^+$$

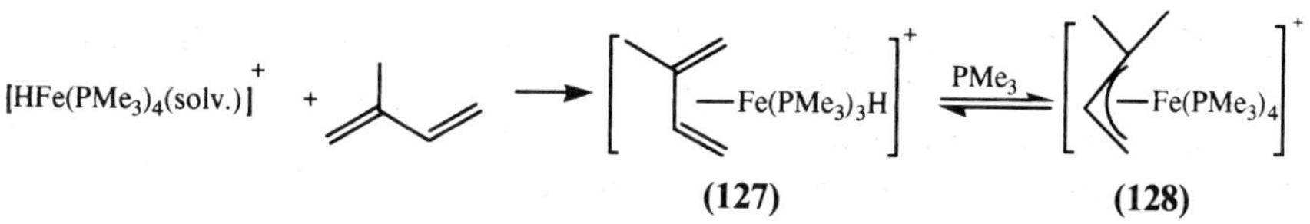

(127) **(128)**

intermediate hydridodiene complex (**127**) (Karsch, 1977). The nitrosyl hydride, $HFe(CO)_3NO$ adds similarly to dienes and in a related process the corresponding acyls, $RCOFe(CO)_3NO$, add, e.g. to butadiene to give the ketone (**129**) (Chaudhari *et al.*, 1967). The remarkable η^3-1-silapropenyl complexes (**130**) result from reaction of $Fe_2(CO)_9$ with the disilabutenes $CH_2{=}CHSiMe_2SiMe_2R$ (Sakurai *et al.*, 1980).

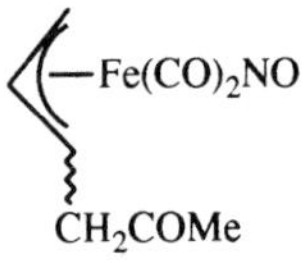

(129) 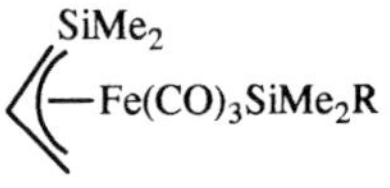 **(130)** 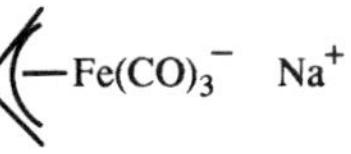 **(131)**

Following reduction of the iodide (**124**; L = CO, X = I) with sodium amalgam, apparently to the salt (**131**), reaction with allyl bromide gives diallyldicarbonyliron (**132**). In analogous fashion, $Na^+[Fe(CO)_3NO]^-$ with allyl halides gives the neutral allyldicarbonylnitrosyliron (**133**) completing the series with the diallyl (**132**) and the dinitrosyl, $Fe(CO)_2(NO)_2$ with allyl or NO groups as alternative 3-electron donors. Reaction of the related

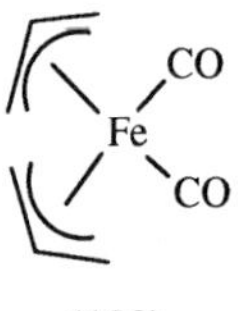

(132) 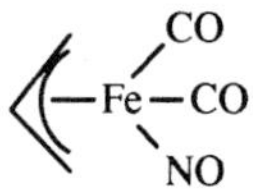 **(133)** 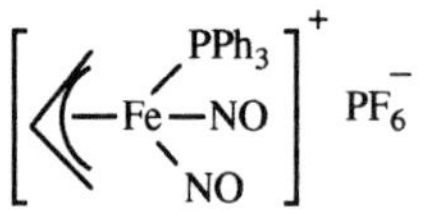 **(134)**

allylnitrosyl complex $C_3H_5Fe(CO)(PPh_3)NO$ with $NOPF_6$ yields the cationic allyldinitrosyl complex (**134**) (Baker and Connelly, 1979). One of the very few close analogues of the diallyl (**132**) is formed by a remarkable rearrangement of an initially generated carbene complex (Scheme 4.59) (Chen *et al.*, 1989).

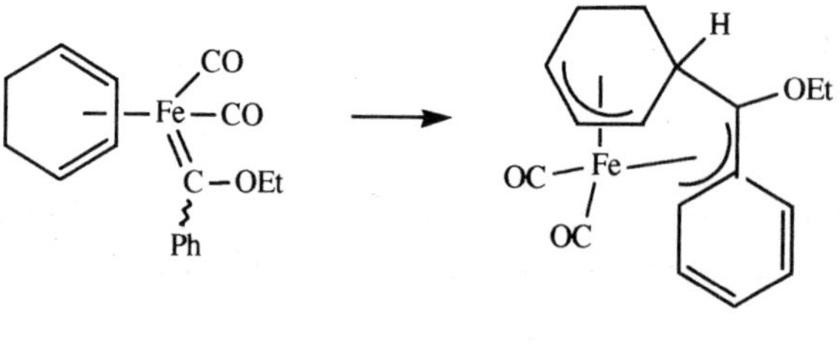

Scheme 4.59

An alternative route to the allylnitrosyl complex (**133**) involves dehalogenation of the halides (**124**, L = CO) by deactivated alumina [or by e.g. zinc metal (Wegner and Delaney, 1976)] which leads to the radical (**135**). When this is produced in an atmosphere of nitric oxide, the nitrosyl (**133**) is obtained in 80% yield (Murdoch, 1965). In the absence of such addends, the paramagnetic allyltricarbonyliron (**135**), which has also been generated electrochemically (Gubin and Denisovich, 1968), is a persistent radical which exists in equilibrium with the dimer (**136**) (Muetterties *et al.*, 1975). The latter

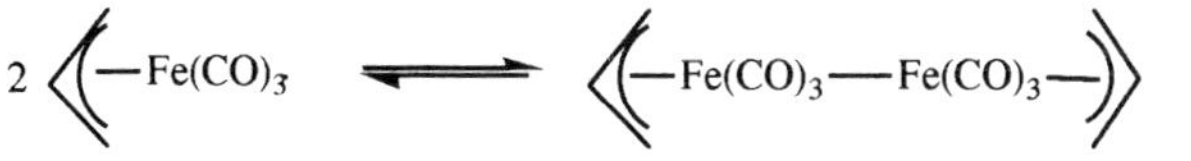

2 ⟨—$Fe(CO)_3$ ⇌ ⟨—$Fe(CO)_3$—$Fe(CO)_3$—⟩

(135) **(136)**

has been obtained in crystalline form, but the weakness of the iron–iron bond is reflected by its extreme length (3.14 Å) (Putnik *et al.*, 1978); for comparison, when the allyl groups are tied as in the hexacarbonyldiiron complex of cycloheptatriene (**137**) the distance is 2.87Å (Cotton *et al.*, 1971).

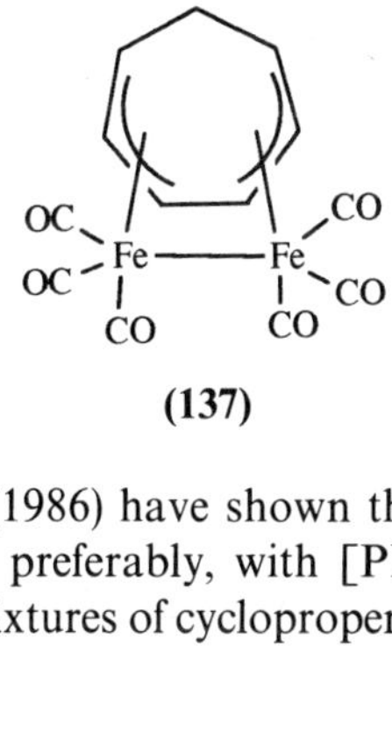

(137)

Hughes and coworkers (1986) have shown that cyclopropenyl salts react with $Na[Fe(CO)_3NO]$ or, preferably, with $[PPN][Fe(CO)_3NO]$ like allyl halides. The products are mixtures of cyclopropenyl (**138**) and oxocyclobutenyl complexes (**139**).

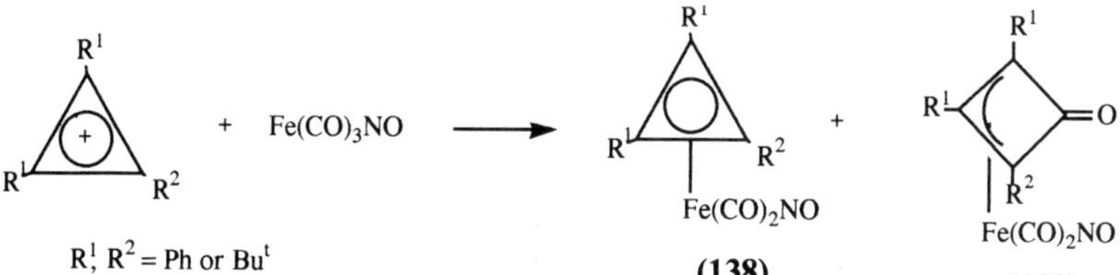

+ $Fe(CO)_3NO$ ⟶ $Fe(CO)_2NO$ + $Fe(CO)_2NO$

R^1, R^2 = Ph or Bu^t

(138) **(139)**

Of potential synthetic interest are the reactions of the nitrosyl (**133**) with carbanions, e.g. Scheme 4.60 (Roustan and Houlihan, 1988), the formation

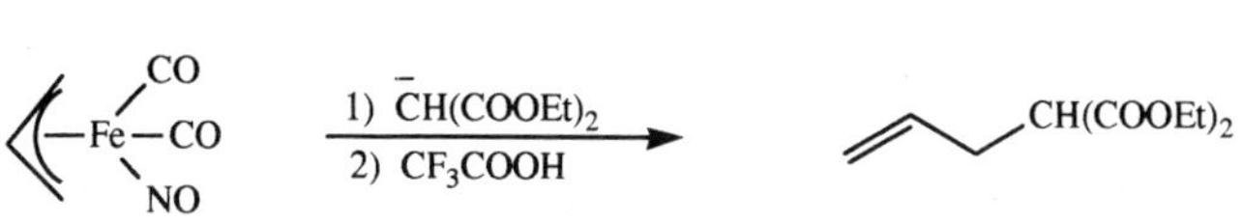

1) $\bar{C}H(COOEt)_2$
2) CF_3COOH

$CH(COOEt)_2$

Scheme 4.60

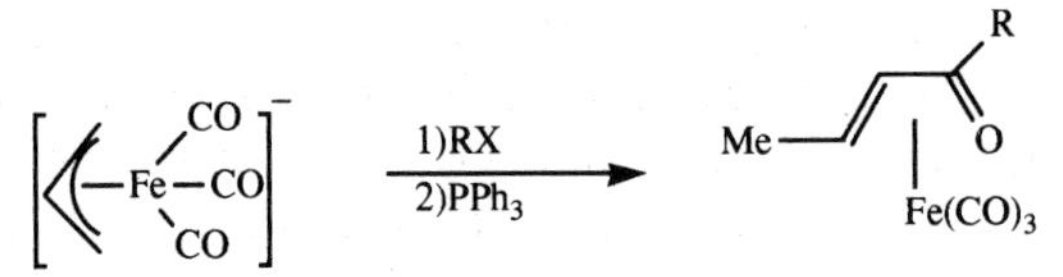

Scheme 4.61

of enone complexes on alkylation of the anionic complex (**131**) (Scheme 4.61) (Brookhart *et al.*, 1989), and the use of the cationic complexes (**123**) as electrophiles in reactions with e.g. allylsilanes (e.g. Scheme 4.62) (Li and Nicholas, 1990) and with highly activated arenes (e.g. Scheme 4.63) (Dieter *et al.*, 1987).

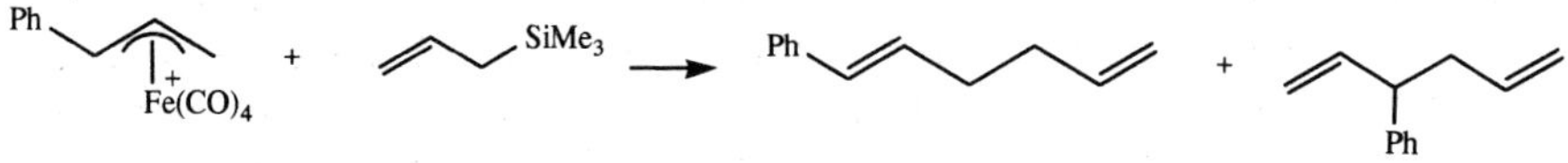

Scheme 4.62

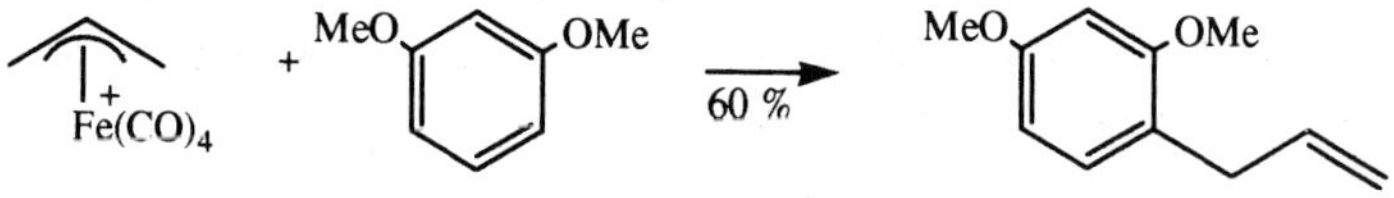

Scheme 4.63

The homoleptic tris-η^3-allyliron (**140**) and its homologues are obtainable by reaction of iron(III) chloride with allylic Grignard reagents at $-80°C$, but are unstable above $-40°C$ (Wilke *et al.*, 1966). Reaction with carbon monoxide results in conversion to $(\eta^3\text{-}C_3H_5)(\eta^1\text{-}C_3H_5)Fe(CO)_3$ which must also be an unobserved intermediate in the above formation of the bis-η^3-complex (**132**).

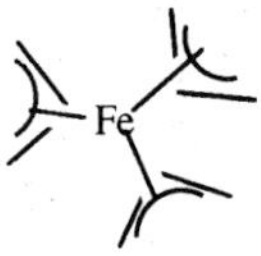

(**140**)

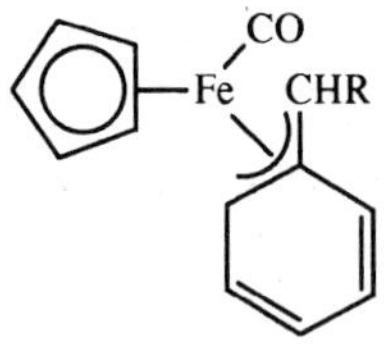

(**141**)

Although η^3-allyl(carbonyl)cyclopentadienyliron complexes have long been known to arise on photolysis of the η^1-allyldicarbonyl compounds (Scheme 4.19) they have received relatively little attention. The kinetics of their formation and *endo* to *exo* isomerisation have been reported (Fish *et al.*, 1976) and benzylic complexes FpCHRPh have been shown to yield analogous η^3-compounds (**141**) on low-temperature photolysis (Brookhart *et al.*, 1989).

4.4.3.2 Ligands which are η^3,η^1 or η^3,η^2 coordinated. There exists a rather large and diverse family of compounds in which an η^3-allyl system is part of a chelating ligand which is simultaneously η^1- or η^2-bonded through other carbon atoms. Among the simpler complexes of this type are the 'ferrilactone' and 'ferrilactam' complexes of the type (**142**) and these are of considerable

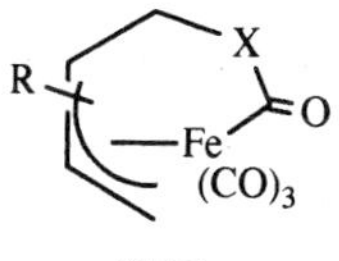

(142)

interest in organic synthesis. The lactone structure ($X = O$) has been confirmed crystallographically (Churchill and Chen, 1976) and the original synthesis (Aumann *et al.*, 1974) from unsaturated epoxides (vinyloxiranes) and $Fe_2(CO)_9$ has remained the major route. Alternatively 1,4-dihydroxy-2-alkenes, used with Lewis acids (Zn halides) and ultrasound react with $Fe_2(CO)_9$ to give moderate but useful yields (Bates *et al.*, 1990) while isomeric 1,2-dihydroxy-3-alkenes are best converted to cyclic sulphites, which then react in good to excellent yields, again with beneficial activation by ultrasound (e.g. Scheme 4.64) (Caruso *et al.*, 1990).

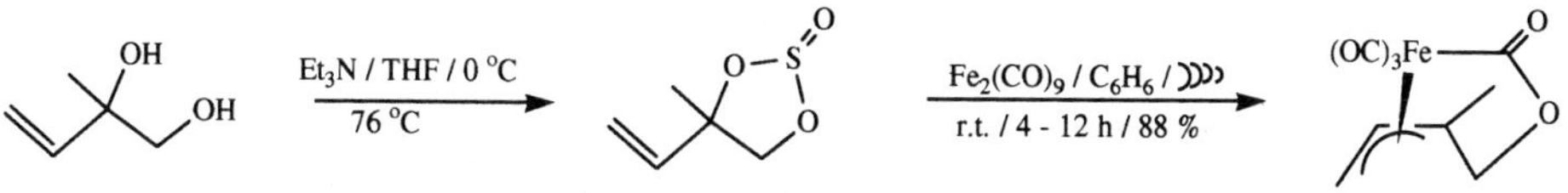

Scheme 4.64

Vinylaziridines react similarly to vinyloxiranes to give the ferrilactams (**142**; $X = NR'$) (Aumann *et al.*, 1974) but these are more commonly made from the lactones (**142**; $X = O$) by reaction with amines in the presence of Lewis acids (typically $ZnCl_2$, Et_2AlCl) (Ley, 1988). A structurally characterised example (**143**) was prepared from (η^2-PhCH=CHCOPh)$Fe(CO)_4$ by successive treatment with boron trifluoride and cyclohexylamine (Nesmeyanov *et al.*, 1978).

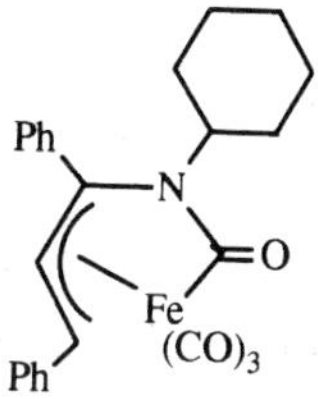

(143)

The value of such complexes in organic synthesis arises from their stereoselective conversion to β- and δ-lactones and lactams. δ-Lactones are obtained as the sole products when the lactone complexes (**142**; X = O) are subjected to carbonylation at high temperature while vinyl-substituted β-lactones are commonly the sole or major products of room temperature ceric oxidation. Whether δ-lactones are also formed under the latter conditions depends on the structure including stereochemistry of the particular iron complex. The analogous oxidation of the lactam complexes yields vinyl-β-lactams with apparent complete selectivity. Examples of all these reactions are collected in Scheme 4.65 (Ley, 1988).

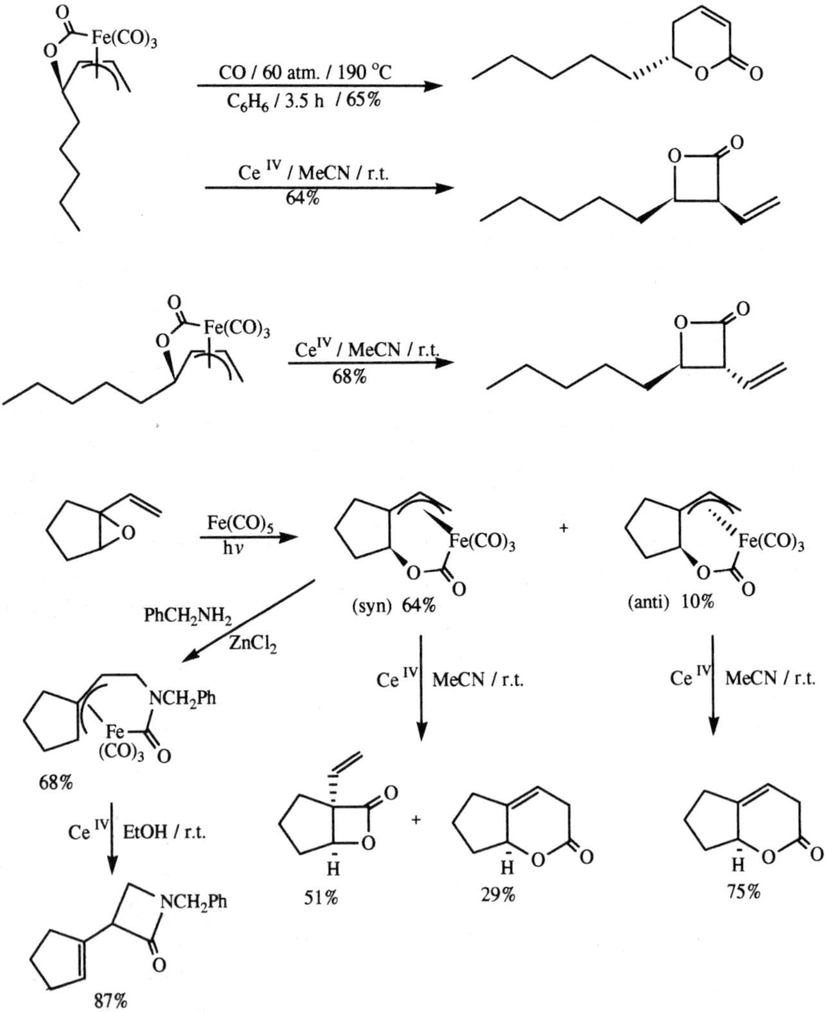

Scheme 4.65

Iron–oxygen linked ferrilactone complexes (**144**) which hydrolyse to β,γ- and/or γ,δ-unsaturated acids are the products of addition of carbon dioxide to (substituted) butadiene(tris-phosphine) complexes (Hoberg *et al.*, 1987).

Other types of η^3-complexes with additional coordination are formed from polyenes or from diene and polyene complexes by a wide variety of reactions of which only a few selected examples can be given here. Thus barbaralone (**145**) opens its 3-membered ring on reaction with $Fe_2(CO)_9$ to yield the η^3,η^1-complex (**146**) (Eisenstadt, 1972) and an exactly analogous process leads

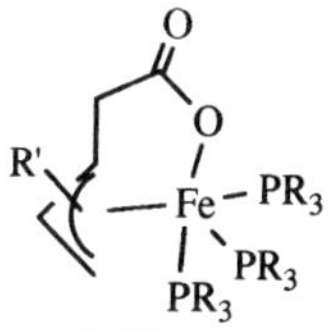

(144) 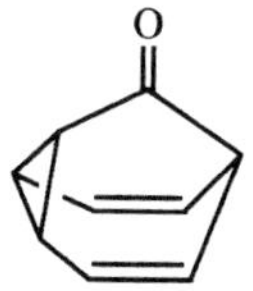 **(145)** 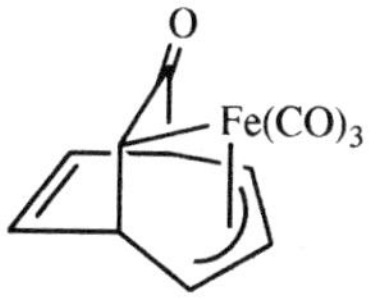 **(146)**

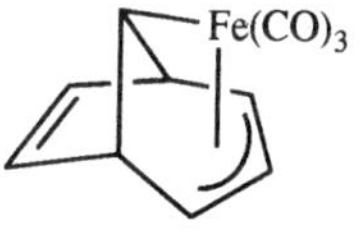

(147) 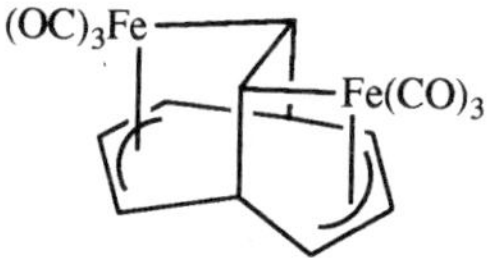 **(148)** 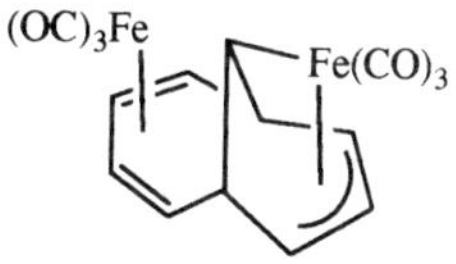 **(149)**

from semibullvalene to the bicyclooctadienediyl complex (**147**) (Moriarty *et al.*, 1971). Bullvalene gives several isomeric products including the bis-$Fe(CO)_3$ complex (**148**) which itself isomerises quantitatively at 120°C to the complex (**149**) (Aumann, 1971). Thermal or photochemical additions of reactive alkynes, alkenes, and other double bonds to diene (including cyclobutadiene) and triene complexes may be exemplified by the formation of complexes (**150**) (Bottrill *et al.*, 1977), (**151**) (Bond and Green, 1972) and (**152**) (Andreetti *et al.*, 1978).

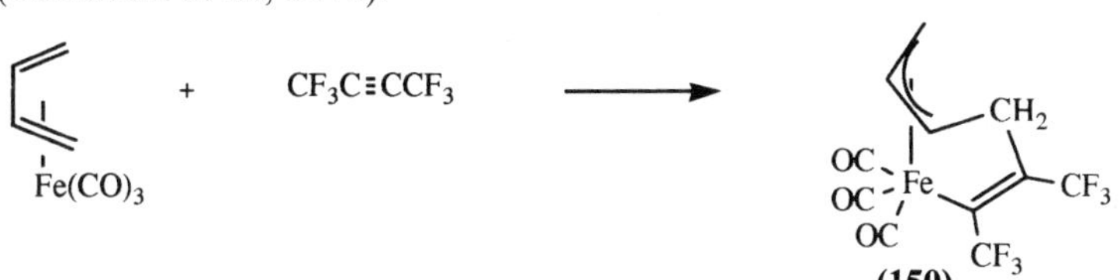

$Fe(CO)_3$(butadiene) + $CF_3C{\equiv}CCF_3$ ⟶ **(150)**

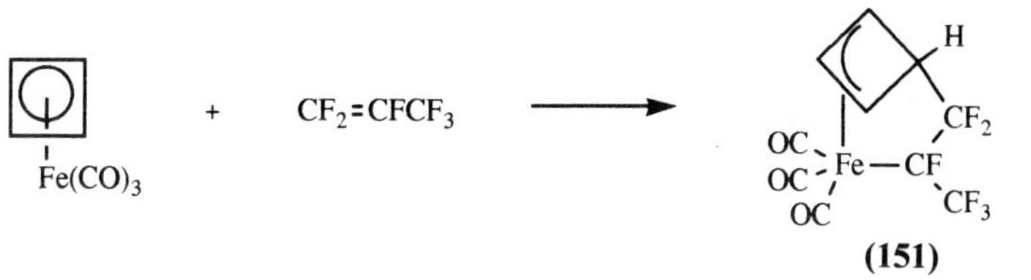

$Fe(CO)_3$(cyclobutadiene) + $CF_2{=}CFCF_3$ ⟶ **(151)**

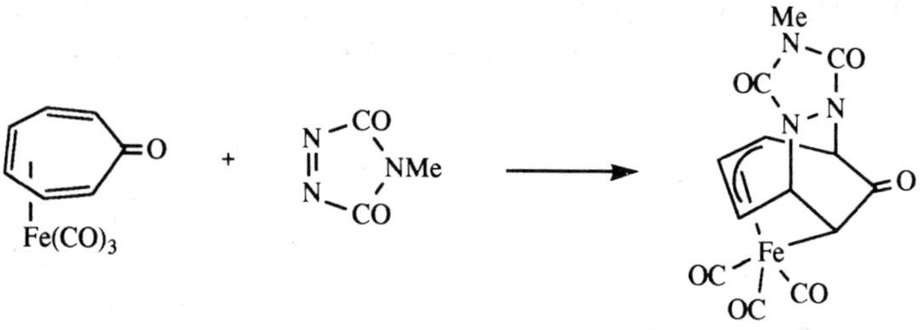

(152)

Aluminium chloride catalysed carbonylation of tricarbonyl(η^4-cyclohexa-1,3-diene)iron (**153**) yields the η^3,η^1-ketone complex (**154**) (cf. Eilbracht *et al.*, 1990), while in analogous fashion, tricarbonyl (η^{1-4}-cycloocta-1,3,5,7-tetraene)-

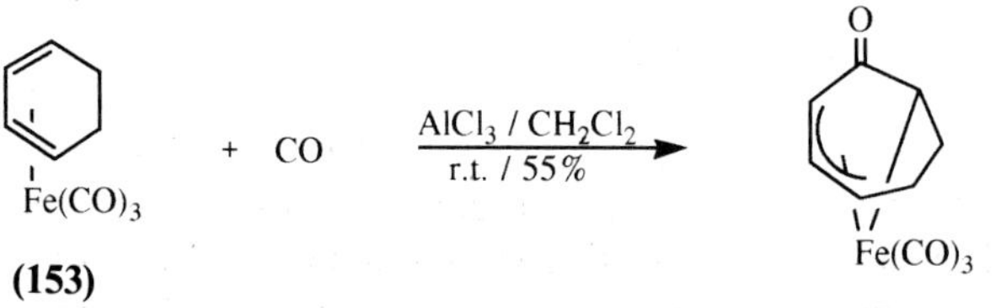

$AlCl_3$ / CH_2Cl_2, r.t. / 55%

(153) **(154)**

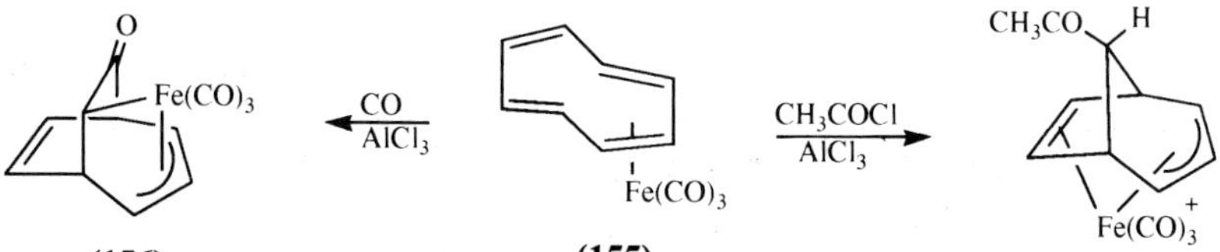

(156) **(155)** **(157)**

iron (**155**) yields complex (**156**) (65%) (Johnson *et al.*, 1978). Acetylation of the tetraene complex (**155**) under Friedel–Crafts conditions gives an example of a cationic η^3,η^2-complex (**157**) (Charles *et al.*, 1977). A simple cation of this type (**158**) is formed by hydride abstraction from tricarbonyl (1,5-cyclooctadiene)-iron. On borohydride reduction the principal product (90%) is the η^3,η^1-complex (**159**) and only 10% reverts to the diene complex (Deeming *et al.*, 1974).

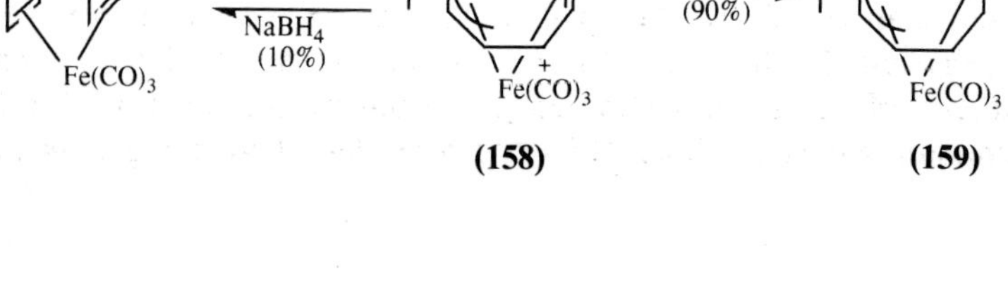

(158) **(159)**

A ferrocene analogue (**161**) with η^3,η^2-ligands has been made by reacting the corresponding allyllithium derivative (**160**) with ferrous chloride (Blümel *et al.*, 1988).

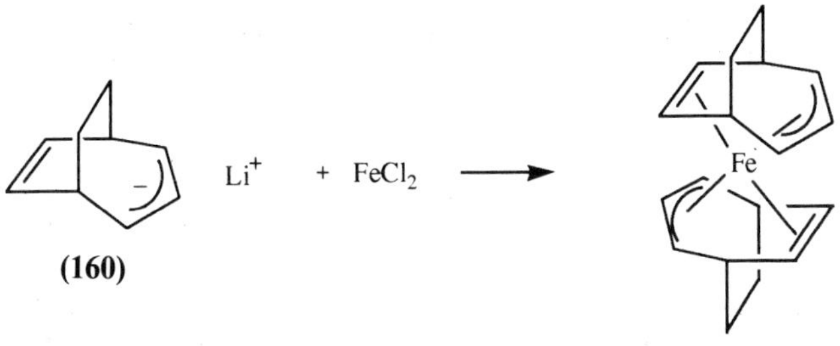

Li^+ + $FeCl_2$ ⟶ Fe

(160) **(161)**

4.4.4 η^4-Carbon–iron complexes

4.4.4.1 η^4-Cyclobutadiene complexes. The original and still probably best (*Organic Syntheses Coll.* **VI**, 310) route to tricarbonylcyclobutadiene-iron (**162**) is the reaction of 3,4-dichlorocyclobutene with $Fe_2(CO)_9$ (Emerson *et al.*, 1965). Alkyl derivatives are readily obtained similarly. Early work on these and aryl derivatives has been fully reviewed (Efraty, 1977). The principal routes to $C_4H_4Fe(CO)_3$ are summarised in Scheme 4.66. The tetraphenyl

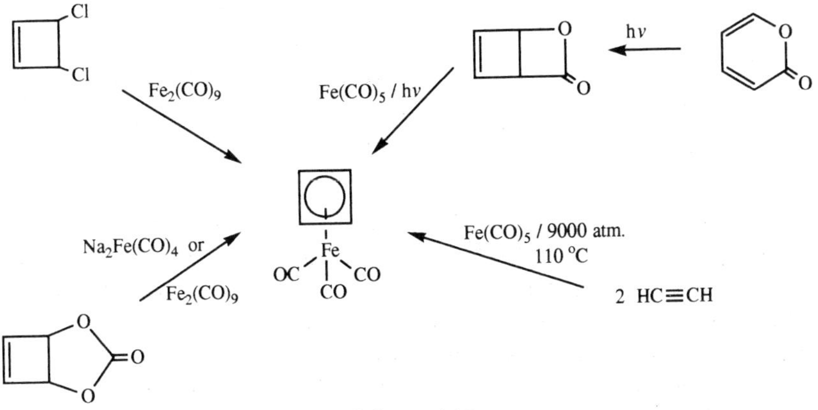

Cl Cl $Fe_2(CO)_9$ $Fe(CO)_5$ / hν O O hν O O

Fe OC CO CO

$Na_2Fe(CO)_4$ or $Fe_2(CO)_9$ $Fe(CO)_5$ / 9000 atm. 110 °C 2 HC≡CH

O O =O

Scheme 4.66

derivative, first isolated as a minor product of the high temperature reaction of diphenylacetylene with pentacarbonyliron, has been obtained *inter al.* by ligand transfer from the palladium and platinum halide complexes $[(C_4Ph_4)MX_2]_2$ (M = Pd, Pt) on reaction with $Fe(CO)_5$. A convenient synthesis of tricarbonyl(tetramethylcyclobutadiene)iron from 2-butyne via aluminium chloride adduct (**163**) has been described (Fongers *et al.*, 1982).

Other highly substituted derivatives include the diamino derivatives (**164**) obtained from $Me_3SiC\equiv CNR_2$ (R = Me, Et, Pr^i) and $Fe(CO)_5$ (King *el al.*, 1987) and the methoxy-substituted cation (**165**) obtained by alkylation

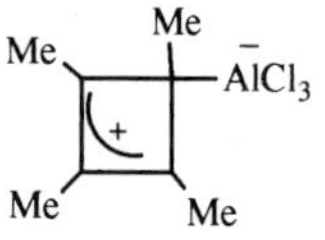

(**163**) 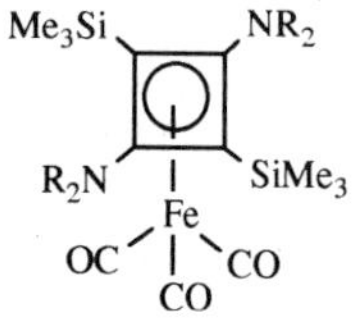 (**164**) 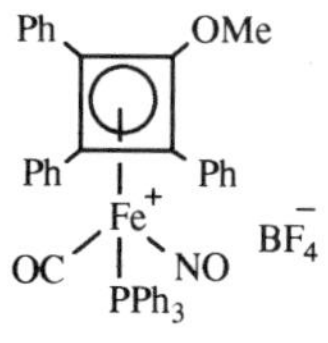 (**165**) 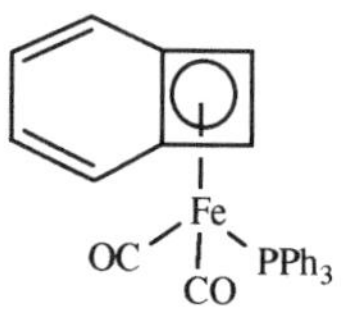 (**166**)

of the corresponding oxocyclobutenyl complex with Me_3OBF_4 (Behrens *et al.*, 1988).

Extensive physical studies include photo-electron (Kostic and Fenske, 1982) and electron transmission spectra (Olthoff *et al.*, 1987) and low-temperature crystallographic study of $C_4H_4Fe(CO)_3$ (Harvey *et al.*, 1988); structures of derivatives include that of the benzocyclobutadiene complex (**166**) (Riley and Davis, 1983).

The neutral tricarbonyl complexes react with $NOPF_6$ to give the cationic nitrosyls (**167**) which, in turn, add certain nucleophiles Y^- to give the η^3-allylic complexes (**168**), e.g. $Y = PR_3^+$ (Choi and Sweigart, 1982) or $Y = 4\text{-}Me_2NC_6H_4$ (by addition to $PhNMe_2$ and loss of a proton) (Calabrese *et al.*, 1983) while

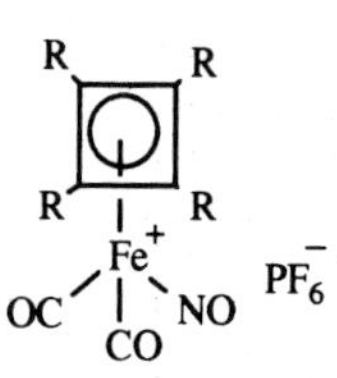

(**167**) 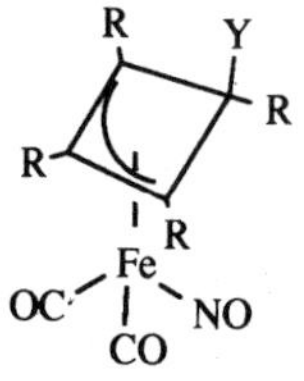 (**168**) 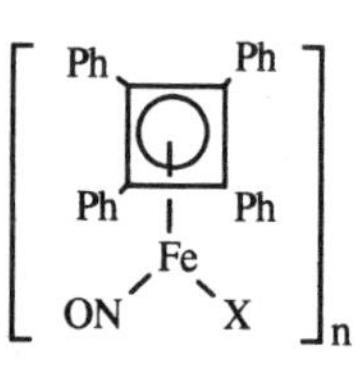 (**169**)

coordinating nucleophiles X^- (X = Cl, NCO, N_3, acac, S_2CNEt_2, etc.) displace both carbonyl groups from the tetraphenyl complex (**167**; R = Ph) to give probably dimeric ($n = 2$) or chelated (e.g. X = acac; $n = 1$) nitrosyls (**169**) (Lalor *et al.*, 1984).

Photolysis causes reversible loss of three carbonyl groups from two molecules of the tricarbonyls. For the product from the di-tert.-butyl-diphenyl-cyclobutadiene complex the triply bridged, Fe–Fe triple-bonded (Fe–Fe = 2.177Å) structure (**170**) was demonstrated by X-ray crystallography, whereas the unsubstituted analogue has been formulated on the basis of its vibrational spectra as having two terminal and one bridging carbonyl group. Photolysis in a matrix involves insertion of carbon monoxide into the C_4H_4 ring resulting in production of the cyclopentadienone complex (**171**) (Rest and Taylor, 1981) and photolysis in the presence of certain addends leads to partial or

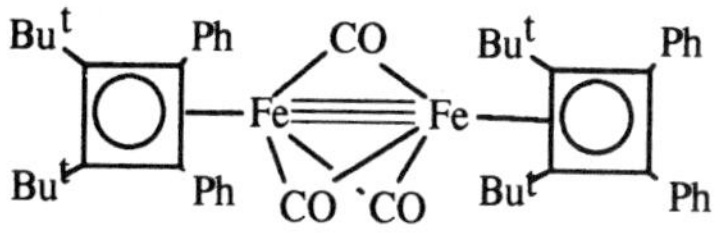

(170)

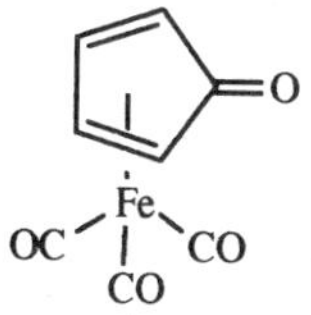

(171)

complete replacement of CO as in the formation of the hydridosilyl complex (**172**) (Hill and Wrighton, 1985) and the azepine complex (**173**). By contrast

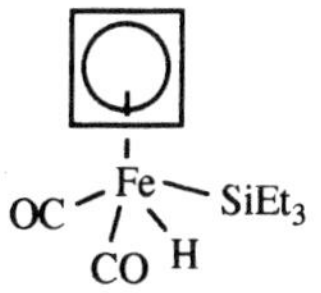

(172)

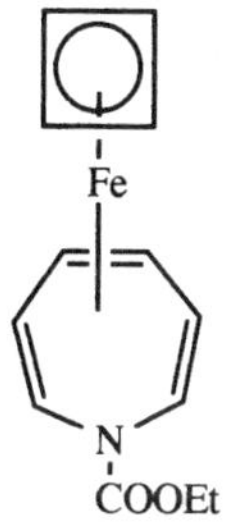

(173)

irradiation with, for example, perfluoroalkenes as mentioned earlier (section 4.4.3) leads to addition to the ring, as does irradiation with unactivated alkynes (Gist and Reeves, 1981) or with cycloheptatriene or oxepin which simultaneously displace one CO group giving the triene-dicarbonyl complexes (**174**); $X=CH_2$ or O). Additions with iron insertion into the four-membered ring occur in reactions with $(C_5H_5)M(CO)_2$ (M = Co or Rh) to yield the binuclear complexes (**175**) (King *et al.*, 1982) and on photolysis with B_5H_9 to yield compound (**176**) (Fehlner, 1978).

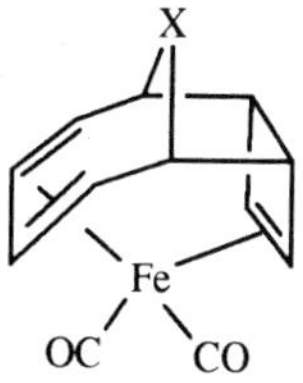

(174)

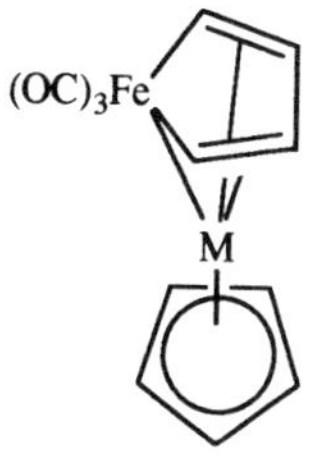

(175)

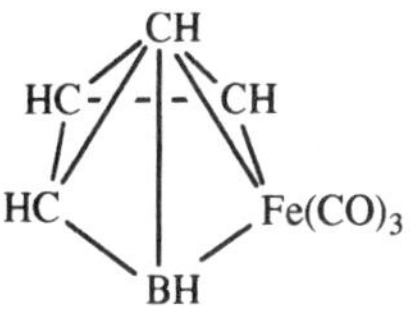

(176)

Oxidative (e.g. Ce^{IV}) cleavage of $C_4H_4Fe(CO)_3$ provides the most convenient source of the transient free cyclobutadiene and has been used synthetically, e.g. in the formation of 'Dewar-benzenes' (bicyclo[2.2.0]-hexadienes) (**177**) by its addition to alkynes ($RC{\equiv}CR'$) or for its capture by

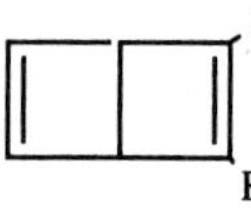

(177)

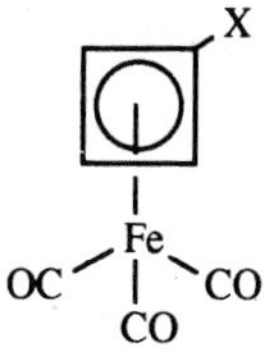

(178)

various other dienophiles or dienes, e.g. Schemes 4.67 (Gree *et al.*, 1980) and 4.68 (Grimme and Köser, 1981).

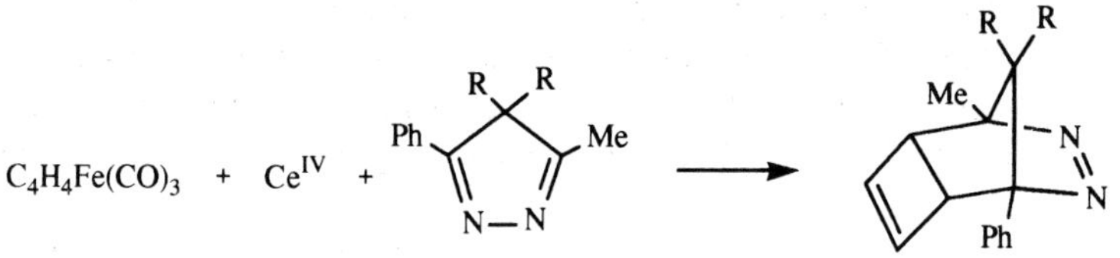

$C_4H_4Fe(CO)_3$ + Ce^{IV} +

Scheme 4.67

$C_4H_4Fe(CO)_3$ + Ce^{IV} +

Scheme 4.68

Electrophilic substitutions of tricarbonylcyclobutadieneiron occur under conditions very similar to those for ferrocene (see Scheme 4.2) yielding functionally substituted complexes (**178**) with *inter al.* acyl (X = COR), formyl (X = CHO), dimethylaminomethyl ($X = CH_2NMe_2$), sulphonic acid ($X = SO_3H$) and metallo (X = HgCl; X = Li) groups and allowing further elaboration of the organic residue, e.g. $CH_2NMe_2 \rightarrow \rightarrow CH_2CH(NH_2)COOH$ (Brunet *et al.*, 1981). Even 1,2-diacetylation has been reported (Dinulescu *et al.*, 1977). Also, as in the ferrocene series, cations α- to the cyclobutadiene ligand (i.e. **178**; $X = \overset{+}{C}RR'$) are stabilised (Reeves, 1981).

4.4.4.2 η^4-1,3-Diene complexes. Butadienetricarbonyliron (**3**) is mentioned in the introduction and its original synthesis by heating the diene with $Fe(CO)_5$ remains the method of choice. Numerous other dienes (as well as trienes and polyenes) yield η^4-diene complexes under similar conditions. The ability of $Fe(CO)_5$ to catalyse double-bond shifts, presumably by a process illustrated by Scheme 4.69, allows the use of 1,4- and 1,5- as well as 1,3-dienes. The ready availability of 1,4-cyclohexadienes by Birch reduction of arenes has made these the commonly preferred precursors of many 1,3-cyclohexadiene complexes. On the other hand such double bond migration introduces ambiguity and may lead to mixtures of products whose structures have to

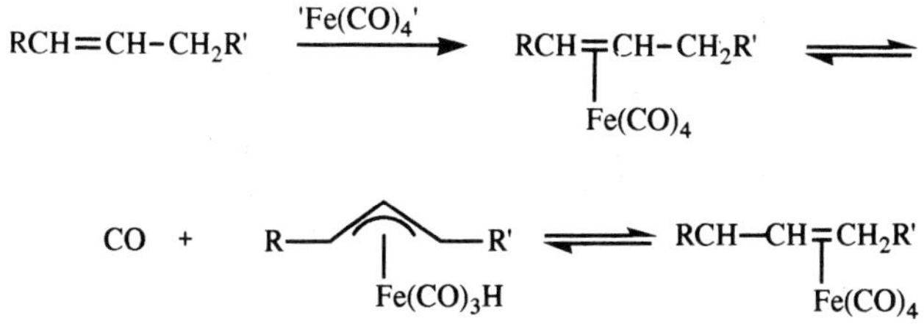

$$RCH{=}CH{-}CH_2R' \xrightarrow{'Fe(CO)_4'} RCH{=}CH{-}CH_2R'(Fe(CO)_4) \rightleftharpoons$$

$$CO + R{-}(\eta^3\text{-allyl})Fe(CO)_3H{-}R' \rightleftharpoons RCH{-}CH{=}CH_2R'(Fe(CO)_4)$$

Scheme 4.69

be determined separately. Thus 1-methoxy-1,4-cyclohexadiene yields mixtures of two of the three possible conjugated diene complexes (**179** and **180**) whereas 1-carbomethoxy-2,5-cyclohexadiene reacts regio- and stereoselectively to give only complex (**181**) which is converted by acids to the thermodynamically more stable isomer (**182**) (Pearson, 1980). Vinylcyclohexenes also afford

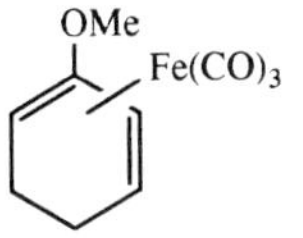

OMe, $Fe(CO)_3$ (**179**) OMe, $Fe(CO)_3$ (**180**) COOMe, $Fe(CO)_3$ (**181**) COOMe, $Fe(CO)_3$ (**182**)

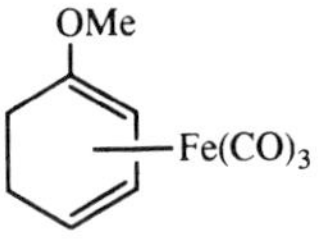

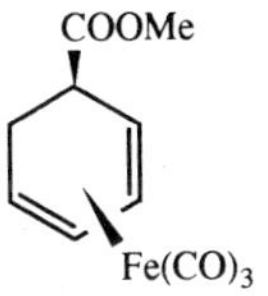

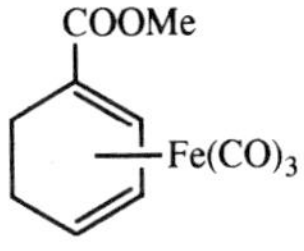

mixtures, convertible with acids to the thermodynamically most stable complexes (Rodriguez *et al.*, 1987). Alkyl migrations, ring opening (e.g. Scheme 4.70) (Sarel, 1978), ring closure (e.g. Scheme 4.71) (Paquette, *et al.*,

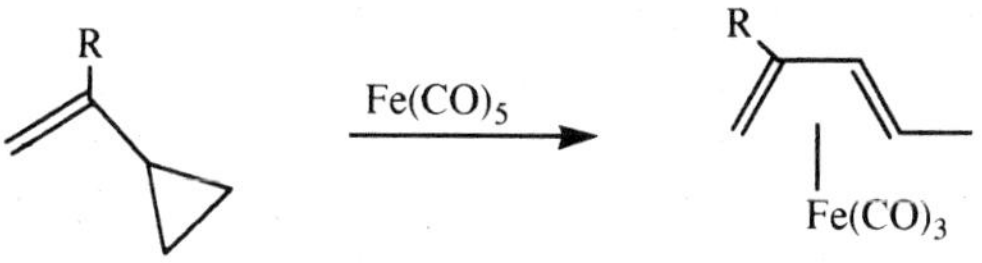

R $\xrightarrow{Fe(CO)_5}$ R, $Fe(CO)_3$

Scheme 4.70

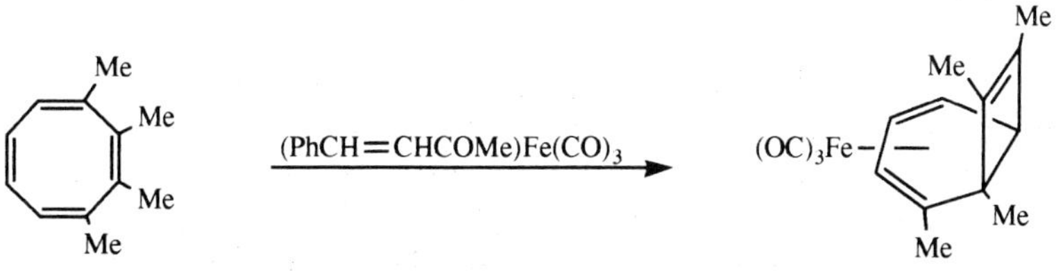

Me, Me, Me, Me $\xrightarrow{(PhCH{=}CHCOMe)Fe(CO)_3}$ $(OC)_3Fe$, Me, Me, Me, Me

Scheme 4.71

1975) and more deep-seated skeletal rearrangements as in the reaction of santonin (**183**) to yield the complex (**184**) (Alper and Keung, 1972) may accompany complexation.

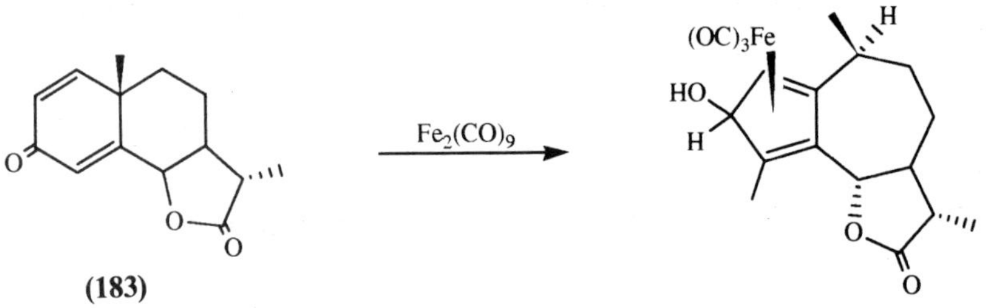

$Fe_2(CO)_9$

(183) (184)

The last two examples also illustrate the use of more reactive sources of the $Fe(CO)_3$ groups and much milder conditions than are required with $Fe(CO)_5$. Other commonly used reagents are $Fe_3(CO)_{12}$ and (alkene) $Fe(CO)_4$ complexes.

Among the many complexes prepared by the above methods are complexes of styrenes (**185**) in which one ring bond is complexed. This results in partial bond fixation and allows a second $Fe(CO)_3$ group to be complexed as in compound (**186**) (Herbstein and Reisner, 1972). The end-ring of linearly-fused polycyclic aromatics and heteroaromatics may be similarly complexed as in example (**187**) (Bauer *et al.*, 1970).

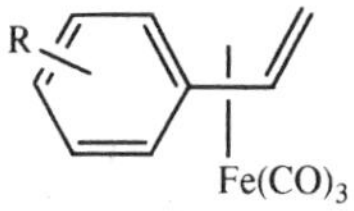

(185) 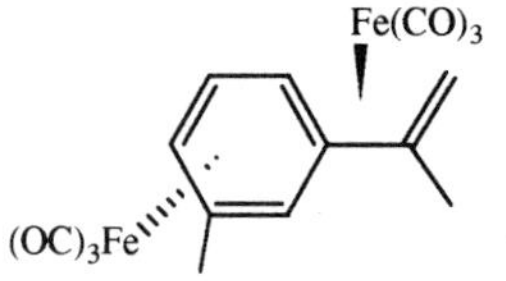 (186) 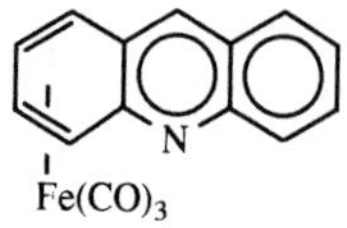 (187)

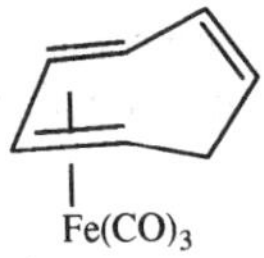

(188) 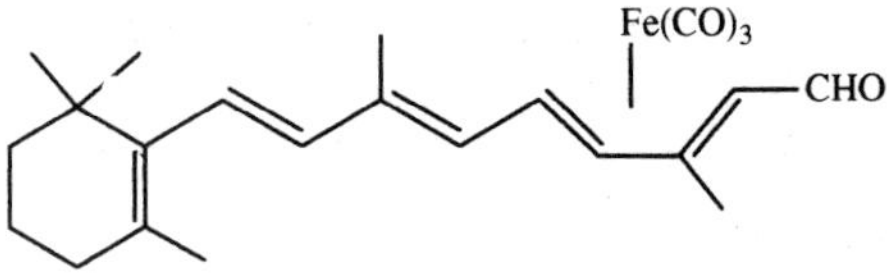 (189)

Trienes, e.g. cycloheptatriene (in **188**) act only as η^6-ligands, leaving one double bond free, while polyenes may yield mono-, bis- or even more highly complexed η^4-$Fe(CO)_3$ derivatives. Thus a tetrakis-$Fe(CO)_3$ complex of β-carotene has been identified (Ichikawa *et al.*, 1967). Some polyenes yield mono-$Fe(CO)_3$ complexes with high specificity as was found for Vitamin A aldehyde which forms only complex (**189**) (Birch and Fitton, 1966; cf. Mason and Robertson, 1970), while others give mixtures, allowing the kinetics of 1,3-migration of the $Fe(CO)_3$ group along a conjugated chain to be studied (Whitlock *et al.*, 1971). Remarkable stereoselectivity, dependent on the polarity of the reaction medium, was displayed in the complexation of ketone (**190**) as shown (Salzer and von Philipsborn, 1979).

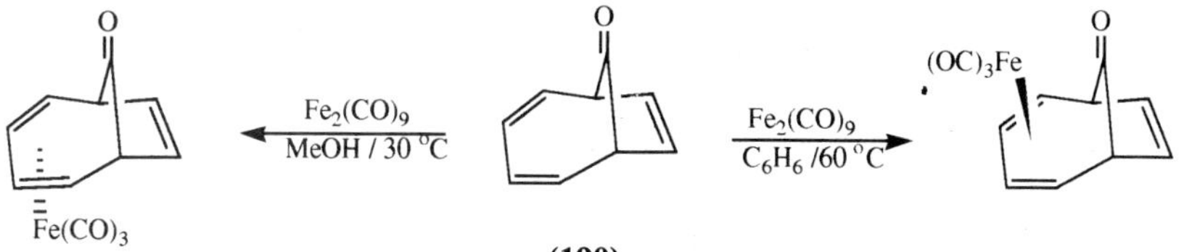

$Fe(CO)_3$ ← $Fe_2(CO)_9$ / MeOH / 30 °C — (**190**) — $Fe_2(CO)_9$ / C_6H_6 /60 °C → $(OC)_3Fe$

Enantioselectivity in the transfer of $Fe(CO)_3$ from an optically active enone to a 1,3-cyclohexadiene to given an optically active complex of the type (**191**) has been demonstrated (Birch *et al.*, 1984). Classical optical resolution of related complexes has been achieved either by replacing one CO ligand by an asymmetric phosphine, e.g. (+)-neomenthyldiphenylphosphine (Howell *et al.*, 1987) to produce a pair of diastereoisomeric complexes, or by HPLC on an active support (Sotokawa *et al.*, 1987).

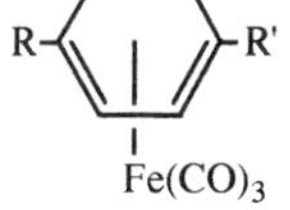

R R' $Fe(CO)_3$ (**191**) $Fe(CO)_3$ (**192**)

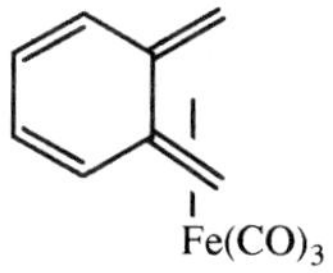

Alternative methods of synthesis involve generation of the diene *in situ*, e.g. by dehydration of allylic alcohols during reaction with $Fe(CO)_5$, while elimination of bromine in the presence of $Fe(CO)_5$ has made possible the synthesis of complex (**192**) from 1,2-bis(bromomethyl)benzene (des Abbayes *et al.*, 1988). Vinylketene complexes have been generated both from enone complexes by Scheme 4.72 (Alcock *et al.*, 1991) and from vinylcarbene complexes by Scheme 4.73 (Mitsudo *et al.*, 1989).

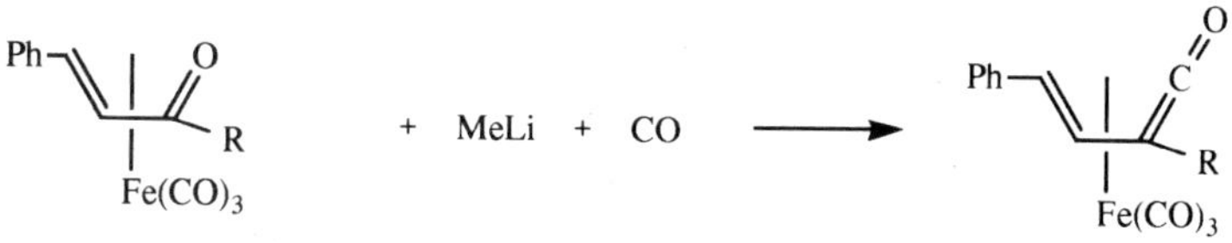

Ph O R $Fe(CO)_3$ + MeLi + CO → Ph O C R $Fe(CO)_3$

Scheme 4.72

COOMe OMe $Fe(CO)_3$ + PPh_3 → O C MeOCO OMe $Fe(CO)_2PPh_3$

Scheme 4.73

Addition of nucleophiles to the actionic cyclopentadienyl complex, $[(C_5H_5)Fe(CO)_3]^+$ has been used to obtain the 5-*exo*-substituted cyclopentadiene complexes (**193**) (Darensbourg, 1972). Analogous additions to cyclohexadienyl complexes are discussed in section 4.4.5.

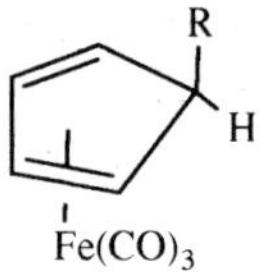

R H $Fe(CO)_3$ $Fe(PF_3)_3$

(193) 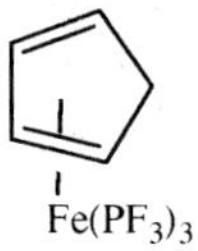 **(194)**

Complexes with other counterligands include the phosphorus trifluoride complex (**194**) obtained by photolysis of $Fe(PF_3)_5$ with cyclopentadiene (Kruck and Knoll, 1972) and the butadiene(phosphine)iron complexes (**195**) and (**196**; $L = PR_3$) obtained according to Scheme 4.74 at $-40°C$ and room

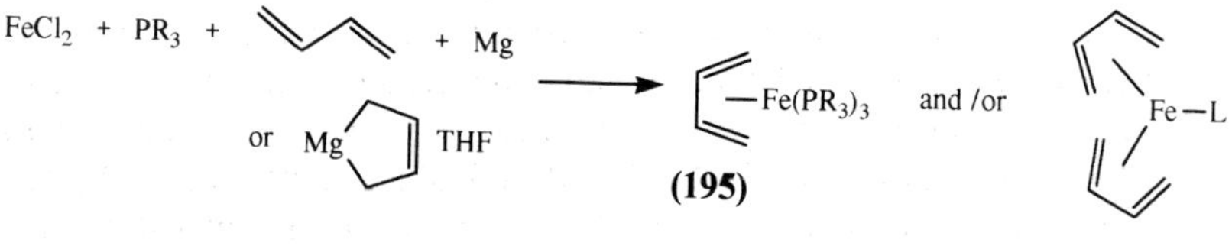

$FeCl_2$ + PR_3 + + Mg or Mg THF → $-Fe(PR_3)_3$ and /or Fe—L

(195) **(196)**

Scheme 4.74

temperature respectively (Hoberg *et al.*, 1987). The carbonyl (**196**; $L = CO$) is the product of photolysing $C_4H_6Fe(CO)_3$ with excess of butadiene.

Skeletal rearrangements of intact diene complexes include the first order conversion of the bicyclo[5.1.0]– (**197**) to the bicyclo[4.2.0.]–complex (**198**)

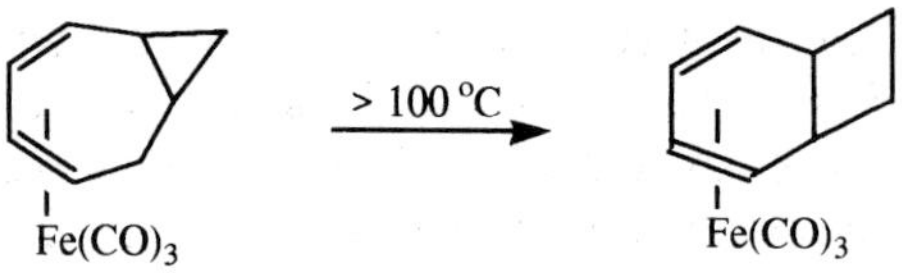

$Fe(CO)_3$ > 100 °C $Fe(CO)_3$

(197) **(198)**

(Brookhart *et al.*, 1974) and the ring opening of the spiro compound (**199**) to give the cyclopentadienyl complex (**200**) (Eilbracht and Dahler, 1977).

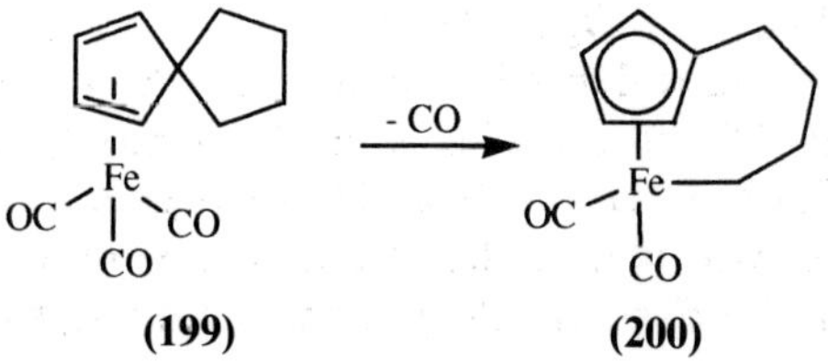

Fe OC CO CO - CO Fe OC CO

(199) **(200)**

The addition of protons to η^4-diene complexes has been discussed in section 4.4.3 as a route to η^3-allylic compounds. Acyl cations add similarly under typical Friedel–Crafts conditions; the initial products, formed stereoselectively, were shown to be η^3-allylic cations of the type (**201**) in which the acyl oxygen donates electrons to the iron atom (Greaves *et al.*, 1974). These, however, readily lose a proton on base treatment or hydrolysis and careful workup allows isolation of the *anti*- (or *endo*) acyl-substituted diene complexes (**202**). The facile acid or base catalysed isomerisation of the

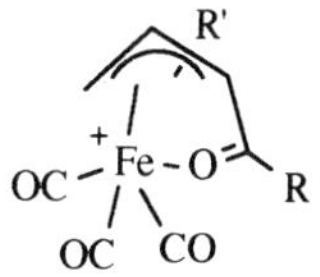

R' OC Fe O R OC CO

(**201**)

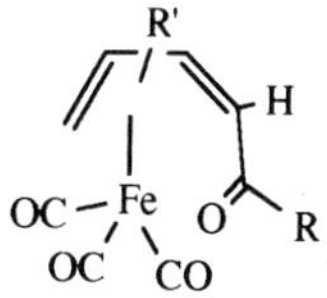

R' H OC Fe O R OC CO

(**202**)

R' R H O OC Fe OC CO

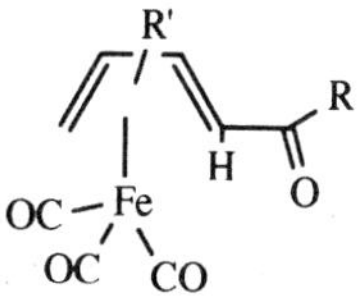

(**203**)

latter leads almost quantitatively to the *syn* (or *exo*) ketones (**203**) so that stereochemically pure samples of both isomers are readily accessible. By further elaboration of the acyl side-chains and decomplexation this has led to efficient stereoselective syntheses of a group of insect pheromones (Knox and Thom, 1981). Directive effects of silyl and other substituents (R′) have been studied with a view to other synthetic applications (Franck-Neumann *et al.*, 1987, 1991).

When uncomplexed double bonds are present these add electrophiles in preference to the complexed bonds. In these and many other reactions of functional substituents, the $Fe(CO)_3$ group acts to protect the diene fragment and this is probably its most important synthetic use. The stability of the diene-$Fe(CO)_3$ systems to acids and bases means that a wide range of addition, substitution, reduction and even some selective oxidation reactions can be carried out using this protective function to avoid polymerisation or to change the site of reaction. Thus complexation of steroidal dienes has allowed modification of isolated double bonds (e.g. Evans *et al.*, 1975) and been used to modify intermediates in prostaglandin syntheses (Corey and Moinet, 1973). That the reactivity of attached groups may be strikingly modified was shown, e.g. in reactions of complexed thebaine (Birch and Fitton, 1969) and the behaviour of complexed tropone (**204**). The latter, unlike the more delocalised free tropone undergoes typical 1,2- and 1,4-additions to the free enone system (Johnson *et al.*, 1971; Childs and Rogerson, 1980; Rigby and Ogbu, 1990).

All such synthetic uses depend of course on the possibility of efficient removal of the $Fe(CO)_3$ group. This is most commonly effected with such oxidants as Ce^{IV}, $CuCl_2$, CrO_3, $FeCl_3$ and H_2O_2 (e.g. Franck-Neumann *et al.*, 1991; Bovicelli and Mincione, 1988; Thompson, 1976; Corey and Moinet, 1973; Semmelhack and Fewkes, 1987) or the more selective Me_3NO which destroys the complex by oxidising CO ligands to CO_2 (Eekhof *et al.*, 1976; Childs and Regerson, 1980). Occasionally vigorous reducing agents, e.g. $LiAlH_4$, have also caused clean decomplexation (e.g. Nesmeyanov *et al.*, 1970).

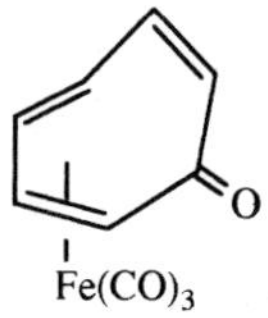

O $Fe(CO)_3$ **(204)** CR_2 $Fe(CO)_3$ **(205)** O $Fe(CO)_3$ **(206)**

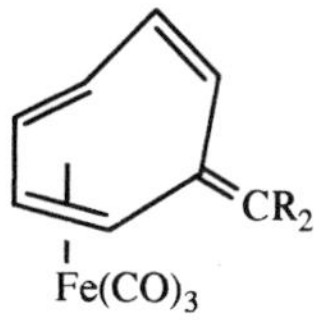

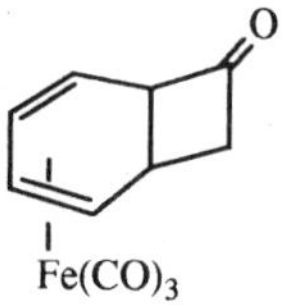

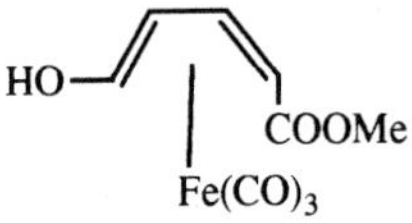

HO COOMe $Fe(CO)_3$ **(207)** O $Fe(CO)_3$ **(208)** OMe + $Fe(CO)_3$ **(209)**

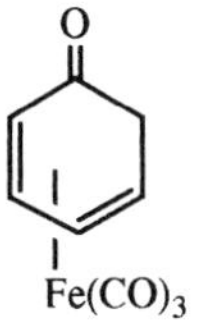

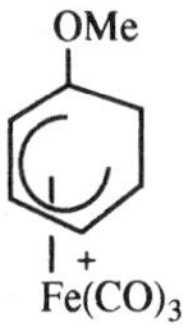

An important feature of diene-iron complexes is the ability to stabilise species which have only transient existencc in the free state. In addition to cyclobutadiene (section 4.4.4.1) this is exemplified by the $Fe(CO)_3$ complexes of cyclopentadienone (**171**) and of heptafulvenes (**205**), two ligands which rapidly polymerise, by complexes of the unstable bicyclic tautomer of cycloocta-2,4,6-trien-1-one (**206**), of dienols (e.g. **207**) (De Puy *et al.*, 1977) and of cyclohexa-2,4-dien-1-one, the keto tautomer of phenol, (**208**), which results on hydride abstraction from the methoxydiene complex (**180**) and the facile hydrolysis of the resultant cation (**209**).

4.4.4.3 η^4-Trimethylenemethane-iron complexes. Emerson and coworkers (1966) first showed that the dichloroisobutene (**210**) reacts with $Fe_2(CO)_9$ to yield the trimethylenemethane complex (**211**), isomeric with butadienetricarbonyliron (**3**). The free ligand must be thought of as a diradical and is only captured in low yield by addition to TCNE or other alkenes when the complex is oxidised in their presence. The structure has been confirmed by gas-phase electron diffraction which shows Fe–C distances of 1.938 Å to the central and 2.123 Å to the outer carbon atoms. The complex adds HCl or Br_2 to give the haloallyltricarbonyliron compounds (**212**; R = H, X = Cl, viz.

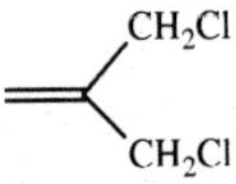

CH_2Cl CH_2Cl **(210)** CH_2 CH_2 CH_2 $Fe(CO)_3$ **(211)** RCH_2 $Fe(CO)_3Cl$ **(212)** $Fe(CO)_3$ **(213)**

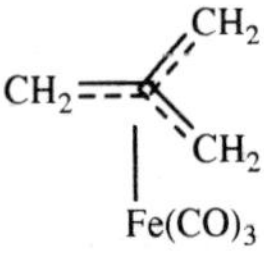

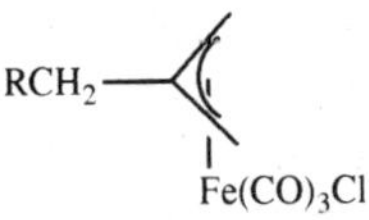

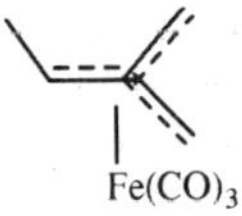

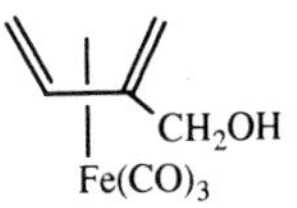

(214) 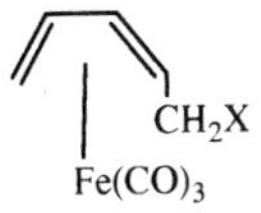 (215) 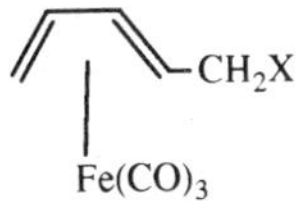 (216) 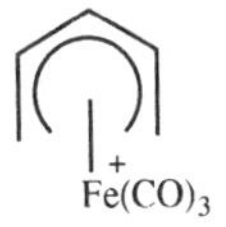 (217)

$R = X = Br$) from which it is regenerated on pyrolysis. Various substituted derivatives have been generated either similarly or from allenes (Roustan *et al.*, 1980; Martina *et al.*, 1982; Aumann and Melchers, 1988) or from vinylcyclopropanes (Noyori *et al.*, 1969; Pinhas and Carpenter, 1980). A remarkable synthesis of the methyl derivative (**213**) in optically active form has been effected by treating the (−) (1*S*, 4*R*)-camphanoyl ester of the dienol complex (**214**) with BF_3 and Et_3SiH (Kappes *et al.*, 1989).

4.4.5 *η^5-Dienyl iron complexes except cyclopentadienyl*

The *anti*- (**215**; X = H) but not the more stable *syn*-penta-1,3-diene complex-(**216**; X = H) loses hydride on treatment with triphenylmethyl tetrafluoroborate (or related salts) to yield the cationic pentadienyl complex (**217**). On the other hand both isomeric alcohols (**215**; X = OH) and (**216**; X = OH) react with strong acids to give the same cation (**217**) (e.g Donaldson, 1990).

In the widely studied cyclohexadiene series alternative oxidants, e.g. Tl^{III} or DDQ/HBF_4 have also been used to generate dienyl cations (Alexander and Stephenson, 1987). Although giving poorer yields in most cases, Tl^{III} shows significantly different regioselectivity and can also cause hydride loss from molecules too hindered to react with Ph_3C^+. Dehydration of the dienol complex (**218**) by HBF_4 was shown to give the salt (**219**) with stereoselective loss of the *endo*-H (Bandara and Birch, 1984).

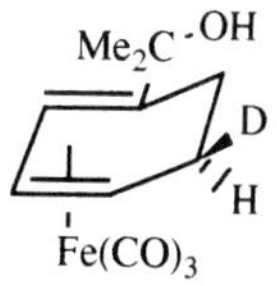

(218) 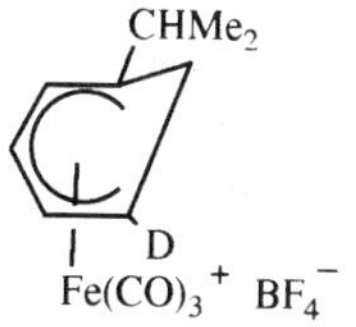 (219)  (220)

Whereas many nucleophiles add to the dienyl moiety of such cationic complexes (see below), iodide displaces CO, converting, for example, the cyclohexadienyl cation $(C_6H_7)Fe(CO)_3^+$, to the neutral iodide (**220**). On treatment with FpNa the latter yields a mixture of the symmetrical (**221**) and mixed dinuclear complexes (**222**). Hydride abstraction (using Ph_3C^+) converts each of these to an unstable cationic benzene complex (**223**); $L = C_6H_7$ or C_5H_5) (Begley *et al.*, 1987). The crystal structure of the open-chain 2,4-dimethyl-pentadienyl analogue of complex (**221**) has been determined (Gedridge *et al.*, 1987).

Dienal complexes, e.g. (**224**) are protonated by FSO_3H to yield initially the *syn*-hydroxymethylenedienyl cations (**225**) which isomerise to the *anti* form (**226**) (Brookhart and Harris, 1972). The tricarbonylcycloheptadienyliron cation (**227**; X = H) is formed both by hydride abstraction from the diene complex $(C_7H_{10})Fe(CO)_3$ and, more conveniently, by the action of acids on tricarbonyl(η^4-cycloheptatriene)iron, (**188**). Use of CF_3COOD or D_2SO_4

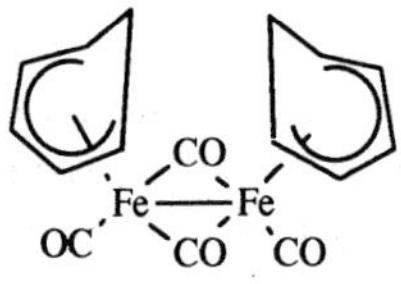

(221)

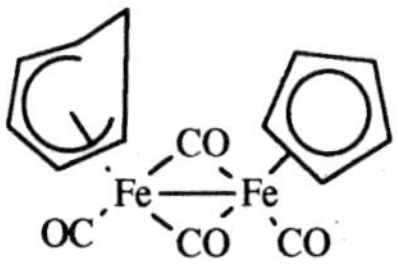

(222)

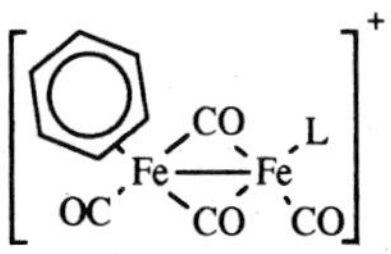

(223)

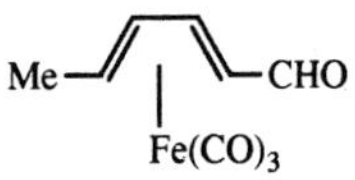

(224)

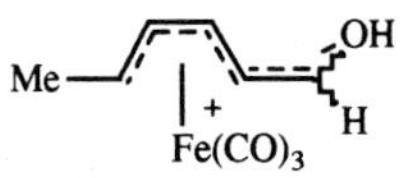

(225)

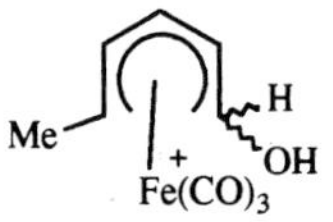

(226)

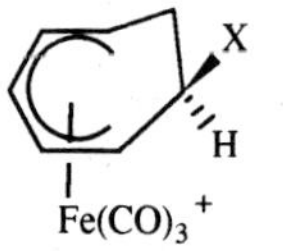

(227)

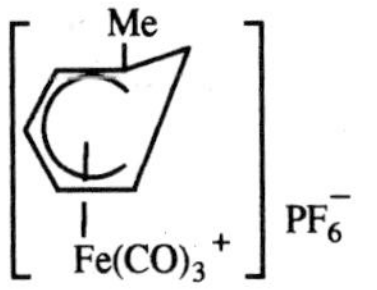

(228)

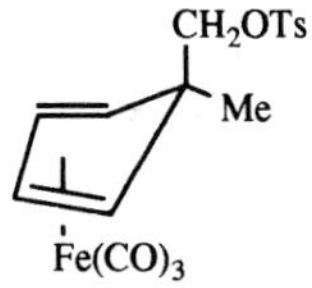

(229)

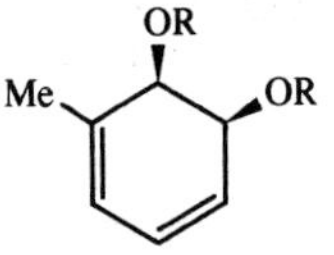

(230)

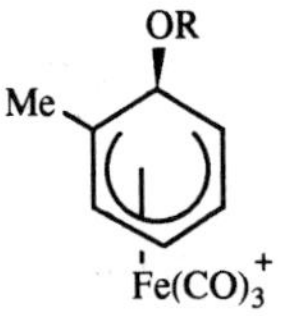

(231)

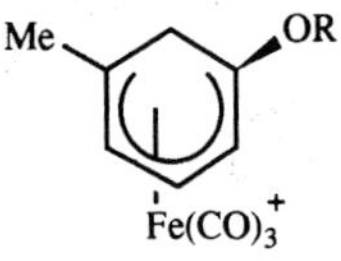

(232)

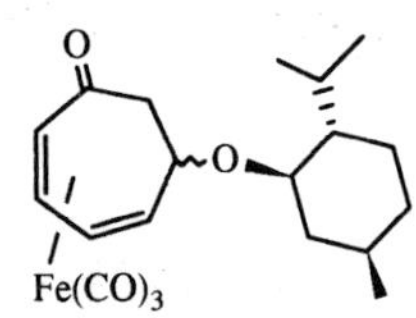

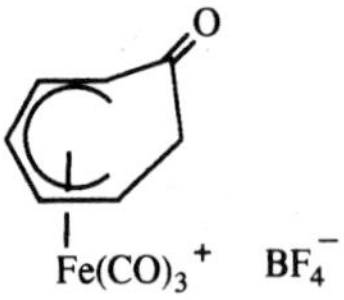

(234) **(233)**

gives exclusively the *exo*-deuteriated product (**227**; X = D) (Brookhart *et al.*, 1977). The 1-methylcyclohexadienyl complex (**228**) bas been obtained in high yield by the acid-catalysed ring-enlargement of the tosylate (**229**) (Herberich and Müller, 1977).

1,2-Dihydroxycyclohexa-3,5-dienes, e.g. the 3-methyl derivative (**230**; R = H) which are obtained in homochiral form by microbial oxidation of the corresponding arenes, have been converted into their $Fe(CO)_3$ complexes and hence with acids or Ph_3C^+ to homochiral hydroxy- or alkoxy-cyclohexadienyl complexes (**231**) and (**232**) (Howard *et al.*, 1989). The optically active (+)-oxocycloheptadienyliron salt (**233**) has been generated from the η^4-tropone complex (**204**) by acid-catalysed addition of 1-menthol to yield separable diastereoisomers of the ether (**234**) followed by treatment with HBF_4/Ac_2O and has been converted to the (+)-enantiomer of the initial complex (**204**) with triethylamine (Morita *et al.*, 1988).

The facile additions of nucleophiles to tricarbonylcyclohexadienyliron cations are exemplified in Scheme 4.75 for the 2-methoxy derivative, which is regarded as a synthon equivalent to a cyclohexenone cation (for reviews see Birch *et al.*, 1981, 1985; Pearson, 1980). The value of such synthetic uses

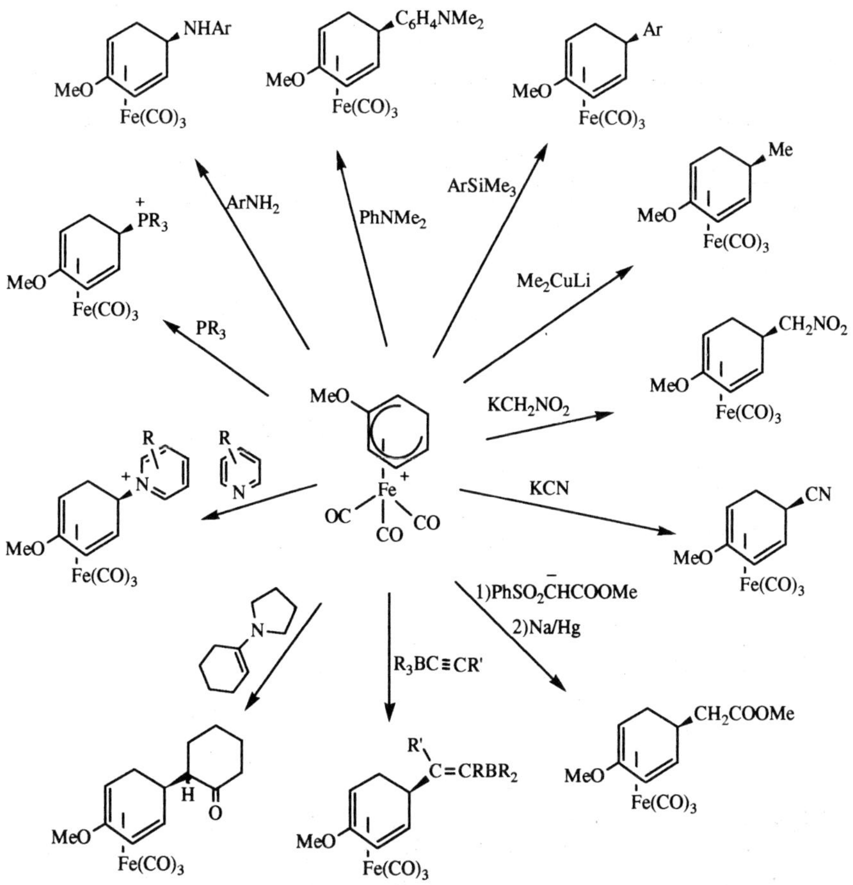

Scheme 4.75

has led to extensive studies of selectivity in the formation of variously substituted cations of this type and of their isomerisation reactions. More complex examples continue to find use in a range of natural product syntheses (e.g. Pearson *et al.*, 1983; Pearson and O'Brien, 1989). Although in all these cases the addition of the nucleophile shows strong *exo* selectivity, when the addition is reversible the kinetically preferred *exo* product may be slowly replaced (in part) by the more slowly formed *endo*-adduct (Burrows *et al.*, 1980) and the choice of solvent may also affect the outcome (Brown *et al.*, 1981).

When nucleophilic addition is strongly hindered it may be replaced by proton loss to give an (η^4-triene)$Fe(CO)_3$ complex (McArdle, 1973; Alper and Huang, 1973). Thus the η^{1-5}-dienyl complex formed from 1,3- or 2,4-cholestadiene (part-structure **235**) yields only the η^{1-4}-cholesta-1,3,5-triene complex (**236**) with a wide range of nucleophiles (Alper, *loc. cit.*). Proton loss may also occur when it leads to additional conjugation (Cowles *et al.*, 1972) or more generally when non-nucleophilic bases are used. In all cases it then provides a route to trienes from diene precursors.

A number of ferrocene analogues with 'open' dienyl systems or one such and one cyclopentadienyl group have been described. They include the bis(di- and tri-methylpentadienyl) complexes (**237**; R = Me, R′ = H or Me) which have been characterised crystallographically as well as the parent complex (**237**; R = R′ = H) and are obtained from the corresponding dienyl anion and $FeCl_2$ (Stahl *et al.*, 1987; Han *et al.*, 1987). If (bis-trialkylphosphine)iron(II) chlorides are used as starting materials, η^5,η^3-bis-dienyl complexes of the type (**238**; R = Me or Et) resulting from incomplete loss of phosphine, and/or the cations (**239**; R = Me or Et; n = 2 or 3) may be isolated (Bleeke *et al.*, 1990). The bis(cyclohexadienyl)- and cyclohexadienylcyclopentadienyl-iron complexes are products of nucleophilic addition to the corresponding arene

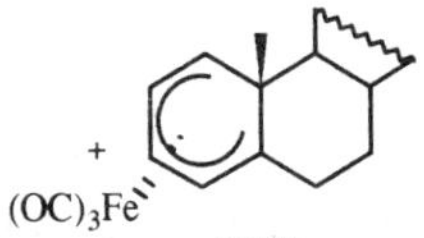

(235) 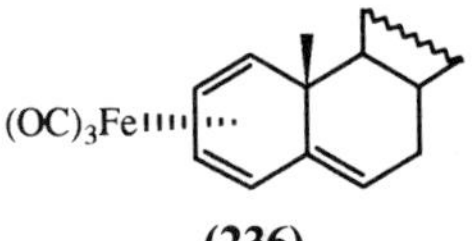 **(236)** 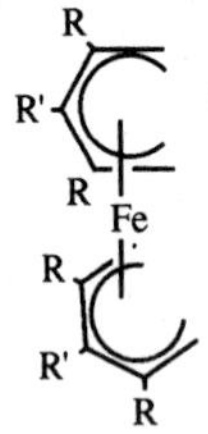 **(237)**

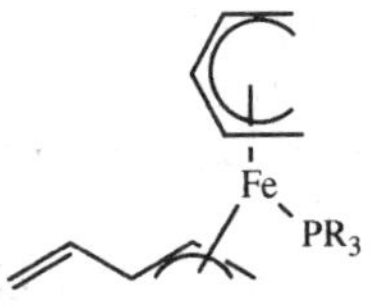

(238) 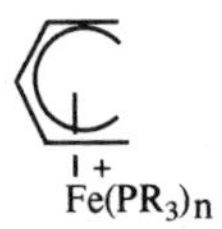 **(239)** 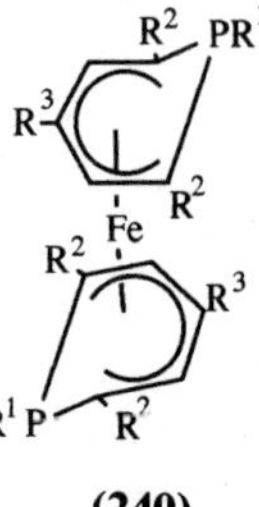 **(240)**

$$\text{(phosphinine oxide)} + Cp_2Fe_2(CO)_4 \xrightarrow{75\%} \textbf{(241)}$$

cations and are therefore discussed in section 4.4.6. Heterocyclic analogues include the phosphinine complexes (**240**) obtained from $FeCl_2$ and the appropriate anion (Dave *et al.*, 1985) and the mixed complex (**241**), obtained as shown (Deschamps *et al.*, 1984). Although commonly written as having completely delocalised 'borabenzene' ligands (for a theoretical study see Kostic and Fenske, 1983), the complexes of the types (**242**–**244**) have at best weak Fe–B bonds, the rings being non-planar with the boron atom bent away from iron, so that they may also be thought of as 6-bora-η^{1-5}-cyclohexadienyl complexes and hence analogues of the compounds discussed above. They are variously obtained from boracyclohexa-1,4-dienes (Ashe *et al.*, 1975, 1979; Herberich *et al.*, 1977, 1986) or from borabenzene-cobalt complexes (Herberich and Leonhard, 1974; Herberich and Carsten, 1978) as shown in Scheme 4.76. Also shown in this scheme is the fact that both borabenzene analogues of ferrocene (**243**, **244**) on oxidation with Ce^{IV}, Ag^+, etc. lose boron with (partial) migration of the boron substituent, and that attempted Friedel–Crafts acetylation yields cationic toluene complexes by replacement of a BR group by the MeC fragment. The parent borabenzene complex (**243**; R = H) is efficiently obtained as a sublimable red solid by lithium aluminium hydride reduction of the methoxy derivative (**243**; R = OMe) (Ashe *et al.*, 1979).

4.4.6 η^6-Arene–iron complexes

The classical Fischer synthesis of η^6-arene complexes from arene and metal halide in the presence of aluminium halide (and optionally aluminium or other reducing metal) yields the bis(arene)iron dications (**245**) when applied to iron(II) chloride (Fischer and Böttcher, 1956). Apart from benzene itself, it appears to have been used only for alkylbenzenes including [2.2]-paracyclophane (Elzinga and Rosenblum, 1982). When applied to allylbromotricarbonyliron it yields (η^6-arene)(η^3-allyl)carbonyliron cations (**246**) (Begley *et al.*, 1989) and with halo-dicarbonylcyclopentadienyliron, FpX, the (η^6-arene)(η^5-cyclopentadienyl)iron cations (**247**) are formed (Coffield *et al.*, 1957). All such cations are conveniently isolated as hexafluorophosphates or tetrafluoroborates; isolation as perchlorates should be avoided as the perchlorate of the benzene complex (**247**; R = H) has caused a serious

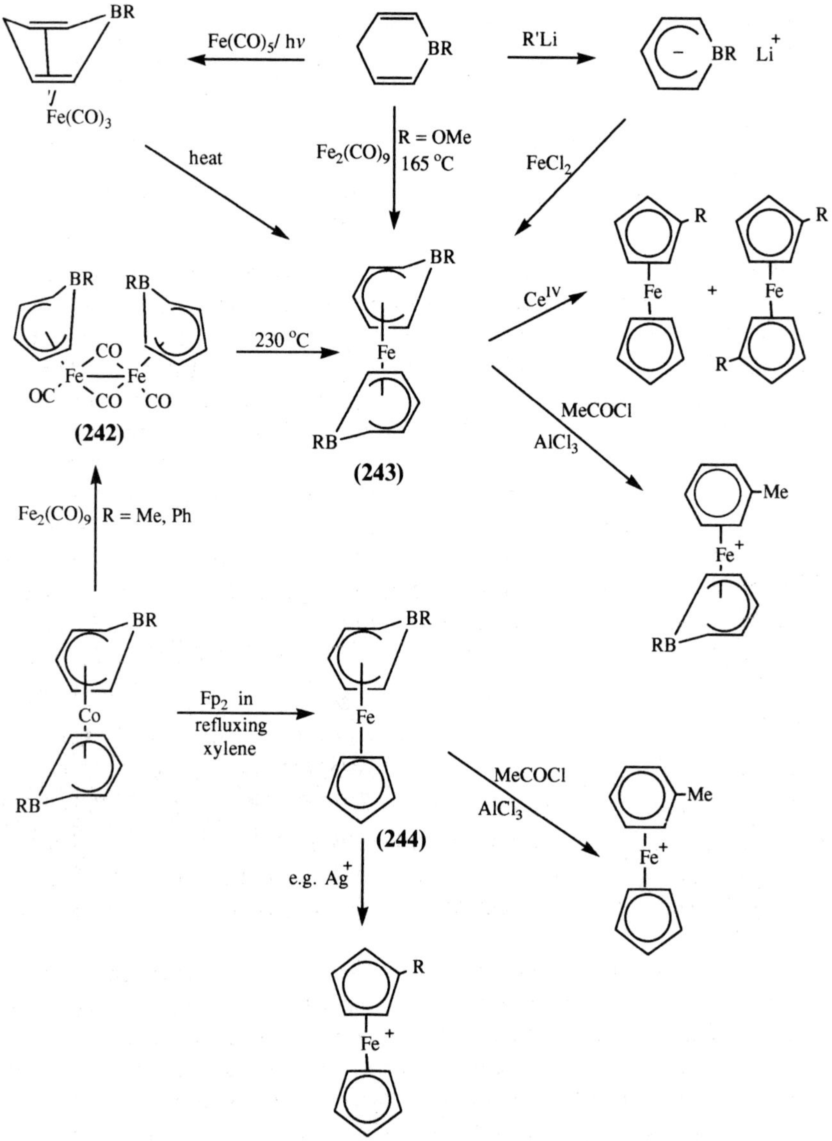

Scheme 4.76

explosion (Denning and Wentwort, 1966). The alternative route of preparing the arene-cyclopentadienyl cations (**247**) from ferrocene under similar conditions (Nesmeyanov *et al.*, 1963) is commonly preferred, if only because of the ready availability of ferrocene, and has been applied to a wide range of substituted arenes as well as substituted ferrocenes. If the latter are monosubstituted, the unsubstituted ring is replaced in preference to one bearing electron

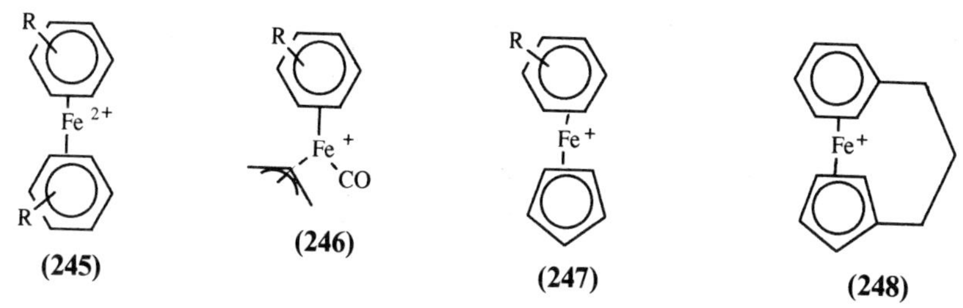

withdrawing groups (e.g. acyl) (see e.g. Roberts *et al.*, 1986). Nevertheless, 1,1′-diacetylferrocene can suffer replacement of both rings at higher temperature, reacting e.g. with mesitylene to give the dication (**245**; R = 1,3,5-Me_3) (Astruc, 1973). Many polycyclic arenes including naphthalenes (e.g. Billig *et al.*, 1988) and coronene (Schmitt *et al.*, 1978) and some heterocycles, e.g. tetramethylthiophene (Bachmann and Singer, 1976) have been used. The reduction (e.g. anthracene to 9,10-dihydroanthracene) sometimes observed during complexation has been attributed to solvent impurities and hence is readily avoidable (Lacoste *et al.*, 1989). The bromide, $(C_5Me_5)Fe(CO)_5Br$, is the preferred starting material for obtaining pentamethylcyclopentadienyl analogues of cation (**247**), but excessive crowding as in the attempt to obtain $(C_5Me_5)Fe(C_5Et_6)^+$ may lead to loss of alkyl groups (Hamon and Astruc, 1989). Intramolecular reaction has been employed to convert the ferrocene derivative $Fe[C_5H_4(CH_2)_3Ph]_2$ to the bridged cation (**248**) (Nesmeyanov *et al.*, 1977).

Arene exchange, e.g. replacing *p*-xylene from the cation (**247**; R = 1,4-Me_2) photochemically by cyclophanes (Schirch and Boekelheide, 1981) or by indoles (Kuhn and Lampe, 1990) has been found useful, but such exchange should proceed much more easily with more labile complexes, e.g. $(C_5H_5)Fe(C_{10}H_8)^+$; it has also been shown to be facilitated by prior reduction of the cations (**247**) to the corresponding neutral compounds (see below) (Nesmeyanov *et al.*, 1973).

The only other significant preparative method is the low temperature reaction of iron vapour with arenes, alone or with other ligands. This yields a variety of initial arene-iron complexes depending on the conditions. At 77 K a mixture of two products is formed from benzene; these have been identified by Mössbauer spectroscopy (Parker *et al.*, 1984) as the bis-η^6-(20 electron) and the η^6,η^4-(18 electron) (**249**) forms of $(C_6H_6)_2Fe$ with the latter in greater amount. Addition of cyclopentadiene to such a mixture at −78°C leads to the complex (**247**; R = H) (Beard *et al.*, 1981), while addition of phosphines, phosphites, or dienes leads to a series of complexes of the type (**250**; e.g. L = $P(OMe)_3$ or L_2 = 1,5-C_8H_{12}) (Ittel and Tolman, 1982). Addition of 1-methylnaphthalene to the bis-toluene complex, $(C_6H_5Me)_2Fe$, has led to a relatively stable 5:3 mixture of the (η^6-toluene)(η^4-methylnaphthalene) complexes (**251**; R = Me, R′ = H and R = H, R′ = Me respectively) which undergo facile replacement of the naphthalene moiety by other ligands

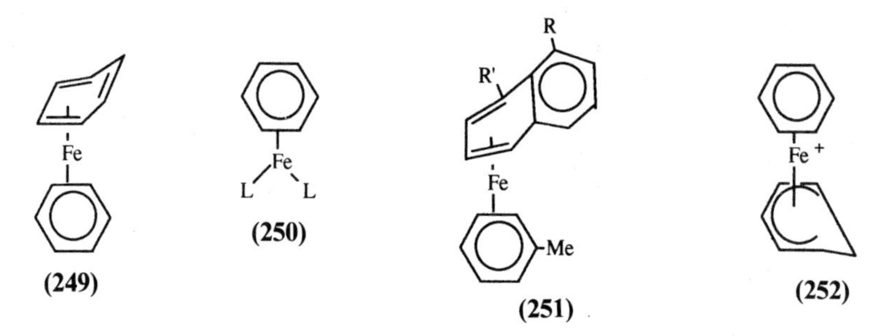

(Schäufele *el al.*, 1989). Replacement of trimethylphosphine from complex (**250**; $L = PMe_3$) by a tridentate phosphine ('P_3') has been shown to lead to a complex of the type $(\eta^4\text{-}C_6H_6)Fe(P_3)$ (Boncella and Green, 1987). A compound of the type (**250**; L_2 = cyclohexa-1,3-diene) was first prepared by reaction of iron(III) chloride and cyclohexadiene with isopropylmagnesium bromide followed by irradiation (Fischer and Müller, 1962). This compound (**250**; $L_2 = C_6H_8$) reacts with Ph_3C^+ at $-50°C$ to yield the (η^6-benzene)(η^5-cyclohexadienyl)iron cation (**252**) (Mandon and Astruc, 1989). Only two dicarbonyls of the type (**250**; L = CO) appear to be known: the hexamethylbenzene complex $(\eta^6\text{-}C_6Me_6)Fe(CO)_2$ being formed in straightforward fashion from $(\eta^6\text{-}C_6Me_6)_2Fe$ and CO at room temperature (Weber and Brintinger, 1977), whereas the structure (**253**) is the remarkable result of the addition of phenyllithium to tricarbonyl(η^4-1,5-cyclooctadiene)iron and isolated as the ethyl (R = Et) or trimethylsilyl ether ($R = SiMe_3$) after alkylation of the intermediate lithium salt (R = Li) (Chen *et al.*, 1987). Reaction of $(C_6Me_6)_2Fe$ with 2-butyne gives a low yield of 'decamethylisoferrocene' (**254**) (Weber and Brintinger, 1977).

All the cationic arene complexes can be reduced to neutral compounds. Reversible reduction of $(C_6Me_6)_2Fe^{2+}$ with either dithionite (Fischer and Röhrscheid, 1962) or, more efficiently, with sodium amalgam (Michaud and Astruc, 1982; Astruc *et al.*, 1990) has been shown to proceed in two steps, giving first the deep purple 19-electron cation $(C_6Me_6)_2Fe^+$ and then the neutral 20-electron complex $(C_6Me_6)_2Fe$, isolated as very air-sensitive black plates. Various arenecyclopentadienyliron cations (**247**) have also been reduced, both with Na/Hg and electrochemically, to neutral, 19-electron arenecyclopentadienyliron(I) compounds (e.g. Nesmeyanov, Solodnikov *et al.*, 1978) and for the naphthalene complex $[(c_{10}H_8)Fe(C_5H_5)]^+$ two-electron reduction to the corresponding anion has been observed (Nesmeyanov *et al.*, 1981). The neutral parent complex $(C_6H_6)Fe(C_5H_5)$ dimerises slowly (15 h) in solution at room temperature giving the di(cyclohexadienyl) compound (**255**) which regenerates the cation (**247**; R = H) on oxidation (O_2, I_2 etc.) (Nesmeyanov *et al.*, 1977); it also reacts with various alkyl halides, RX, to give a mixture ($\sim$1:1) of the same cation and the *exo*-substituted cyclohexadienyl compounds (**256**). When the green, neutral hexamethylbenzene complex is oxidised by O_2, the

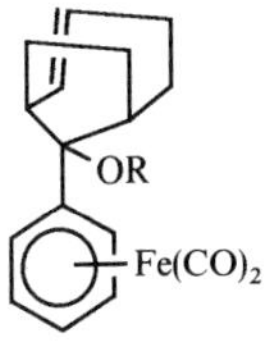

(253) 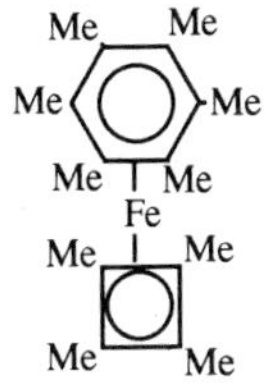 (254) 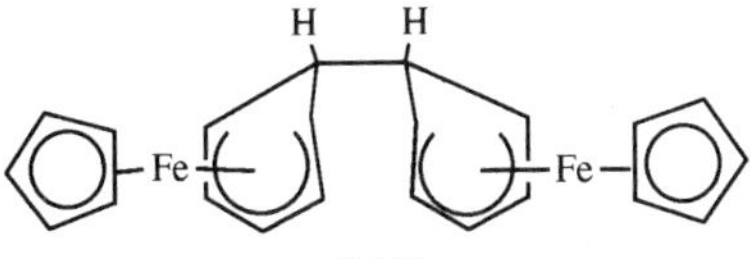 (255)

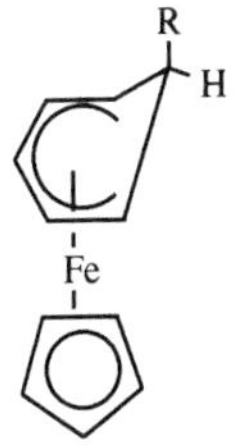

(256) 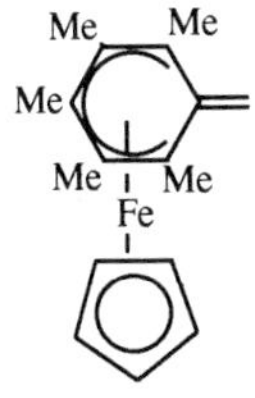 (257) 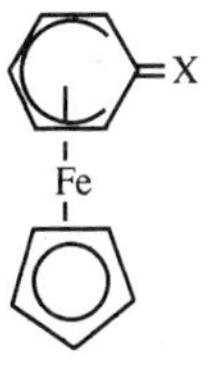 (258) 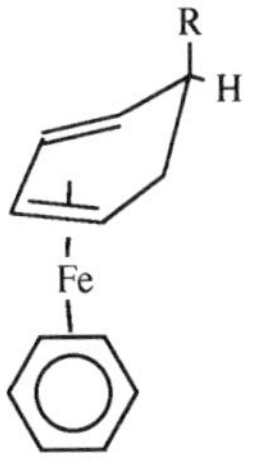 (259) 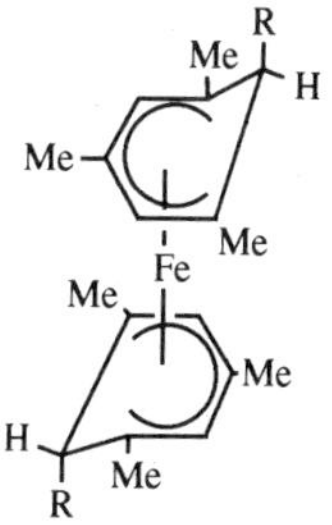 (260)

corresponding cation is only a minor product, while the principal product is the methylenecyclohexadienyl compound (**257**), also readily formed from the cation with base ($KOBu^t$) (Astruc *et al.*, 1979); since this product adds iodomethane to give the (ethylpentamethylbenzene)cyclopentadienyliron cation, it is possible, with excess base and alkyl halide, to convert the hexamethyl- to the hexa-ethyl complex, or, under phase-transfer conditions, to higher alkylbenzene complexes (Moulines and Astruc, 1989). Deprotonation of less highly substituted complexes (**247**; R = CHR'_2, R = OH, or R = NHR′) to give methylene (e.g. **258**; X = CMe_2) (Helling and Gill, 1984), oxo (**258**; X = O) (Helling and Hendrickson, 1979) or imino compounds (**258**; X = NR′) (Lee *et al.*, 1981; Michaud and Astruc, 1982) occurs similarly.

The high oxidative stability of the cations of type (**247**) is illustrated by the possibility of effecting side-chain oxidation with permanganate (e.g. **247**; R = CH_2R' to R = COR′) (Lee *et al.*, 1983). As expected by analogy with other cationic complexes discussed in this chapter, both the mono- (**247**) and dicationic complexes (**245**) add nucleophiles to give neutral complexes. However, the aromatic nature of the hydrocarbon ligand restricts such additions to good nucleophiles which add irreversibly: in practice only hydride (as $NaBH_4$ or $LiAlH_4$) and carbanions have been so added. When the arene ring contains halogen substituents these are strongly activated towards nucleophilic substitution (S_N1) and are readily replaced by such nucleophiles as RO^-, RS^-, CN^-, N_3^-, amines, malonate, etc. However, the faster attack at unsubstituted positions ensures that H^- or alkyl$^-$ still add as the major pathway. Thus even the *p*-dichlorobenzene complex (**247**; R = 1,4-Cl_2) adds organolithium compounds to the 2-position (Cambie *et al.*, 1989, 1991).

Additions to the dications (**245**) must first yield derivatives of the cationic (arene)(cyclohexadienyl)iron (**252**). According to the Davies, Green, Mingos rules (Davies *et al.*, 1978) this should preferentially add to the 'even' i.e. the η^6-arene group. In practice, the tendency to retain the aromatic nature of this group appears too strong and complex (**252**) [obtained from $(C_6H_6)_2Fe^{2+}$ (**245**; R = H) with $NaBH_4$ at 0°C] adds H^-, CN^-, $CH(COOEt)_2^-$, etc. all to the η^5-dienyl group giving the *exo*-substituted cyclohexa-1,3-diene complexes (**259**) (Madonik *et al.*, 1984) and only the bis-mesitylene complex (**245**; R = 1,3,5-Me_3), which cannot give a cyclohexadiene derivative without addition to a substituted carbon, adds carbanions, R^-, to give bis(cyclohexadienyl) complexes (**260**) (Helling and Braitsch, 1970).

In the nucleophilic additions to arenecyclopentadienyliron cations (**247**) the directive effects of arene substituents have been studied. It is found that strongly electron donating groups (e.g. NMe_2) direct addends to the *meta*-position (Khand *et al.*, 1969) while electron withdrawing groups (Cl, NO_2, COR, etc.) direct strongly to the *ortho*-position (e.g. Zhang *et al.*, 1988). The products of such additions (**256**) do not usually react with Ph_3C^+ which shows strong selectivity for abstraction of unhindered *exo*-hydrogens, but treatment with *N*-bromosuccinimide leads to the corresponding arene complexes (**247**).

In addition to F and Cl, nucleophiles have been shown to displace NO_2: e.g. aniline displaces NO_2 preferentially from chloronitrobenzene complexes and both Cl and NO_2 are displaced by e.g. stabilised carbon nucleophiles even when hindered by two flanking methyl groups (Abd-el-Aziz, 1988, 1989). Pyrolysis (*c.* 200°C) has been employed to liberate the substituted arenes from the resultant complexes. Other methods of decomplexation, e.g. by displacement by such donor ligands as PhCN, Me_2S, $P(OMe)_3$ and various phosphines have more often been applied as preparative routes for cations $[(C_5H_5)FeL_3]^+$ without isolation of the liberated arene, but should be effective in cases where pyrolysis is unsuitable. Both stepwise and complete substitution of two chlorines from the *p*-dichloro (Pearson *et al.*, 1989) and *o*-dichlorobenzene complexes have found synthetic use, in the latter case leading to novel heterocycles (e.g. Sutherland *et al.*, 1988). Finally it should be noted that nucleophilic substitution in the five-membered ring of $[(ClC_5H_4)Fe(C_6H_6)]^+$ occurs very much less readily than in its isomer (**247**; R = Cl) (Nesmeyanov *et al.*, 1967).

4.4.7 η^6-*Triene complexes*

Few such complexes are known. One (**173**) has been mentioned earlier. Several result when $FeCl_3$ is reduced by isopropylmagnesium bromide in the presence of trienes (or trienes + dienes) under photolysis: e.g. (η^4-cyclohepta-1,3-diene)(η^6-cyclohepta-1,3,5-triene)iron (Fischer and Müller, 1963) has been made in this way from a mixture of the two hydrocarbons, whereas cycloocta-1,3,5-triene yields the complex (**261**) (Müller and Fischer, 1966; Huttner and Bejenke, 1974). A carborane(cyclooctatriene)iron complex is mentioned below. The ring-exchange in the η^6,π^4-bis(cyclooctatetraene) complex (**262**) has been studied by ^{13}C NMR (Mann, 1978).

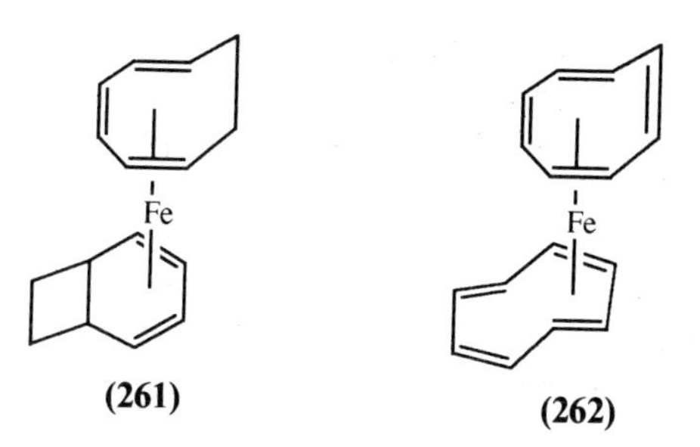

(261) **(262)**

4.5 Miscellaneous complexes

4.5.1 Complexes derived from alkynes and iron carbonyls

Simple η^2-alkyne complexes of the type discussed in section 4.4.2.2 are rarely obtained from reactions of alkynes with carbonyls bearing no other ligands, although $C_2(SiMe_3)_2$ forms $(Me_3SiC{\equiv}CSiMe_3)Fe(CO)_4$ (Pannell and Crawford, 1973) and the phospha-alkyne $Ph_2PC \equiv CBu^t$ reacts with $Fe_2(CO)_9$ forming the η^2-complex (**263**) (Carty *et al.*, 1978). Even binuclear complexes in which the alkyne acts as a bridging η^4-ligand [so strongly favoured in the cobalt series where nearly all alkynes initially form complexes $(RC{\equiv}CR')Co_2(CO)_6$] are only known with iron for very bulky acetylenes, $C_2Bu^t_2$ forming both the mono- (**264**) and bis-alkyne complex (**265**). Their formulation as having Fe–Fe double bonds is supported by the bond distances of 2.316 Å in the former (Cotton *et al.*, 1976) and 2.215 Å in the latter (Nicholas *et al.*, 1971) and 2.225 Å in a close analogue (Schmitt and Ziegler, 1973). A related tetrakis(alkyne)di-iron complex with the structure (**266**) and a rather longer Fe–Fe bond (2.465 Å) has been obtained from $C_2(SiMe_3)_2$ and $(\eta^6\text{-}PhMe)Fe(\eta^2\text{-}C_2H_4)_2$ (Schäufele *et al.*, 1988).

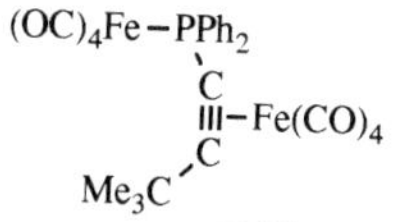

(263) 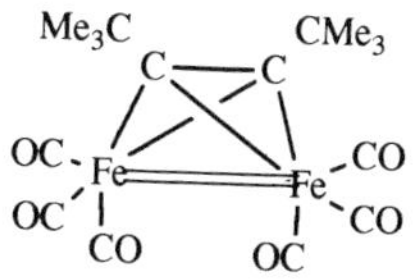 **(264)** 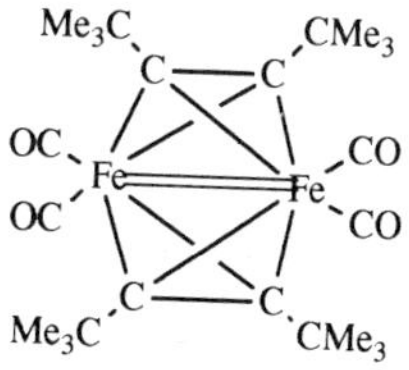 **(265)**

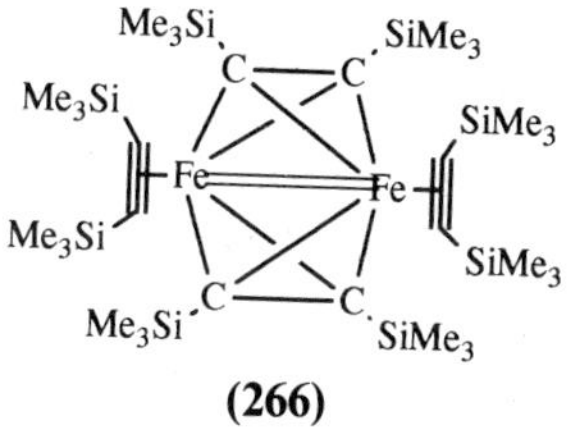

(266) 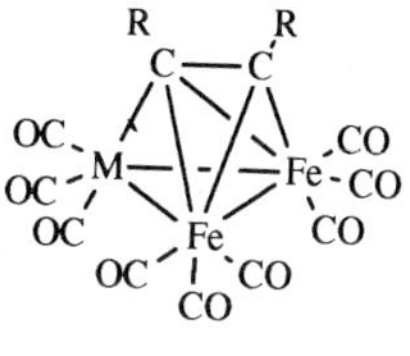 **(267)**

Trinuclear complexes containing a single alkyne unit linked as shown in structure (**267**) (Blount *et al.*, 1966) are known for several alkynes RC≡CR including 3-hexyne which forms the complex (**267**; M = Fe, R = Et) in up to 35% yield when briefly refluxed with $Fe_3(CO)_{12}$ in hexane (Carty *et al.*, 1988) and the mixed Fe_2Ru complex (**267**; M = Ru, R = Ph) similarly (Busetti *et al.*, 1984). A change in bonding pattern accompanies electrochemical reduction of the complexes (**267**) to dianions (Osella *et al.*, 1986). However, the majority of reactions of alkynes with iron carbonyls involve more than one alkyne unit and lead to the formation of complex organic ligands, frequently with incorporation of CO. Because of the variety of structural types which may result from a single reaction they have not been included in earlier sections, but some typical examples are given here. The complexity of the changes undergone by the alkynes in some of these processes is well illustrated by a product isolated from reaction of $Fe_3(CO)_{12}$ with propyne, which was shown crystallographically to have incorporated four molecules of the latter in forming the structure (**268**) (Aime *et al.*, 1977). A related complex from 1-butyne has been described as one of *c.* 30 products (Sappa *et al.*, 1978)! An equally remarkable structure (**269**) (King, 1962) belongs to one of the products from $Fe_3(CO)_{12}$ and phenylacetylene; other products obtained with this alkyne include the triphenyltropone complex (**270**) (Smith and Dahl, 1962), probably formed via compound (**269**), as well as (2,5-diphenylcyclopentadienone)tricarbonyliron and incompletely identified products with up to five alkyne units per iron atom (Hübel and Braye, 1959; Sappa *et al.*, 1983).

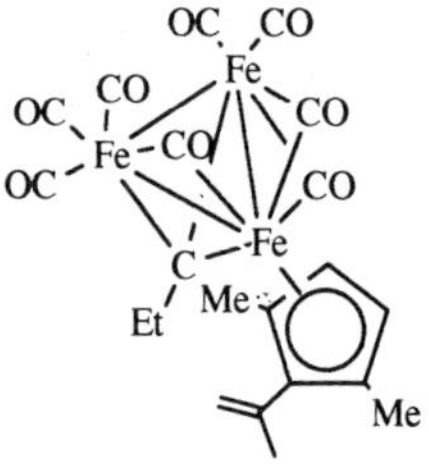

(268) 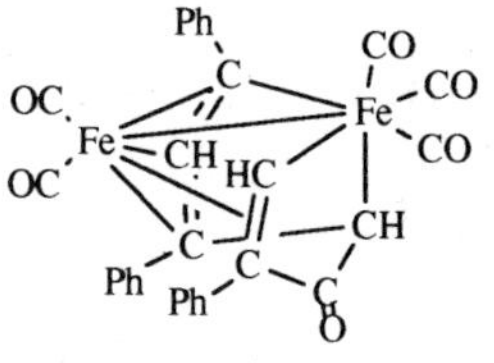 **(269)** 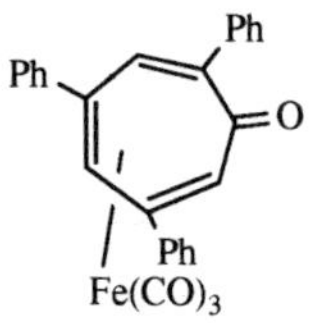 **(270)**

Acetylene itself yields *inter al.* (Weiss *et al.*, 1962) the unsubstituted cyclopentadienone (**171**) and tropone (**204**) complexes, the maleyl complex (**271**), as well as complexes having the structures (**272**) (Meunier-Piret *et al.*, 1965) and (**273**; R = H, X = CH_2) (Piret *et al.*, 1965) and the ferracyclopentadiene complex (**274**; $R^1 = R^2$ = H). The cyclopentadienyl complex (**272**) is probably formed by cyclisation and rearrangement of compound (**273**; R = H, X = CH_2). Related to the latter are a series of ketonic complexes (**273**; X = O) which are generally major products from internal alkynes. Before the symmetry of these compounds was established by Mössbauer spectroscopy (Greatrex *et al.*, 1968) and the structure of the tetramethyl derivative (**273**; R = Me, X = O) established

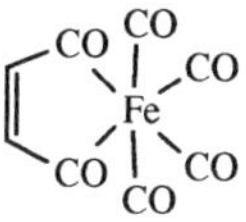

(271) 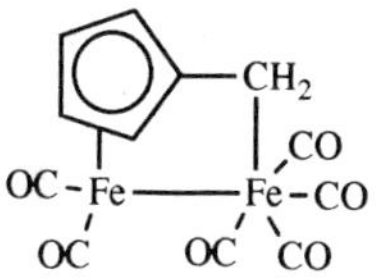 (272) 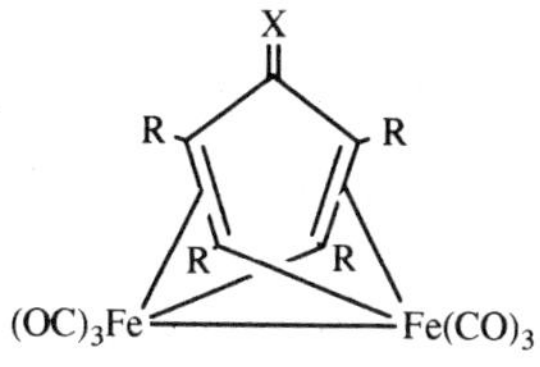 (273)

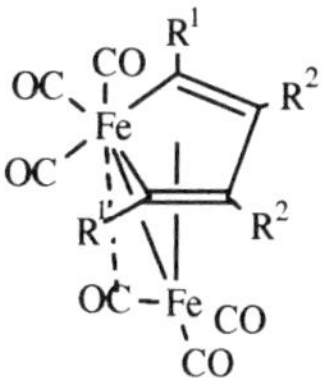

(274) 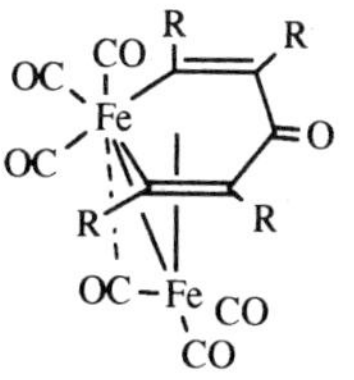 (275) 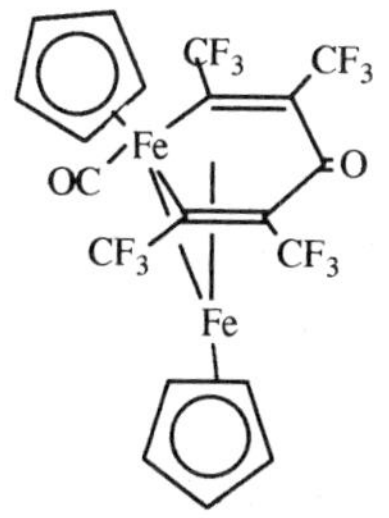 (276)

by X-ray crystallography (Piron *et al.*, 1969), they were thought to have the structures (**275**) by analogy with complexes (**274**). Formation of such a structure is now invoked to account for the fluxionality of these compounds (Aime *et al.*, 1986) and a hexafluorobut-2-yne derived cyclopentadienyliron complex (**276**) has this type of structure in the crystal (Davidson *et al.*, 1975).

Structure (**274**) is another type which is formed in substantial amount from most acetylenes; whereas normally incorporating two alkyne units (e.g. $R^1 = Ph, R^2 = H$), tolane, PhC≡CPh, also yields complex (**277**) from a single

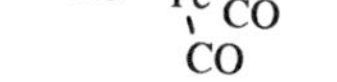

(278) 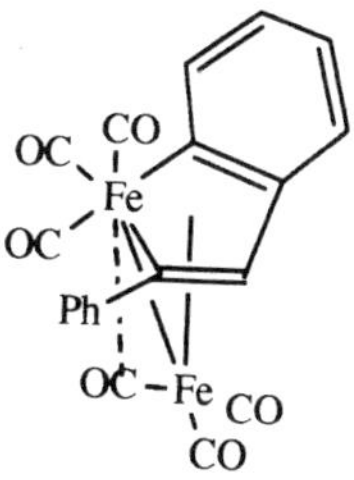 (277)

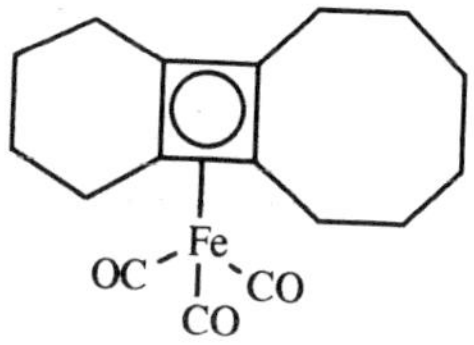

(279) 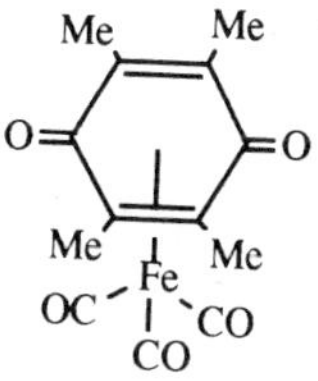 (280)

molecule. These ferracyclopentadienes (**274**) can become the sole products when (benzylideneacetone) tricarbonyliron is the source of the $Fe(CO)_3$ group (King and Ackermann, 1973). The trinuclear complexes (**278**) which are rare examples of metal coordination on both sides of a hydrocarbon fragment (e.g. (**278**); R = Ph) (Dodge and Schomaker, 1965) have been shown to be readily formed from the trinuclear complexes (**267**) (Blount *et al.*, 1966; Busetti *et al.*, 1984); it appears to be unknown whether they can also arise from the structurally related binuclear compounds (**274**). Moreover, although complex (**274**; $R^1 = R^2 = Ph$) has been converted, by treatment with e.g. Ph_2SiCl_2, to the cyclobutadiene complex $[C_4Ph_4Fe(CO)_3]$, it is not a precursor of the latter under the conditions where both are formed from tolane. Formation of cyclobutadiene complexes is apparently especially favoured in the case of very large ring diacetylenes; thus, cyclotetradeca-1,7-diyne yields complex (**279**) (King and Haiduc, 1972). Whereas photolysis of but-2-yne with $Fe(CO)_5$ yields chiefly the duroquinone complex (**280**) (Sternberg *et al.*, 1958), dicyclopropylacetylene on such photolysis gives chiefly the complexes (**273**; $R = C_3H_5$) and (**274**; $R = C_3H_5$) along with an analogue of compound (**269**) (Victor *et al.*, 1977). Several complexes of the type (**274**) have been obtained from starting materials other than alkynes, including a vinylcyclopropene (Dettlaf *et al.*, 1978); indeed, the parent compound (**274**; $R^1 = R^2 = H$) is probably best obtained from thiophene which forms this product both by reaction with $Fe_3(CO)_{12}$ (Dettlaf and Weiss, 1976) and with iron vapour at −196°C followed by carbon monoxide (Chivers and Timms, 1976).

Dihydroxy-substituted complexes (**274**; $R^1 = OH$) are the sole products isolated when acetylene or substituted acetylenes react with alkaline solutions believed to contain $HFe_2(CO)_8^-$ as the reactive species. The product (**274**; $R^1 = OH$, $R^2 = Me$) was the first to be characterised crystallographically (Hock and Mills, 1961); the presence of a semibridging carbonyl group noted in that case has been confirmed by all subsequent X-ray studies of compounds of the type (**274**) (e.g. Dettlaf *et al.*, 1978). The trinuclear anion, $HFe_3(CO)_{11}^-$ is reported to react with acetylene at room temperature to give an anion $[Fe_2(CO)_7CH{=}CH_2]^-$ (Lourdichi and Mathieu, 1982), conceivably an intermediate in the formation of compound (**274**; $R^1 = OH$, $R^2 = H$); the latter has been oxidised to compound (**271**).

4.5.2 *Complexes of heterodienes*

Heterodienes such as cinnamaldehyde, benzylideneacetone and chalcone or the corresponding imines (Schiff bases) react readily with $Fe_2(CO)_9$ to give complexes of the type (**281**; X = O, NPh, etc., R = H, Me, Ph, etc.). They are more labile than the η^4(carbon)-diene compounds and indeed are used in the formation of the latter and other hydrocarbon complexes by replacement of the heterodiene moiety as exemplified in Scheme 4.71. The oxo-complexes form via η^2-alkene complexes, e.g. (η^2-PhCH=CHCOR)

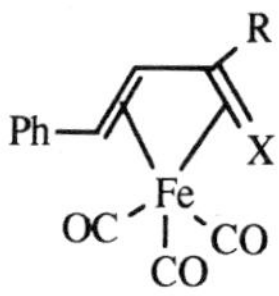

(281) 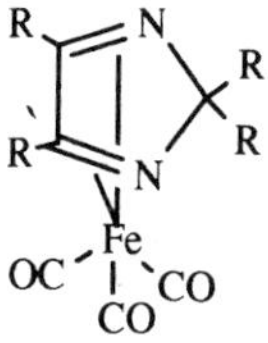 (282) 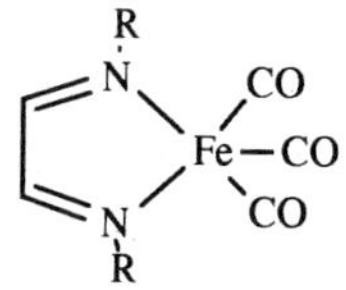 (283)

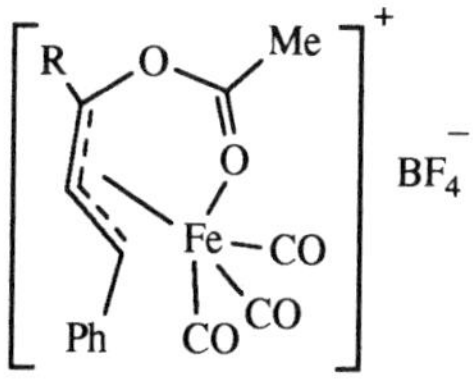

(284) 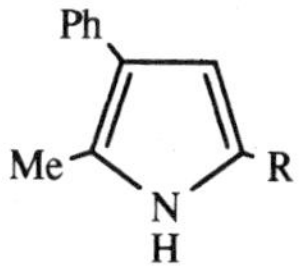 (285)

$Fe(CO)_4$ (Cardaci, 1975), but the azadiene complexes form via N-donor complexes (Cardaci and Bellachioma, 1976). Moreover, whereas cyclic diazadienes yield complexes such as (**282**) (tom Dieck and Bock, 1968), the open-chain analogues tend to give *N*,*N*-bonded complexes (**283**) (Lebfritz and tom Dieck, 1976; Otsuka *et al.*, 1967). Treatment of the oxo complexes (**281**; X = O) with acetylium tetrafluoroborate yields the salts (**284**) in direct analogy to the acetylation of simple diene complexes to give cations (**201**). Methanol reverses the process, whereas aniline converts the salts (**284**) to the imine complexes (**281**; X = NPh) (Nesmeyanov *et al.*, 1974). A more convenient preparation of the latter uses the reaction of aniline with the (η^2-enone)$Fe(CO)_4$ complexes (Semmelhack and Cheng, 1990). Recent interest in these complexes has centred on their synthetic potential. Conversion to vinylketene complexes has been mentioned (Scheme 4.72). Reaction with organolithium (R′Li) or related reagents converts the oxo-complexes (e.g. (**281**; R = Me, X = O) to 1,4-diketones, $R'COCHPhCH_2COMe$ (Danks *et al.*, 1988; Kitahara *et al.*, 1988) while the anils (**281**, X = NPh) react (at −78°C) with methyllithium to give pyrroles (**285**) (Danks and Thomas, 1990) or with secondary or tertiary alkyllithiums (R′Li) to give allylamines PhCH=CHCRR′NHPh (Cheng, 1988).

4.5.3 *Complexes with carboranes and other boracyclic ligands*

Various carboranes, but most notably the dianion $[B_9C_2H_{11}]^{2-}$, behave like cyclopentadienide in the formation of metal complexes. The ferrocenium analogues (**286**) and (**287**), the dianionic form of the latter (a ferrocene analogue) as well as analogues of carbonylcyclopentadienyliron compounds (see Lee *et al.*, 1991) were among the first such complexes to be prepared and structurally characterised. Early work on these and complexes with other

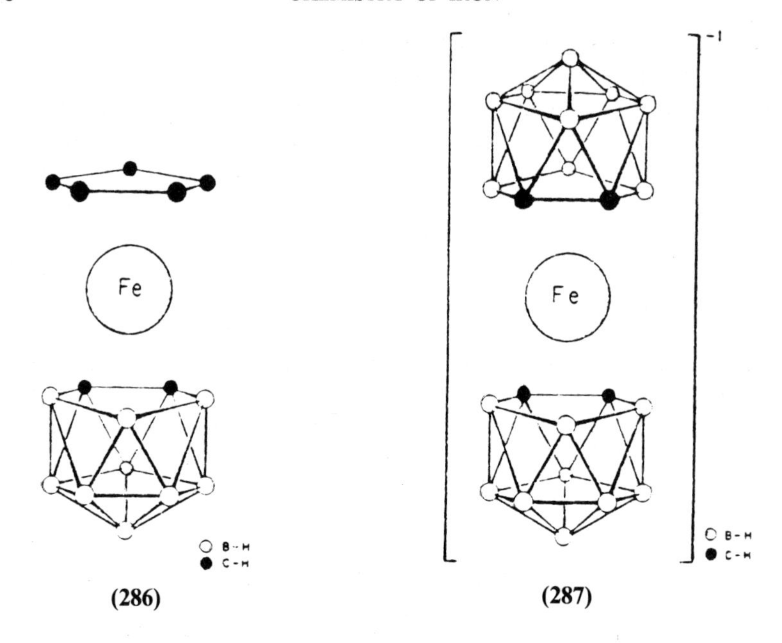

(286) (287)

carborane ligands has been fully reviewed (Grimes, 1978; Hawthorne, 1975). Unlike the compound (**176**) mentioned earlier, they are generally derived from ligands with several boron and one or two carbon atoms, e.g. $(B_3C_2H_7)^-$, $(B_{10}C_2H_{12})^-$, $(B_9CH_{10})^{2-}$, $(B_{10}CH_{11})^{3-}$, and include C–Me as well as C–H derivatives, e.g. $(B_4H_5C_2Me_2)^-$ which reacts with $FeCl_2$ to give the hydrido complex $(B_4H_4C_2Me_2)_2FeH_2$. More recent work has concentrated on smaller boracycles. Thus reaction of a mixture of $C_8H_8^{2-}$ and $Et_2C_2B_4H_5^-$ with $FeCl_2$ yields complex (**288**). This, with benzene in the presence of aluminium chloride gives the complex (**289**) which loses a boron atom when treated with tetramethylethylenediamine to give the *nido*-1,2,3-(η^6-C_6H_6) $Fe(Et_2C_2B_3H_5)$ (**290**). The latter when treated with KH followed by $FeCl_2$ and O_2 gives a complex $(C_6H_6)Fe(Et_4C_4B_6H_6)$ (Swisher *et al.*, 1983). Treatment of com-

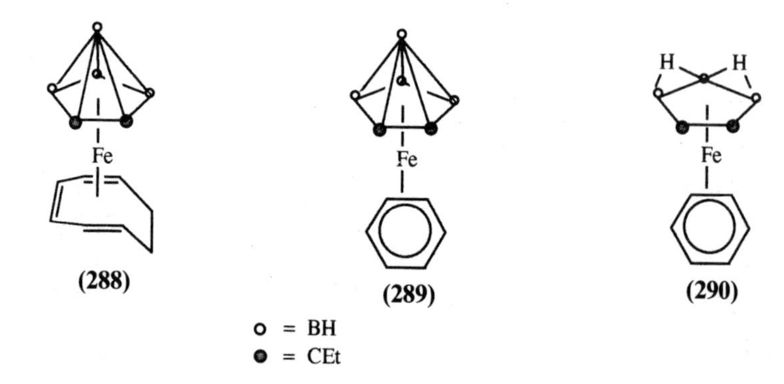

(288) (289) (290)

pound (**290**) with KH and iodomethane leads to specific methylation on the central boron atom (Davis *et al.*, 1990).

The diborathiacyclopentene (**291**; R = Me, X = S) reacts on heating with $Fe(CO)_5$ to yield complex (**292**) whereas irradiation produces chiefly the *bis*(thiadiborolene) complex, $(Et_2C_2B_2Me_2S)_2Fe(CO)$. The latter with Fp_2 produces a 'triple-decker' $(CpFe)_2(Et_2C_2B_2Me_2S)$ (Edwin *et al.*, 1985). The mixed iron–manganese triple-decker (**293**) is obtained in one step from the thiadiborolene (**291**; R = Me, X = S) with $Fp_2 + Mn_2(CO)_{10}$ and the iron–cobalt triple-decker (**294**) was obtained similarly using the diboracyclopentene (**291**; R = Et, X = CHMe) with a mixture of $CpCo(CO)_2$ and Fp_2 (Siebert and Bochman, 1977). The iron–manganese complex (**293**) is stable enough to have C_5H_5 replaced by C_6H_6 in the presence of aluminium chloride to give a cationic benzene–iron–manganese complex which in turn loses benzene at 140°C (vac.) to give an $FeMn_2$ tetradecker (Siebert *et al.*,

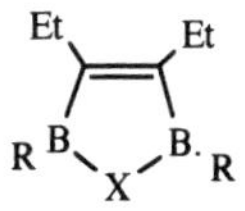

(**291**) 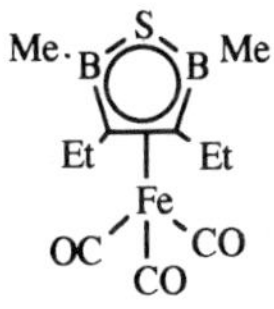 (**292**) 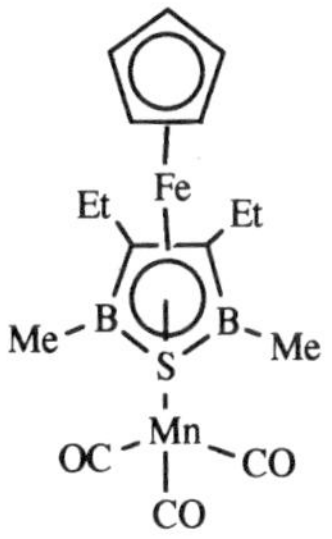 (**293**)

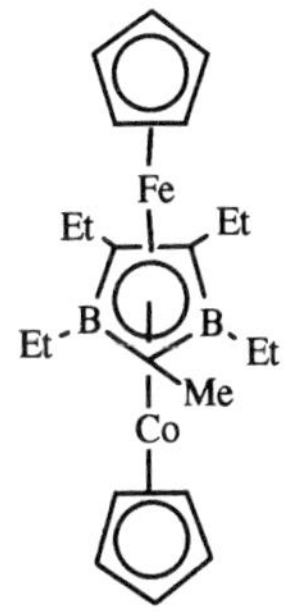

(**294**) 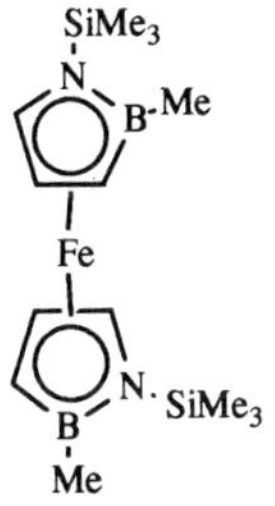 (**295**) 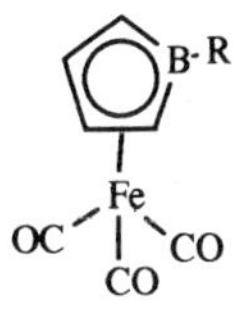 (**296**) 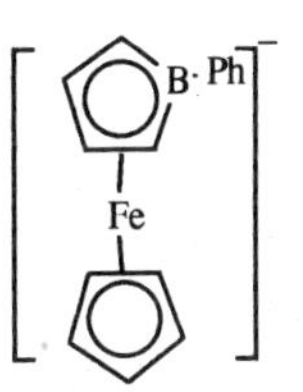 (**297**)

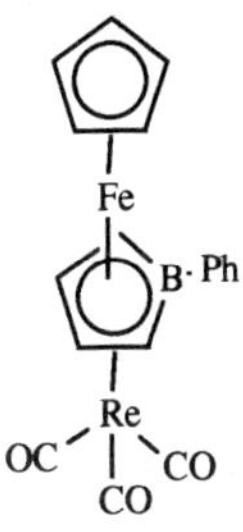

(**298**) 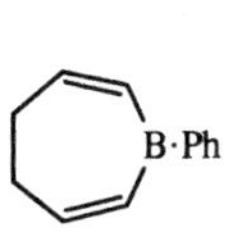 (**299**) 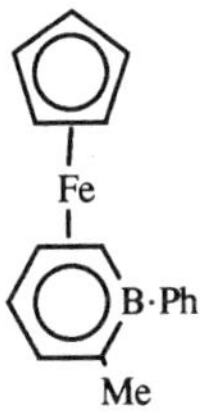 (**300**) 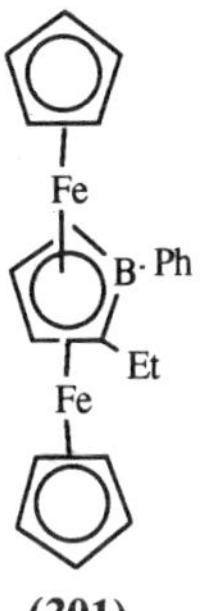 (**301**)

1978; for a review of polydecker complexes see Siebert, 1988). The azaborolinyliron complex (**295**) and its 'trans' isomer have been obtained by the metal vapour method (Amirkhalili *et al.*, 1982) and the analogues of cyclopentadienyliron carbonyls have been prepared in the analogous Bu^t-N (in place of Me_3Si-N) series (Schmid and Rohling, 1989). The simple borole complexes (**296**) are formed efficiently by irradiating the corresponding 4,5- or 2,5-dihydroboroles with $Fe(CO)_5$ (Herberich *et al.*, 1986). The related ferrocene analogues are anions, e.g. compound (**297**), which reacts with $BrRe(NCMe)_2(CO)_3$ to give the triple-decker (**298**) (Herberich *et al.*, 1989a; cf. Herberich *et al.*, 1989b). The seven-membered ring precursor (**299**) suffers ring contraction on heating with metal carbonyls; the products in the case of Fp_2 include both 5- and 6-membered boron rings in the complexes (**300**) and (**301**) (Herberich *et al.*, 1983).

4.5.4 *Complexes with other heterocycles*

Cyclopentadienylpyrrolyliron (azaferrocene) (**89**) has been mentioned. 1,1′-Diazaferrocenes require bulky substituents flanking the basic nitrogen atom for stabilisation as in the tetra-tert.-butyl derivative (**302**) (Kuhn *et al.*, 1991). By contrast, stable complexes of polyphosphorus and -arsenic heterocycles are known in considerable number. They include the diphosphacyclobutadiene complexes (**303**) and (**304**) formed together with compound (**305**;

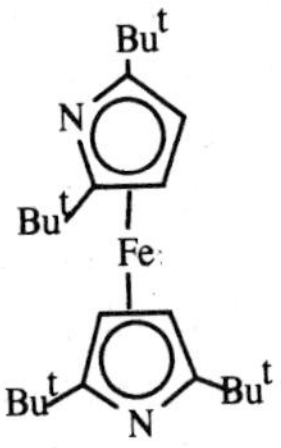

(302) 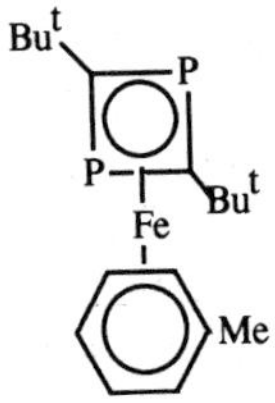 **(303)** 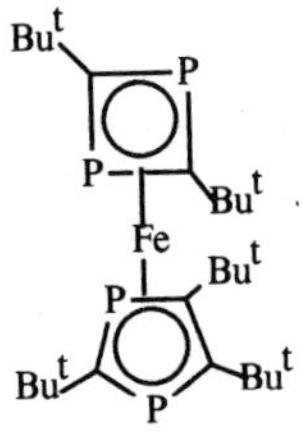 **(304)** 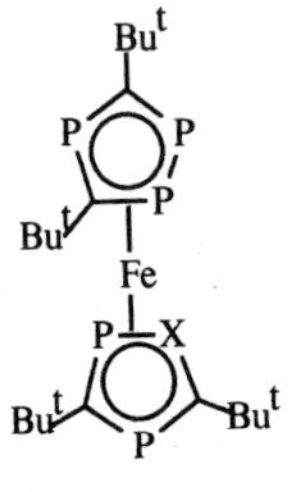 **(305)**

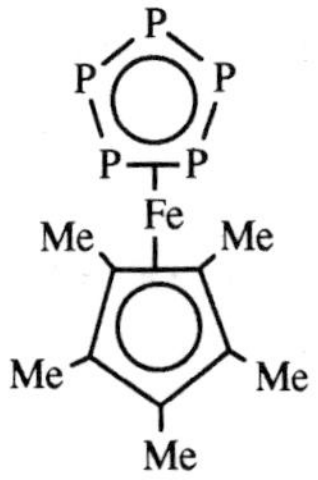

(306) 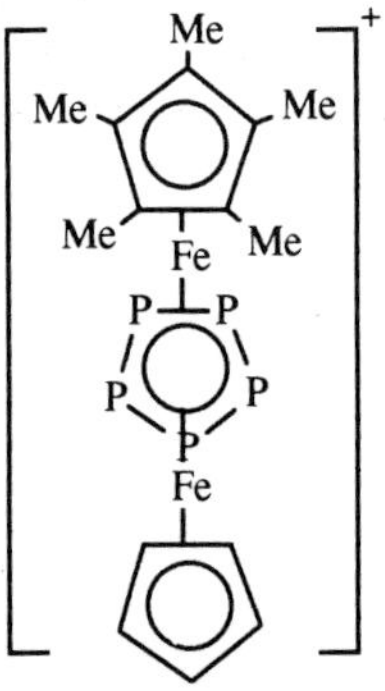 **(307)** 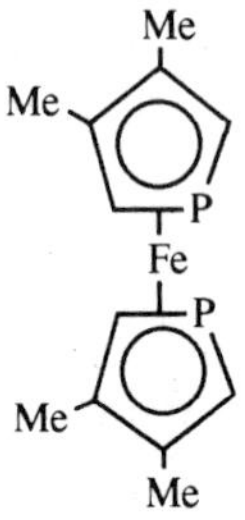 **(308)**

$X = CBu^t$) when the η^4-naphthalene complex (**251**) is treated with $Bu^tC\equiv P$ (Driess *et al.*, 1987). The anion $(C_2Bu^t{}_2P_3)^-$ reacts with $FeCl_2$ to yield complex (**305**; $X = P$) or, when mixed with Cp^- to give $(C_5H_5)Fe(C_2P_3Bu^t_2)$ (Bartsch *et al.*, 1988). LiP_5 together with $(C_5Me_5)Li$ have been similarly converted to compound (**306**) (Baudler *et. al.*, 1988) which, on irradiation with $[(C_5H_5)Fe(C_6H_6)]PF_6$ yields the cationic triple-decker (**307**) (Scherer *et al.*, 1989). The As_5 analogue of the latter has been obtained similarly (Scherer *et al.*, 1990).

Simpler mono-and 1,1′-diphosphaferrocenes, e.g. the tetramethyl derivative (**308**) have been well studied and their chemistry has been reviewed (Mathey, 1990). In common with the above poly-phosphorus compounds they can form P-donor complexes. They behave like ferrocene in being oxidisable to cations (e.g. Roberts *et al.*, 1987) as well as in reacting with arenes in the presence of $AlCl_3$ to replace one phosphacyclopentadienyl ligand to form cationic arene complexes (Roberts and Wells, 1987).

4.6 Practical applications of organo-iron compounds

The utility of many of the complexes as intermediates in laboratory syntheses has been stressed throughout this chapter. The author has no information to indicate that any of these methods are currently in use on a production scale. Use of ferrocene as an anti-knock compound in motor-spirits was tried immediately after its discovery. However, although it has a substantial effect it cannot be used for this purpose because no way has been found to remove the iron oxide which forms and which rapidly causes the engines to seize up. However, extensive trials have shown that ferrocene has beneficial effects in controlling combustion of other fuels. This is believed to have been made use of in rocket propellants in the former USSR and possibly elsewhere. Medicinal use of ferrocene derivatives as a source of iron has also been studied and the salt (**309**) whose effectiveness was first reported by Nesmeyanov *et al.* (1973) has probably found some practical use in the former USSR to prevent or treat anaemia in animals and/or humans.

The one well-established use of ferrocene derivatives is in sensors which

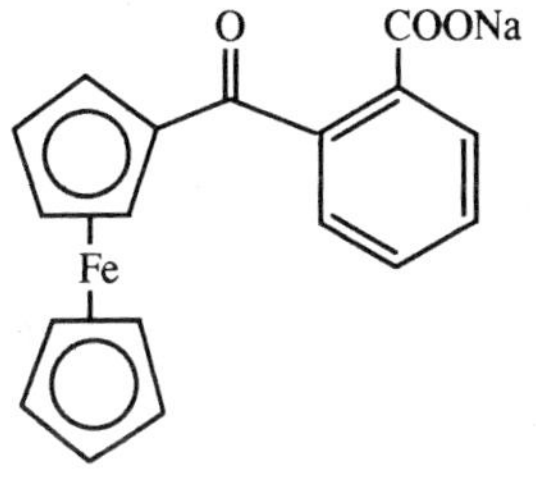

(309)

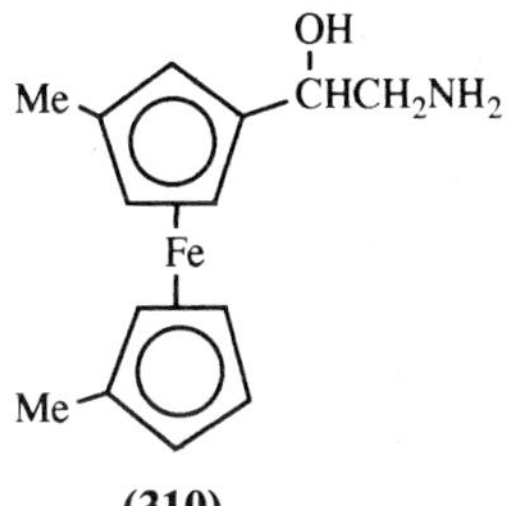

(310)

rely on their redox properties. In the 'ExacTech' pen-meter, now widely used to monitor glucose levels in diabetics, a ferrocene derivative, probably the amino-alcohol (**310**) mediates electron transfer between glucose and the enzyme glucose oxidase with the oxidised (ferrocenium) form replacing oxygen as a cofactor for the enzyme. This use, pioneered by H.A.O. Hill and his coworkers is being extended to various other sensors, e.g. for cholesterol estimation (see e.g. Green and Hill, 1986; Frew and Hill, 1987, 1988; Armstrong *et al.*, 1988; di Gleria *et al.*, 1989; Kyvik *et al.*, 1990).

Further reading

For individual compounds the reader is referred to comprehensive information in *Gmelin* (supplement, Fe) and for more selective coverage to the *Dictionary of Organometallic Compounds* (Chapman & Hall). For synthesis and reactions of organo-iron compounds the most extensive treatment will be found in *Comprehensive Organometallic Chemistry* (Eds. G. Wilkinson, F.G.A. Stone and E.W. Abel) Vol. 4, chapter 31 and Vol. 8, chapters 58, 59 (Pergamon Press, 1982) and for σ-organo-iron compounds in the chapter by A. Segnitz in *Houben-Weyl, Methoden der Organischen Chemie* Vol. 13/9a (1986) pp. 175–525; much more selective treatment of π-complexes is included in part 1 of Vol. E18.

Monographs on *Organic Chemistry of Iron* (Eds. E.A. Koerner von Gustorf, F.-W. Grevels and I. Fischer), 2 volumes (Academic Press, 1978 and 1981) and on *Chemistry of the Iron Group Metallocenes* by M. Rosenblum (Wiley, 1965) cover selected aspects and the Annual Survey volumes of the *J. Organomet. Chem.* review progress regularly.

The following references concentrate on the more recent and less readily accessible information; they are intended to lead the user back to earlier work and therefore tend to ignore precedence. References to *Angew. Chem.* relate to the International Edition in English.

References

des Abbayes, H. *et al.* (1988) *Organometallics* **7**, 2293.
Abd-el-Aziz, A.S. (1988) *J. Organomet. Chem.* **348**, 95.
Abd-el-Aziz, A.S. (1989) *Can. J. Chem.* **67**, 1618.
Adams, R.D., Chen, G. and Wang, J.G. (1989) *Polyhedron* **8**, 2521.
Aime, S. *et al.* (1976) *J. Chem. Soc., Chem. Commun.* 370.
Aime, S. *et al.* (1977) *J. Chem. Soc., Dalton Trans.* 227.
Aime, S. *et al.* (1986) *Organometallics* **5**, 1829.
Aktogu, N., Felkin, H. and Davies, S.G. (1982) *J. Chem. Soc., Chem. Commun.* 1301.
Aktogu, N. *et al.* (1984) *J. Organomet. Chem.* **262**, 49.
Alcock, N.W., Richards, C.J. and Thomas, S.E. (1991) *Organometallics* **10**, 231.
Alexander, R.P. and Stephenson, G.R. (1987) *J. Chem. Soc., Dalton Trans.* 885.
Allcock, H.R. *et al.* (1984) *J. Am. Chem. Soc.* **106**, 2337.
Almenningen, A. *et al.* (1979) *J. Organomet. Chem.* **173**, 293.
Alper, H. and Huang, C.-C. (1973) *J. Organomet. Chem.* **50**, 213.

Alper, H. and Keung, E. C.-H. (1972) *J. Am. Chem. Soc.* **94**, 2144.
Altbach, M.I. *et al.* (1986) *J. Organomet. Chem.* **306**, 375.
Amirkhalili, S., Hoehner, U. and Schmid, G. (1982) *Angew. Chem.* **21**, 68.
Andreetti, G.D., Bocelli, G. and Sgarabotto, P. (1978) *J. Organomet. Chem.* **150**, 85.
Appleton, T.D. *et al.* (1985) *J. Organomet. Chem.* **279**, 5.
Armstrong, F.A., Hill, H.A.O. and Walton, N.J. (1988) *Accounts Chem. Research* **21**, 407.
Ashe, A.J. *et al.* (1975) *J. Am. Chem. Soc.* **97**, 6865.
Ashe, A.J., Butler, W. and Sandford, H.F. (1979) *J. Am. Chem. Soc.* **101**, 7066.
Astruc, D. (1973) *Tetrahedron Lett.* 3437.
Astruc, D. *et al.* (1979) *J. Am. Chem. Soc.* **101**, 2240, 5445.
Astruc, D. *et al.* (1990) *Organometallics* **9**, 2155.
Aumann, R. (1971) *Angew. Chem.* **10**, 189, 190.
Aumann, R., Fröhlich, K. and Ring, H. (1974) *Angew Chem.* **13**, 275.
Aumann, R. and Melchers, H.-D. (1988) *J. Organomet. Chem.* **355**, 357.
Bachmann, P. and Singer, H. (1976) *Z. Naturforsch.* B. **31**B, 525.
Baer, H.H. and Hanna, H.R. (1982) *Carbohydr. Res* **102**, 169.
Bagenova, T.A. *et al.* (1981) *J. Organomet. Chem.* **222**, C1.
Baird, G.J. and Davies, S.G. (1983) *J. Organomet. Chem.* **248**, C1.
Baker, P.K. and Connelly, N.G. (1979) *J. Organomet. Chem.* **178**, C33.
Balch, A.L. and Miller, J. (1972) *J. Am. Chem. Soc.* **94**, 417.
Bandara, B.M.R. and Birch, A.J. (1984) *J. Organomet. Chem.* **265**, C6.
Barrett, A.G. *et al.* (1988) *Organometallics* **7**, 2553.
Bartsch, R., Hitchcock, P.B. and Nixon, J.F. (1988) *J. Organomet. Chem.* **340**, C37.
Bates, D.J., Rosenblum, M. and Samuel, S.B. (1981) *J. Organomet. Chem.* **209**, C55.
Bates, R.W. *et al.* (1990) *Tetrahedron* **46**, 4063.
Baudler, M. *et al.* (1988) *Angew. Chem.* **27**, 280.
Bauer, R.A., Fischer, E.O. and Kreiter, C.G. (1970) *J. Organomet. Chem.* **24**, 737.
Beard, L.K., Silvon, M.P. and Skell, P.S. (1981) *J. Organomet. Chem.* **209**, 245.
Beckman, D.E. and Jacobson, R.A. (1979) *J. Organomet. Chem.* **179**, 187.
Begley, M.J., Puntambekar, S.G. and Wright, A.H. (1987) *J. Organomet. Chem.* **329**, C7.
Begley, M.J., Puntambekar, S.G. and Wright, A.H. (1989) *J. Organomet. Chem.* **362**, C11.
Behrens, U. *et al.* (1988) *J. Organomet. Chem.* **348**, 379.
Bellerby, J.M. and Mays, M.J. (1976) *J. Organomet. Chem.* **117**, C21.
Benoit, A. *et al.* (1982) *J. Organomet. Chem.* **233**, C51.
Billig, W. *et al.* (1988) *J. Organomet. Chem.* **338**, 227.
Birch, A.J. and Fitton, H. (1966) *J. Chem. Soc., Part C*, 2060.
Birch, A.J. and Fitton, H. (1969) *Australian J. Chem.* **22**, 971.
Birch, A.J. *et al.* (1981) *Tetrahedron* **37** (Suppl. No 1) 289.
Birch, A.J. and Kelly, L.F. (1985) *J. Organomet. Chem.* **285**, 267.
Birch, A.J., Raverty, W.D. and Stephenson, G.R. (1984) *Organometallics* **3**, 1075.
Birk, R. *et al.* (1988) *Chem. Ber.* **121**, 471; *J. Organomet. Chem.* **345**, 321.
Blaha, J.P. *et al.* (1985) *J. Am. Chem. Soc.* **107**, 4561.
Bleeke, J.R. *et al.* (1990) *J. Am. Chem. Soc.* **112**, 6539.
Blount, J.F. *et al.* (1966) *J. Am. Chem. Soc.* **88,** *292.*
Blümel, J., Köhler, F.H. and Müller, G. (1988) *J. Am. Chem. Soc.* **110**, 4846.
Bly, R.S., Silverman, G.S. and Bly, R.K. (1988) *J. Am. Chem. Soc.* **110**, 7730.
Boland-Lussier, B.E. and Hughes, R.P. (1982) *Organometallics* **1**, 628.
Boncella, J.M. and Green, M.L.H. (1987) *J. Organomet. Chem.* **325**, 217.
Bond, A. and Green, M. (1972) *J. Chem. Soc., Dalton Trans.* 763.
Bottrill, M. *et al.* (1977) *J. Chem. Soc., Dalton Trans.* 1252.
Bovicelli, P. and Mincione, E. (1988) *Synth. Commun.* **18**, 2037.
Brandt, S. and Helquist, P. (1979) *J. Am. Chem. Soc.* **101**, 6473.
Brookhart, M., Buck, R.C. and Danielson, E. (1989) *J. Am. Chem. Soc.* **111**, 567.
Brookhart, M., Demond, R.E. and Lewis, B.F. (1974) *J. Organomet. Chem.* **72**, 239.
Brookhart, M. *et al.* (1980) *J. Am. Chem. Soc.* **102**, 7802.
Brookhart, M. *et al.* (1983) *J. Am. Chem. Soc.* **105**, 6721.
Brookhart, M. and Harris, D.L. (1972) *J. Organomet. Chem.* **42**, 441.
Brookhart, M., Karel, K.J. and Nance, L.E. (1977) *J. Organomet. Chem.* **140**, 203.
Brookhart, M. *et al.* (1991) *J. Am. Chem. Soc.* **113**, 927.

Brookhart, M., Studabaker, W.B. and Husk, G.R. (1985) *Organometallics* **4**, 943.
Brookhart, M., Tucker, J.R. and Husk, G.R. (1983) *J. Am. Chem. Soc.* **105**, 258.
Brookhart, M., Whitesides, T.H. and Crockett, J.M. (1976) *Inorg. Chem.* **15**, 1550.
Brookhart, M., Yoon, J. and Noh, S.K. (1989) *J. Am. Chem. Soc.* **111**, 4117.
Broussier, R. *et al.* (1990) *Inorg. Chem.* **29**, 1817.
Brown, D.A., Glass, W.K. and Hussein, F.M. (1981) *J. Organomet. Chem.* **218**, C15.
Brun, P. *et al.* (1983) *J. Chem. Soc., Dalton Trans.* 1357.
Brunet, J.C. *et al.* (1981) *J. Organomet. Chem.* **216**, 73.
Bryan, R.F. *et al.* (1970) *J. Chem. Soc., Part A*, 3064, 3068.
Burrows, A.L. *et al.* (1980) *J. Chem. Soc., Dalton Trans.* 1135.
Busetti, V. *et al.* (1984) *Organometallics* **3**, 1510.
Butler, I.R. *et al.* (1983) *Organometallics,* **2**, 128.
Caballero, C. *et al.* (1989) *J. Organomet. Chem.* **371**, 329.
Calabrese, J.C. *et al.* (1983) *Organometallics* **2**, 226.
Calderazzo, F. *et al.* (1980) *J. Organomet. Chem.* **191**, 217.
Calderon, J.L. *et al.* (1974) *J. Organomet. Chem.* **64**, C16.
Cambie, R.C. *et al.* (1989) *J. Organomet. Chem.* **359**, C 14.
Cambie, R.C. *et al.* (1991) *J. Organomet. Chem.* **409**, 385.
Cardaci, G. (1975) *J. Am. Chem. Soc.* **97**, 1412.
Cardaci, G. and Bellachioma, G. (1976) *J. Chem. Soc., Dalton Trans* 1735.
Carty, A.J., Taylor, N.J. and Sappa, E. (1988) *Organometallics* **7**, 405.
Carty, A.J., Smith, W.F. and Taylor, N.J. (1978) *J. Organomet.* Chem. **146**, C1.
Caruso, M., Knight, J.G. and Ley, S.V. (1990) *Synlett* 224.
Casey, C.P. *et al.* (1982) *J. Am. Chem. Soc.* **104**, 3761.
Casey, C.P. *et al.* (1986) *Organometallics* **5**, 199, 1873, 1879.
Casey, C.P. *et al.* (1988) *Organometallics* **7**, 670.
Casey, C.P., Gable, K.P. and Roddick, D.M. (1990) *Organometallics* **9**, 221.
Casey, C.P., Vosejpka, P.C. and Crocker, M. (1990) *J. Organomet.* Chem. **394**, 339.
Charles, A.D. *et al.* (1977) *J. Organomet. Chem.* **128**, C31.
Chaudhari, F.M., Knox, G.R. and Pauson, P.L. (1967) *J. Chem. Soc., Part C*, 2255.
Chen, J. *et al.* (1989) *Sci China Ser. B* **32**, 129 [*Chem. Abstr.* 1990, **112**, 139421; cf 1988, **108**, 94730].
Chen, J., Yin, J. and Xu, W. (1987) *Huaxue Xuebao* **45**, 1140 [*Chem. Abstr.* 1988, **108**, 190529].
Cheng, C.H. (1988) *J. Chin. Chem. Soc. (Taipei)* **35**, 261.
Childs, R.F. and Rogerson, C.V. (1980) *J. Am. Chem. Soc.* **102**, 4159.
Chivers, T. and Timms, P.L. (1976) *J. Organomet. Chem.* **118**, C37.
Choi, H.S. and Sweigart, D.A. (1982) *Organometallics* **1**, 60.
Churchill, M. R. (1967) *Inorg. Chem.* **6**, 185.
Churchill, M.R. and Chen, K.-N. (1976) *Inorg. Chem.* **15**, 788.
Churchill, M.R. and Wormald, J. (1974) *J. Chem. Soc., Dalton Trans.* 2410.
Cociolos, P., Laviron, E. and Guilard, R. (1982) *J. Organomet. Chem.* **228**, C39.
Coffield, T.H., Sandel, V. and Closson, R.D. (1957) *J. Am. Chem. Soc.* **79**, 5826.
Collman, J.P. (1975) *Accounts Chem. Research* **8**, 342.
Collman, J.P. *et al.* (1977) *J. Am. Chem. Soc.* **99**, 7381.
Conder, H.L. and Darensbourg, M.Y. (1974) *Inorg. Chem.* **13**, 506.
Connolly, J.W. (1984) *Organometallics* **3**, 1333.
Corey, E.J. and Moinet, G. (1973) *J. Am. Chem. Soc.* **95**, 7185.
Cotton, F.A., De Boer, G.B. and Marks, T.J. (1971) *J. Am. Chem. Soc.* **93**, 5069.
Cotton, F.A. *et al.* (1977) *J. Am. Chem. Soc.* **99**, 3293.
Cotton, F.A., Jamerson, J.D. and Stults, B.R. (1976) *J. Am. Chem. Soc.* **98**, 1774.
Cotton, F.A. and Troup, J.M. (1974) *J. Am. Chem. Soc.* **96**, 4422.
Cowan, D.O. and Le Vanda, C. (1972) *J. Am. Chem. Soc.* **94**, 9271.
Cowles, R.J.H. *et al.* (1972) *J. Chem. Soc., Dalton Trans.* 1768.
Crawford, S.S., Firestein, G. and Kaesz, H.D. (1975) *J. Organomet. Chem.* **91**, C57.
Cunningham, A.F. (1991) *J. Am. Chem. Soc.* **113**, 4864.
Cutler, A. *et al.* (1976) *J. Am. Chem. Soc.* **98**, 3495.
Danks, T.N., Rakshit, D. and Thomas, S.E. (1988) *J. Chem. Soc., Perkin Trans. 1* 2091.
Danks, T.N. And Thomas, S.E. (1990) *J. Chem. Soc., Perkin Trans. 1* 761.
Darensbourg, M.Y. (1972) *J. Organomet. Chem.* **38**, 133.

Dave, T. *et al.* (1985) *Organometallics* **4**, 1565.
Davidson, J.L. *et al.* (1975) *J. Chem. Soc., Chem. Commun.* 286.
Davies, S.G. (1988) *Pure Appl. Chem.* **60**, 13.
Davies, S.G. (1990) *Aldrich Chimica Acta* **23**, 31.
Davies, S.G., Green, M.L.H. and Mingos, D.M.P. (1978) *Tetrahedron* **34**, 3047.
Davis, J.H., Attwood, M.D. and Grimes, R.N. (1990) *Organometallics* **9**, 1171.
Dawkins, G.M. *et al.* (1983) *J. Chem. Soc., Dalton Trans.* 499.
Deeming, A.J. *et al.* (1974) *J. Chem. Soc., Dalton Trans.* 2093.
Delbaere, L.T.J., Kruczynski, L.S. and McBride, D.W. (1973) *J. Chem. Soc., Dalton Trans.* 307.
Denning, R.G. and Wentworth, R.A.D. (1966) *J. Organomet. Chem.* **5**, 292.
De Puy, C.H., Parton, R.L. and Jones, T. (1977) *J. Am. Chem. Soc.* **99**, 4070.
Deschamps, E. *et al.* (1984) *Organometallics* **3**, 1144.
Dettlaf, G. *et al.* (1978) *J. Organomet. Chem.* **152**, 203.
Dettlaf, G. and Weiss, E. (1976) *J. Organomet. Chem.* **108**, 213.
Dieter, J.W., Li, Z. and Nicholas, K.M. (1987) *Tetrahedron Lett.* **28**, 5415.
Dieter, J. and Nicholas, K.M. (1981) *J. Organomet. Chem.* **212**, 107.
Dinulescu, I.G., Georgescu, E.G. and Avram, M. (1977) *J. Organomet. Chem.* **127**, 193.
Dodge, R.P. and Schomaker, V. (1965) *J. Organomet. Chem.* **3**, 274.
Doedens, R.J. (1968) *Inorg. Chem.* **7**, 2323.
Doedens, R.J. and Ibers, J.A. (1969) *Inorg. Chem.* **8**, 2709.
Donaldson, W.A. (1990) *J. Organomet. Chem.* **395**, 187.
Driess, M. *et al.* (1987) *J. Organomet. Chem.* **334**, C35.
Edwin, J., Siebert, W. and Krüger, C. (1985) *J. Organomet. Chem.* **282**, 297.
Eekhof, J.H., Hogeveen, H. and Kellogg, R.M. (1976) *J. Chem. Soc., Chem. Commun.* 657.
Efratry, A. (1977) *Chem. Reviews* **77**, 691.
Eilbracht, P. and Dahler, P. (1977) *J. Organomet. Chem.* **135**, C23.
Eilbracht, P. *et al.* (1990) *Chem. Ber.* **123**, 1071, 1079, 1089.
Eisenstadt, A. (1972) *Tetrahedron Lett.* 2005.
Elder, M. and Hall, D. (1969) *Inorg. Chem.* **8**, 1424.
Elzinga, J. and Hogeveen, H. (1977) *J. Chem. Soc., Chem. Commun.* 705.
Elzinga, J. and Rosenblum, M. (1982) *Tetrahedron Lett.* **23**, 1535.
Emerson, G.F. *et al.* (1966) *J. Am. Chem. Soc.* **88**, 3172.
Emerson, G.F., Watts, L. and Pettit, R. (1965) *J. Am. Chem. Soc.* **87**, 3253.
Evans, G., Johnson, B.F.G. and Lewis, J. (1975) *J. Organomet. Chem.* **102**, 507.
Faller, J.W. and Adams, M.A. (1979) *J. Organomet. Chem.* **170**, 71.
Fehlhammer, P., Mayr, A. and Kehr, W. (1980) *J. Organomet. Chem.* **197**, 327.
Fehlner, T.P. (1978) *J. Am. Chem. Soc.* **100**, 3250.
Fischer, E.O. and Böttcher, R. (1956) *Chem. Ber.* **89**, 2397.
Fischer, E.O. *et al.* (1972) *Chem. Ber.* **105**, 162, 588.
Fischer, E.O. and Müller, J. (1962) *Z. Naturforsch. B* **17**b, 776.
Fischer, E.O. and Müller, J. (1963) *J. Organomet. Chem.* **1**, 89.
Fischer, E.O. and Röhrscheid, F. (1962) *Z. Naturforsch B* **17**b, 483.
Fish, R.W. *et al.* (1976) *J. Organomet. Chem.* **105**, 101.
Fjare, D.E. and Gladfelter, W.L. (1981) *Inorg. Chem.* **20**, 3533.
Fongers, K.S., Hogeveen, H. and Kingsma, R.F. (1982) *Synthesis* 839.
Franck-Neumann, M. *et al.* (1991) *Synlett* 331.
Franck-Neumann, M., Sedrati, M. and Mokhi, M. (1987) *J. Organomet. Chem.* **326**, 389.
Frew, J.E. and Hill, H.A.O. (1987) *Anal. Chem.* **59**, 933A.
Frew, J.E. and Hill, H.A.O. (1988) *Eur. J. Biochem.* **172**, 261.
Frey, M., Jenny, T.A. and Stoeckli-Evans, H. (1990) *Organometallics* **9**, 1806.
Gedridge, R.W. *et al.* (1987) *J. Organomet. Chem.* **331**, 73.
Geoffroy, G.L. (1980) *Accounts Chem. Research* **13**, 469.
Gibson, D.H. and Erwin, D.K. (1975) *J. Organomet. Chem.* **86**, C31.
Gibson, D.H. and Vonnahme, R.L. (1972) *J. Am. Chem. Soc.* **94**, 5090.
Gibson, D.H. and Vonnahme, R.L. (1974) *J. Organomet. Chem.* **70**, C33.
Gist, A.V. and Reeves, P.C. (1981) *J. Organomet. Chem.* **215**, 221.
di Gleria, K. Hill, H.A.O. and Chambers, J.A. (1989) *J. Electroanal. Chem.* **267**, 83.
Glidewell, C. and McKechnie, J.S. (1987) *J. Organomet. Chem.* **321**, C21.

de Graaf, W. *et al.* (1989) *J. Organomet. Chem.* **378**, 115.
Greatrex, R., Greenwood, N.N. and Pauson, P.L. (1968) *J. Organomet. Chem.* **13**, 533.
Greaves, E.O. *et al.* (1974) *J. Chem. Soc. Chem. Commun.* 257.
Gree, R. Park, H. and Paquette, L.A. (1980) *J. Am. Chem. Soc.* **102**, 4397.
Green, M., Stone, F.G.A. and Underhill, M. (1975) *J. Chem. Soc., Dalton Trans.* 939.
Green, M.J. and Hill, H.A.O. (1986) *J. Chem. Soc., Faraday Trans. 1* **82**, 1237.
Grimes, R.N. (1978) *Accounts Chem. Research* **11,** 420.
Grimme, W. and Köser, H.G. (1981) *J. Am. Chem. Soc.* **103**, 5919.
Grössmann, U. *et al.* (1991) *J. Organomet. Chem.* **408**, 203.
Grossel, M.C. *et al.* (1991) *Organometallics* **10**, 851.
Gubin, S.P. and Denisovich, L.I. (1968) *J. Organomet. Chem.* **15**, 471.
Hamon, J.-R. and Astruc, D. (1989) *Organometallics* **8**, 2243.
Han, J.C., Hutchinson, J.P. and Ernst, R.D. (1987) *J. Organomet. Chem.* **321**, 389.
Harvey, P.D. *et al.* (1988) *Inorg. Chem.* **27**, 57.
Hawthorne, M.F. (1975) *J. Organomet. Chem.* **100**, 97.
Hayashi, T. *et al.* (1979) *Tetrahedron Lett.* 425.
Hayashi, T. *et al.* (1980) *Tetrahedron Lett.* **21**, 1871.
Hayashi, T. *et al.* (1988) *J. Org. Chem.* **53**, 113.
Hayashi, T. *et al.* (1989) *J. Chem. Soc., Chem. Commun.* 495; *J. Am. Chem. Soc.* **111**, 6301.
Hayashi, T. *et al.* (1990) *Tetrahedron Lett.* **31**, 1743.
Helling, J.F. and Braitsch, D.M. (1970) *J. Am. Chem. Soc.* **92**, 7207
Helling, J.F. and Gill, U.Ș. (1984) *J. Organomet. Chem.* **264**, 353.
Helling, J.F. and Hendrickson, W.A. (1979) *J. Organomet. Chem.* **168**, 87.
Helquist, P. (1991) in Liebeskind, L.S. (Ed.) *Adv. in Metal-Org. Chem.* **2**, 143.
Hendrickson, D.N., Sohn, Y.S. and Gray, H.B. (1971) *Inorg. Chem.* **10**, 1559.
Herberhold, M. and Leonhard, K. (1974) *J. Organomet. Chem.* **78**, 253.
Herberich, G.E. and Bauer, F. (1977) *Chem. Ber.* **110**, 1167.
Herberich, G.E., Becker, H.J. and Greiss, G. (1974) *Chem. Ber.* **107**, 3780.
Herberich, G.E. and Carsten, K. (1978) *J. Organomet. Chem.* **144**, C1
Herberich, G.E., Dunne, B.J. and Hesner, B. (1989a) *Angew. Chem.* **28**, 737.
Herberich, G.E. *et al.* (1983) *J. Organomet. Chem.* **246**, 141.
Herberich, G.E. *et al.* (1986) *J. Organomet. Chem.* **308**, 153.
Herberich, G.E., Hessner, B. and Köffer, D.P.J. (1989b) *J. Organomet. Chem.* **362**, 243.
Herberich, G.E. and Müller, H. (1971) *Chem. Ber.* **104**, 2781.
Herberich, G.E. and Raabe, E. (1986) *J. Organomet. Chem.* **309**, 143.
Herbstein, F.H. and Reisner, M.G. (1972) *J. Chem. Soc., Chem. Commun.* 1077.
Hill, R.H. and Wrighton, M.S. (1985) *Organometallics* **4,** 413.
Hisatome, M. *et al.* (1990) *Organometallics* **9**, 497.
Hitchcock, P.B. and Mason, R. (1967) *J. Chem. Soc., Chem. Commun.* 242.
Hoberg, H. *et al.* (1987) *J. Organomet. Chem.* **320**, 325.
Hoberg, H. and Jenni, K. (1987) *J. Organomet. Chem.* **322**, 193.
Hock, A.A. and Mills, O.S. (1961) *Acta Cryst.* **14**, 139.
Hooker, R.H., Mahmoud, K.A. and Rest, A.J. (1983) *J. Chem. Soc., Chem. Commun.* 1022.
Howard, P.W., Stephenson, G.R. and Taylor, S.C. (1989) *J. Organomet. Chem.* **370**, 97.
Howell, J.A.S. *et al.* (1987) *J. Organomet. Chem.* **319**, C45.
Hriljac, J.A., Swepston, P.N. and Shriver, D.F. (1985) *Organometallics* **4**, 158.
Hsieh, A.T.T. and Mays, M.J. (1972) *J. Organomet. Chem.* **37**, C53.
Hübel, W. and Braye, E.H. (1959) *J. Inorg. Nucl. Chem.* **10**, 250.
Hughes, R.P. *et al.* (1986) *Organometallics* **5**, 789, 797.
Huttner, G. and Bejenke, V. (1974) *Chem. Ber.* **107**, 156.
Huttner, G. and Gartzke, W. (1972) *Chem. Ber.* **105**, 2714.
Ichikawa, M. Tsutsui, M. and Vohwinkel, F. (1967) *Z. Naturforsch. B* **22**b, 376.
Ikariya, T. and Yamamoto, A. (1976) *J. Organomet. Chem.* **118**, 65.
Impastato, F.J. and Ihrman, K.G. (1961) *J. Am. Chem. Soc.* **83**, 3726.
Ito, Y. *et al.* (1989) *Tetrahedron Lett.* **30**, 4681.
Ittel, S.D. and Tolman, C.A. (1982) *Organometallics* **1**, 1432.
Jacob, M. and Weiss, E. (1977) *J. Organomet. Chem.* **131**, 263.
Jacob, M. and Weiss, E. (1978) *J. Organomet. Chem.* **153**, 31.

Johnson, B.F.G. *et al.* (1971) *Chem. Commun.* 177.
Johnson, B.F.G., Karlin, K.D. and Lewis, J. (1978) *J. Organomet. Chem.* **145**, C23.
Joshi, K.K., Pauson, P.L. and Stubbs, W.H. (1963) *J. Organomet. Chem.* **1**, 51.
Jutzi, P. and Mix, A. (1990) *Chem. Ber.* **123**, 1043.
Kappes, D., *et al.* (1989) *Angew. Chem.* **28**, 1657.
Karsch, H.H. (1977) *Chem. Ber.* **110**, 2222.
Karsch, H.H. (1977) *Chem. Ber.* **110**, 2699.
Khand, I.U., Lanez, T. and Pauson, P.L. (1989) *J. Chem. Soc. Perkin Trans. 1* 2075.
Khand, I.U., Pauson, P.L. and Watts, W.E. (1969) *J. Chem. Soc. Part C* 2024.
King, G.S.D. (1962) *Acta Cryst.* **15**, 243.
King, M. *et al.* (1982) *Organometallics* **1**, 1718.
King, R.B. and Ackermann, M.N. (1973) *J. Organomet. Chem.* **60**, C57.
King, R.B. and Haiduc, I. (1972) *J. Am. Chem. Soc.* **94**, 4044.
King, R.B. *et al.* (1987) *J. Organomet. Chem.* **330**, 115.
Kitahara, H. *et al.* (1988) *Bull. Chem. Soc. Jpn.* **61**, 3362.
Knox, G.R. and Thom, I.G. (1981) *J. Chem. Soc., Chem. Commun.* 373.
Kolis, J.W. *et al.* (1982) *J. Am. Chem. Soc.* **104**, 6134.
Kostic, N.M. and Fenske, R.F. (1982) *Chem. Phys. Lett.* **90**, 306.
Kostic, N.M. and Fenske, R.F. (1983) *Organometallics* **2**, 1319.
Kremer, K.A.M. *et. al.* (1982) *J. Am. Chem. Soc.* **104**, 6119.
Kremer, K.A.M. and Helquist, P. (1985) *J. Organomet. Chem.* **285**, 231.
Kruck, T. and Knoll, L. (1972) *Chem. Ber.* **105**, 3783.
Kruck, T. and Knoll, L. (1973) *Z. Naturforsch. B* **28**b, 34.
Küpper, F.W. (1968) *J. Organomet. Chem.* **13**, 219.
Kuhn, N. *et al.* (1991) *Chem. Ber.* **124**, 89.
Kuhn, N. and Lampe, E.-M. (1990) *J. Organomet. Chem.* **385**, C9.
Kyvik, K.O. *et al.* (1990) *Diabetics Research and Clinical Practice* **10**, 85.
Lacoste, M. *et al.* (1989) *Organometallics* **8**, 2233.
Lalor, F.J. *et al.* (1984) *J. Chem. Soc., Dalton Trans.* 245.
Lang, H. *et al.* (1991) *J. Organomet. Chem.* **409**, C7.
Lau, W., Huffman, J.C. and Kochi, J.K. (1982) *Organometallics* **1**, 155.
Lebfritz, D. and tom Dieck, H. (1976) *J. Organomet. Chem.* **105**, 255.
Lee, C.C. *et al.* (1983) *J. Organomet. Chem.* **247**, 71.
Lee C.C., Gill, U.S. and Sutherland, R.G. (1981) *J. Organomet. Chem.* **206**, 89.
Lee, S.S., Knobler, C.B. and Hawthorne, M.F. (1991) *Organometallics* **10**, 1054.
Lentzner, H.L. and Watts, W.E. (1971) *Tetrahedron* **27**, 4343.
Levitre, S.A., Tso, C.C. and Cutler, A.R. (1986) *J. Organomet. Chem.* **308**, 253.
Ley, S.V. (1988) *Phil. Trans. R. Soc. London A* **326**, 633.
Li, Z. and Nicholas, K.M. (1990) *J. Organomet. Chem.* **402**, 105.
Liebeskind, L.S. and Welker, M.E. (1983) *Organometallics* **2**, 194.
Longoni, G. Manassero, M. and Sansoni, M. (1980) *J. Am. Chem. Soc.* **102**, 3242.
Lotz, S., van Rooyen, P.H. and van Dyk, M.M. (1987) *Organometallics* **6**, 499.
Lourdichi, M. and Mathieu, R. (1982) *Nouv. J. Chim.* **6**, 231.
McArdle, P. (1973) *J. Chem. Soc., Chem. Commun.* 482.
Madonik, A.M. *et al.* (1984) *J. Am. Chem. Soc.* **106**, 3381.
Mandon, D. and Astruc, D. (1989) *Organometallics* **8**, 2372.
Mann, B.E. (1978) *J. Chem. Soc., Dalton Trans. 1761.*
Marquarding, D. *et al.* (1970) *J. Am. Chem. Soc.* **92**, 5389.
Marquarding, D. *et al.* (1977) *J. Chem. Research* (S) 182, (M) 0915.
Martina, D., Brion, F. and De Cian, A. (1982) *Tetrahedron Lett.* **23**, 857.
Mason, R. and Robertson, G.B. (1970) *J. Chem. Soc. Part A* 1229.
Mathey, F. (1990) *J. Organomet. Chem.* **400**, 149.
Meunier-Piret, J., Piret, P. and Van Meersche, M. (1965) *Acta Cryst.* **19**, 85.
Michaud, P. and Astruc, D. (1982) *J. Chem. Soc., Chem. Commun.* 416.
Michaud, P. *et al.* (1982) *J. Chem. Soc., Chem. Commun.* 1383.
Miller, J.S. *et al.* (1987) *J. Am. Chem. Soc.* **109**, 769.
Mitsudo, T. *et al.* (1989) *Organometallics* **8**, 368.
Moriarty, R.M., Yeh, C.-L. and Ramey, K.L. (1971) *J. Am. Chem. Soc.* **93**, 6709.

Morita, N. *et al.* (1988) *J. Organomet. Chem.* **339**, C1.
Moulines, F. and Astruc, D. (1989) *J. Chem. Soc., Chem. Commun.* 614.
Müller, J. and Fischer, R.D. (1966) *J. Organomet. Chem.* **5**, 275.
Mueller-Westerhoff, U.T. and Eilbracht, P. (1972) *J. Am. Chem. Soc.* **94**, 9272.
Mueller-Westerhoff, U.T. and Eilbracht, P. (1973) *Tetrahedron Lett.* 1855.
Muetterties, E.L. and Rathke, J.W. (1974) *J. Chem. Soc., Chem. Commun.* 850.
Muetterties, E.L., Sosinsky, B.A. and Zameraev, K.I. (1975) *J. Am. Chem. Soc.* **97**, 5299.
Murdoch, H.D. (1965) *Z. Naturforsch. B* **20**b, 179.
Nesmeyanov, A.N., Anisimov, K.N. and Magomedov, G.K. (1970) *Izvest. Akad. Nauk SSSR* 715.
Nesmeyanov, A.N. and Bogomolova, L.G. (1973) *Brit. Pat.* 1320046 [*Chem. Abstracts* **79**, 078966].
Nesmeyanov, A.N. *et al.* (1973) *J. Organomet. Chem.* **61**, 329.
Nesmeyanov, A.N. *et al.* (1974) *J. Organomet. Chem.* **71**, 271.
Nesmeyanov, A.N. *et al.* (1981) *J. Organomet. Chem.* **210**, 103.
Nesmeyanov, A.N., Gubin, S.P. and Rubezhov, A.Z. (1969) *J. Organomet. Chem.* **16**, 163.
Nesmeyanov, A.N., Rybinskaya, M.I. *et al.* (1978) *J. Organomet. Chem.* **149**, 177.
Nesmeyanov, A.N., Solodnikov, S.P. *et al.* (1978) *J. Organomet. Chem.* **148**, C5.
Nesmeyanov, A.N., Tolstaya, M.V. *et al.* (1977) *J. Organomet. Chem.* **142**, 89.
Nesmeyanov, A.N., Vol'kenau, N.A. and Bolesova, I.N. (1963) *Tetrahedron Lett.* 1725; *Doklady Akad. Nauk SSSR* **149**, 615.
Nesmeyanov, A.N., Vol'kenau, N.A. and Isaeva, L.S. (1967) *Doklady Akad. Nauk SSSR* **176**, 106.
Nesmeyanov, A.N., Vol'kenau, N.A. and Petrakova, V.A. (1977) *J. Organomet. Chem.* **136**, 363.
Nicholas, K. *et al.* (1971) *Chem. Commun.* 608.
Noyori, R., Nishimura, T. and Takaya, H. (1969) *Chem. Commun.* 89.
Olthoff, J.K. *et al.* (1987) *J. Chem. Phys.* **87**, 7001.
Ortaggi, G. and Paolesse, R. (1988) *J. Organomet. Chem.* **346**, 219.
Osborne, A.G., Whiteley, R.H. and Meads, R.E. (1980) *J. Organomet. Chem.* **193**, 345.
Osella, D. *et al.* (1986) *Organometallics* **5**, 1247.
Otsuka, S., Yoshida, T. and Nakamura, A. (1967) *Inorg. Chem.* **6**, 20.
Pannell, K.H. and Crawford, G.M. (1973) *J. Coord. Chem.* **2**, 251.
Paquette, L.A., Photis, J.M. and Ewing, G.D. (1975) *J. Am. Chem. Soc.* **97**, 3538.
Parker, S.F. and Peden, C.H.F. (1984) *J. Organomet. Chem.* **272**, 411.
Pearson, A.J. (1980) *Accounts Chem. Research* **13**, 463.
Pearson, A.J. and O'Brien, M.K. (1989) *J. Org. Chem.* **54**, 4663.
Pearson, A.J., Park, J.G. *et al.* (1989) *J. Chem. Soc., Chem. Commun.* 1363.
Pearson, A.J., Rees, D.C. and Thornber, C.W. (1983) *J. Chem. Soc., Perkin Trans. 1* 619.
Pinhas, A.R. and Carpenter, B.K. (1980) *J. Chem. Soc., Chem. Commun.* 17.
Piret, P. *et al.* (1965) *Acta Cryst.* **19**, 78.
Piron, J. *et al.* (1969) *Bull. Soc. Chim. Belges* **78**, 121.
Powell, H.M. and Ewens, R.V.G. (1939) *J. Chem. Soc.* 286.
Putnik, C.F. *et al.* (1978) *J. Am. Chem. Soc.* **100**, 4107.
Rausch, M.D. and Ciappenelli, D.J. (1967) *J. Organomet. Chem.* **10**, 127.
Rebiere, F., Samuel, O. and Kagan, H.B. (1990) *Tetrahedron Lett.* **31**, 3121.
Reeves, P.C. (1981) *J. Organomet. Chem.* **215**, 215.
Reger, D.L. and Swift, C.A. (1984) *Organometallics* **3**, 876.
Reihlen, H. *et al.* (1930) *Justus Liebigs Ann. Chem.* **482**, 161.
Rest, A.J. and Taylor, D.J. (1981) *J. Chem. Soc., Chem. Commun.* 489.
Rigby, J.H. and Ogbu, C.O. (1990) *Tetrahedron Lett.* **31**, 3385.
Riley, P.E. and Davis, R.E. (1983) *Organometallics* **2**, 286.
Roberts, R.M.G., Silver, J. and Wells, A.S. (1987) *Inorg. Chim. Acta* **126**, 61.
Roberts, R.M.G. and Wells, A.S. (1986) *J. Organomet. Chem.* **317**, 233.
Roberts, R.M.G. and Wells, A.S. (1987) *Inorg. Chim. Acta* **126**, 67.
Rodriguez, J., Brun, P. and Waegell, B. (1987) *J. Organomet. Chem.* **333**, C25.
Rosenblum, M. and Abbate, F.W. (1966) *J. Am. Chem. Soc.* **88**, 4178.
Rosenblum, M., Santer, J.O. and Howells, W.G. (1963) *J. Am. Chem. Soc.* **85**, 1450.
Röttinger, E. and Vahrenkamp, H. (1978) *Angew. Chem.* **17**, 273.
Roustan, J.L., Guinot, A. and Cadiot, P. (1980) *J. Organomet. Chem.* **194**, 357.
Roustan, J.-L.A. and Houlihan, F. (1988) *J. Organomet. Chem.* **353**, 215.

Sakurai, H., Kamiyama, Y. and Nakadaira, Y. (1980) *J. Organomet. Chem.* **184**, 13.
Salzer, A. and von Philipsborn, W. (1979) *J. Organomet. Chem.* **170**, 63.
Sappa, E., Tiripicchio, A. and Braunstein, P. (1983) *Chem. Reviews* **83**, 203.
Sappa, E., Tiripicchio, A. and Lanfredi, A.M.M. (1978) *J. Chem. Soc., Dalton Trans.* 552.
Sarel, S. (1978) *Accounts Chem. Research* **11**, 204.
Sato, M. *et al.* (1982) *Tetrahedron Lett.* **23**, 185.
Schäufele, H. *et al.* (1989) *Organometallics* **8**, 396.
Schäufele, H., Pritzkow, H. and Zenneck, U. (1988) *Angew. Chem.* **27**, 1519.
Scherer, O.J., Blath, C. and Wolmershauser, G. (1990) *J. Organometal. Chem.* **387**, C21.
Scherer, O.J., Brück, T. and Wolmershauser, G. (1989) *Chem. Ber.* **122**, 2049.
Schirch, P.F.T. and Boekelheide, V. (1981) *J. Am. Chem. Soc.* **103**, 6873.
Schmid, G. and Rohling, T. (1989) *J. Organomet. Chem.* **375**, 21.
Schmidt, E.K.G. and Thiel, C.H. (1981) *J. Organomet. Chem.* **220**, 87.
Schmitt, G. *et al.* (1978) *J. Organomet. Chem.* **152**, 315.
Schmitt, H.-J. and Ziegler, M.L. (1973) *Z. Naturforsch. B* **28**b, 508.
Seidel, W. and Lattermann, K.J. (1982) *Z. Anorg. Allg. Chem.* **488**, 69.
Seiler, P. and Dunitz, J.D. (1982) *Acta Cryst. B* **38**, 1741.
Semmelhack, M.F. and Cheng, C.H. (1990) *J. Organomet. Chem.* **393**, 237.
Semmelhack, M.F. and Fewkes, E.J. (1987) *Tetrahedron Lett.* **28**, 1497.
Semmelhack, M.F. and Park, J. (1986) *Organometallics* **5**, 2550.
Semmelhack, M.F. and Tamura, R. (1983) *J. Am. Chem. Soc.* **105**, 4099.
Seyferth, D. and Withers, H.P. (1982) *Organometallics* **1**, 1275.
Siebert, W. and Bochman, M. (1977) *Angew. Chem.* **16**, 857.
Siebert, W. *et al.* (1978) *Angew. Chem.* **17**, 527.
Siebert, W. (1988) *Pure and Appl. Chem.* **60**, 1345.
Slocum, D.W., Marchal, R.L. and Jones, W.E. (1974) *J. Chem. Soc., Chem. Commun.* 967.
Smith, D.L. and Dahl, L.F. (1962) *J. Am. Chem. Soc.* **84**, 1743.
Sokolov, V.I., Troitskaya, L.L. and Reutov, O.A. (1979) *J. Organomet. Chem.* **182**, 537.
Sotokawa, H. *et al.* (1987) *Tetrahedron Lett.* **28**, 5872.
Stahl, L. *et al.* (1987) *J. Organomet. Chem.* **326**, 257.
Stanley, K. and McBride, D.W. (1976) *Can. J. Chem.* **54**, 1700.
Sternberg, H.W., Markby, R. and Wender, I. (1958) *J. Am. Chem. Soc.* **80**, 1009.
Struchkov, Yu. T. *et al.* (1978) *J. Organomet. Chem.* **378**, 115.
Sutherland, R.G. *et al.* (1988) *J. Heterocycl. Chem.* **25**, 1107, 1911.
Swisher, R.G., Sinn, E. and Grimes, R.N. (1983) *Organometallics* **2**, 506.
Tahiri, A. *et al.* (1990) *J. Organomet. Chem.* **387**, C47.
Taube, R. and Drevs, H. (1977) *Z. Anorg. Allg. Chem.* **429**, 5.
Thompson, D.J. (1976) *J. Organomet. Chem.* **108**, 381.
Thomson, J.B. (1959) *Tetrahedron Lett.* (6) 26.
Togni, A. and Pastor, S.D. (1990) *J. Org. Chem.* **55**, 1649.
tom Dieck, H. and Bock, H. (1968) *Chem. Commun.* 678.
tom Dieck, H. and Orlopp, A. (1975) *Angew. Chem.* **14**, 251.
Treichel, P.M., Dean, W.K. and Douglas, W.M. (1972) *Inorg. Chem.* **11**, 1609.
Vahrenkamp, H. (1990) *J. Organomet. Chem.* **400**, 107.
Victor, R., Usieli, V. and Sarel, S. (1977) *J. Organomet. Chem.* **129**, 387.
Watts, W.E. (1967) *Organomet. Chem. Rev.* **2**, 231.
Watts, W.E. (1979) *J. Organomet. Chem. Libr.* **7**, 399.
Watts, W.E. (1981) *J. Organomet. Chem.* **220**, 165.
Weber, S.R. and Brintinger, H.H. (1977) *J. Organomet. Chem.* **127**, 45.
Wegner, P.A. and Delaney, M.S. (1976) *Inorg. Chem.* **15**, 1918.
Weiss, E., Hübel, W. and Merényi, R. (1962) *Chem. Ber.* **95**, 1155.
Whitlock, H.W., Reich, C. and Woessner, W.D. (1971) *J. Am. Chem. Soc.* **93**, 2483.
Wilke, G. *et al.* (1966) *Angew. Chem.* **5**, 151.
Zakrewski, J. and Giannotti, C. (1990) *J. Organomet. Chem.* **388**, 175.
Zhang, C.H. *et al.* (1988) *J. Organomet. Chem.* **346**, 67.

5 Spectroscopic methods for the study of iron chemistry

B.W. FITZSIMMONS

5.1 Introduction

Mössbauer spectroscopy is the most powerful tool available in studies of iron compounds or materials. It is the recoilless nuclear resonant absorption which dates back to 1957 when Mössbauer made his first observation of recoilless resonant absorption in ^{191}Ir (Mössbauer, 1958). He explained the experiment he had carried out and in a further paper described suitable equipment for scanning the resonance (Mössbauer, 1959). His particular genius rests with his ability to carry out meticulous experimental work using equipment he had built combined with a complete mastery of the necessary background theory and the gift of communication of the results to a wide readership.

The importance of the experiment lies in its provision of a means of measuring some of the weak electron–nuclear interactions. The effect is specific to a particular atomic nucleus and can only be observed in the solid state.

At the turn of the century Wood (1954) demonstrated resonance in atomic systems. He used a sodium light and focused the yellow radiation on an evacuated glass bulb containing metallic sodium and observed a faint yellow glow. From a quantum mechanical point of view, the characteristic light emitted by the sodium atoms is the result of an electronic transition between the ground state and the excited state of the sodium atom followed by de-excitation back to the ground state.

The energy difference between the two states is related by $\Delta E = hf$ where h is Planck's constant and f is the frequency of the emitted light. The resonant absorption process takes place because the incident photon has the right energy to raise an atom of sodium vapour to an excited state. This is a resonance of the first kind. It parallels nuclear resonant absorption. The decay of radioactive nuclides produces a daughter nucleus which is in an excited state. The latter de-excites by emitting a cascade of γ-ray photons until it reaches a stable ground state. This is obviously an analogue of electronic de-excitation but occurs at much higher energies. The search for the experimental realization of this concept was goaded by the belief that the emitted γ-ray should be good source of monochromatic radiation. The uncertainty in the lifetime of the excited state is given by its mean life τ and the uncertainty in its energy is given by the width of the energy distribution at half-height Γ. They are related by $\Gamma\tau \geqslant h/2\pi$ where h is Planck's constant. τ is related to the half-life of the state by $\tau = \ln 2 \times t_{1/2}$. If Γ is given in electron volts and $\tau_{1/2}$ in seconds then $\Gamma = 4.562 \times 10^{-9}/t_{1/2}$.

Typically, $t_{1/2} = 10^{-7}$s, and $\Gamma = 4.562 \times 10^{-9}$eV. If the energy of the excited state is 45.6 keV, the emitted γ-ray will have an intrinsic resolution of 1 part in 10^{13}. This should be compared with atomic line spectra with resolution of 1 in 10^{8}. There are, however, ways in which the energy of the emitted γ-ray can be lowered; in particular, by the effects of nuclear recoil and of thermal energy. The recoil energy is given by:

$$E_{\mathrm{R}} = E_{\gamma}^{2}/2Mc^{2} \tag{5.1}$$

where M is the mass of the nuclear, E is the energy of the excited nuclear energy and c is the velocity of light. E_{R} depends on the mass of the nucleus and the energy of the γ-ray.

The finite line width arises from the finite lifetime which the nucleus spends in the excited state. In 1956–1957 Mössbauer was studying the scattering of γ-rays from Ir^{191} by Ir and Pt and found an increase in scattering in Ir at low temperatures which was contrary to classical predictions. His interpretation was published and Mössbauer effect research began.

To understand the new idea which he brought to bear, three different cases should be distinguished:

1. If E_{R} is greater than the binding energy of the atom, then that atom will be dislodged from its site.

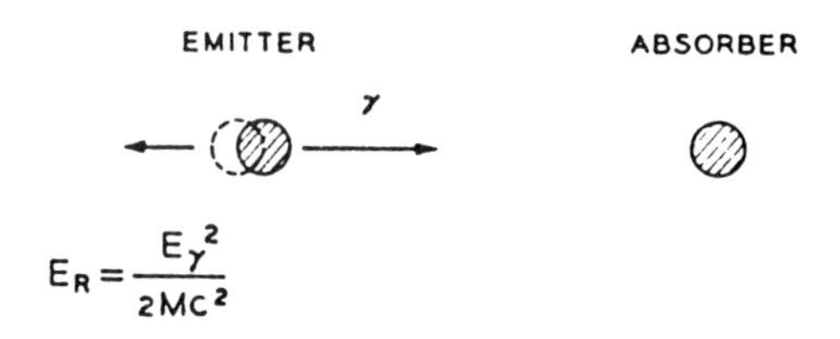

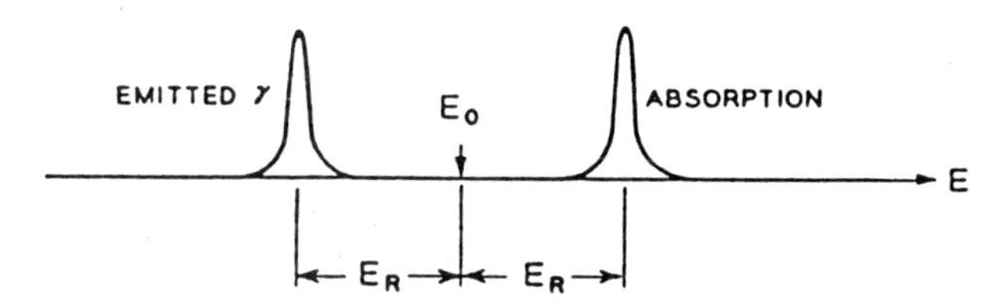

Figure 5.1 Resonant absorption is not possible if the recoil-energy loss exceeds the linewidth. (From Wertheim, 1964.)

2. If E_R is less than the binding energy but greater than the energy of the lattice vibrations, that is, the phonon energy of the solid, then the solid will warm up with the atom remaining in its site (see Figure 5.1)
3. If E_R is less than the phonon energy, then a new situation arises as the lattice is a quantized system which cannot be energized in an arbitrary way.

This last case is the one which accounts for the unexpected increase in the scattering of γ-rays at low temperatures as first observed by Mössbauer. It may be readily understood in terms of an Einstein lattice. A lattice containing n atoms has $3n$ vibrational modes each having the same vibration frequency ω. The solid can be characterized by the quantum numbers of its oscillators. The changes in its state correspond to an increase or decrease in one or more of its quantum numbers. These correspond to the absorption or the emission of quanta of energy $h\omega/2\pi$. The fraction of recoil-free events f is the quantity which determines the intensity of the Mössbauer absorption. This is illustrated in Figure 5.2.

When an average is taken over many emission processes, the energy transferred per event is exactly the free-atom recoil E_R. Thus:

$$E_R = (1 - f)h\omega/2\pi \tag{5.2}$$

$$f = (1 - E_R)h\omega/2\pi \tag{5.3}$$

In zero-phonon emission, the whole crystal rather than a single nucleus

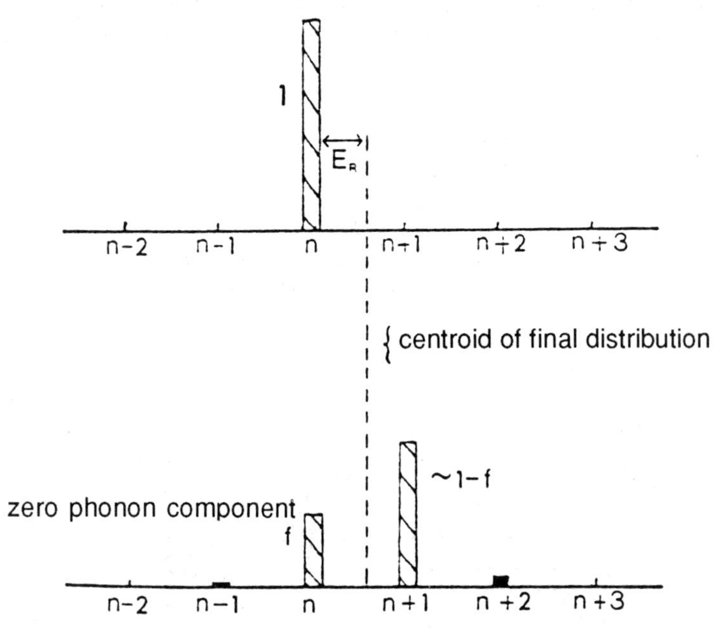

Figure 5.2 The recoil free fraction in the Einstein model. The fraction, f, of the decays does not produce a change in the quantum state of the lattice. (Adapted from Wertheim, 1964.)

recoils. Equation 5.1 contains the reciprocal mass $1/M$. If this is increased to that of a crystallite containing 10^{15} atoms, then the recoil energy becomes very small and much less than Γ. Hence zero-phonon transitions are referred to as recoil-free. The Mössbauer effect than is the resonant emission and subsequent absorption of γ-rays in a solid matrix without lowering by recoil or thermal broadening and it gives an energy distribution dictated by the Heisenberg uncertainty principle. Only these events give rise to a Mössbauer effect.

A general expression for the fraction of elastic or zero phonon processes is:

$$f = \exp[-\pi^2\langle x^2\rangle/\lambda^2] = \exp[-k^2\langle x^2\rangle] \qquad (5.4)$$

where λ is the wavelength of the γ quantum, $k = 2\pi/\lambda = E/(h/2\pi)c$, and $\langle x^2\rangle$ is the component of the mean square vibrational amplitude of the emitting nucleus in the direction of the γ-ray. In order to get a value of f close to unity, $k^2\langle x^2\rangle \ll 1$ which in turn requires that the root mean square displacement of the nucleus be small compared with the wavelenth of the γ-ray. There is a decrease in f with rise in temperature as thermal energy will increase the amplitude of vibration of the nucleus.

The definition of energy of a recoilless γ-ray is about 1 in 10^{12}. This corresponds to the Doppler energy shift produced by small movements. If source and absorbers are in relative motion with a velocity v then the energy of the γ-ray is shifted by an amount $v/c.E\gamma$. If v is zero, the emission and absorption profiles overlap and absorption is at a maximum. Any increase or decrease in v can only decrease the overlap. It follows that a record of

(a)

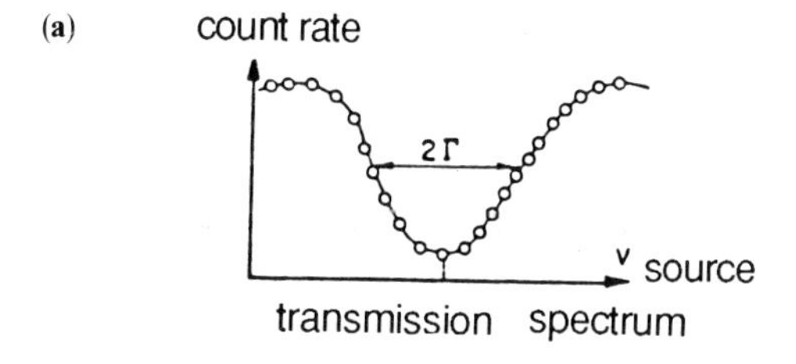

(b)

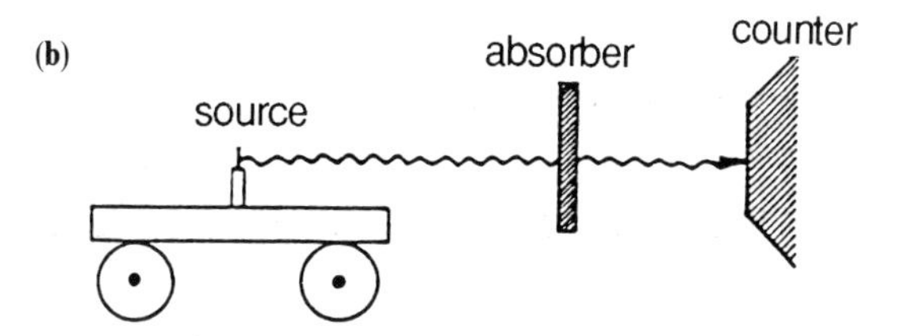

Figure 5.3 Arrangement to observe the Mössbauer effect. (From May, 1971.)

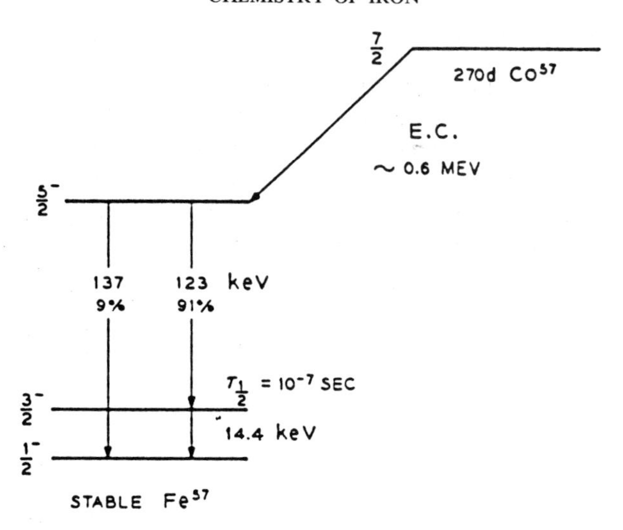

Figure 5.4 The decay of ^{57}Co to ^{57}Fe(E.C. denotes electron capture). (From Wertheim, 1964.)

transmission as a function of v will show an absorption spectrum. The Doppler scanning was originally devised by Mössbaucr and is the method used today to obtain a spectrum. Iron-57 Mössbauer spectroscopy depends upon the fact that ^{57}Fe occurs in nature with 2.1% abundance. The source of radiation in Mössbauer's resonance experiment is the artificial isotope ^{57}Co which goes to the ground state according to the scheme shown in Figure 5.3.

With a ^{57}Co source and any absorber containing some ^{57}Fe as absorber, a plot of transmitted intensity as a function of source velocity constitutes a Mössbauer spectrum. A notional set-up and the resultant spectrum for this experiment is shown in Figure 5.4. The line shape is Lorentzian. The present day equipment is microprocessor based and utilizes an electromechanical device to vibrate the source and a proportional counter as the heart of the detecting system.

5.2 Hyperfine interactions

Mössbauer spectroscopy measures hyperfine interactions which are a function of a nuclear part and an atomic part which relates to chemical and solid state information. The atomic part is a complex function of the electronic energy levels of the atom. There are three main hyperfine interactions:

(1) The chemical isomer shift, which is the electric monopole interaction.
(2) The quadrupole splitting, which is the electric quadruple interaction.
(3) The magnetic hyperfine splitting, which is the magetic dipole interaction.

5.2.1 *Chemical isomer shift (δ)*

Because the nuclear radii of the ground and excited states are different there is a change between the electronic charge at the nucleus (s-electrons) and the nuclear charge itself during the decay. This changes the energy separation by an amount which depends on the electronic charge density at the nucleus and is hence different in source and absorber. This gives a shift of the absorption peak away from zero velocity.

$$\delta = \frac{Ze^2R^2}{5\varepsilon_0}\frac{\Delta R}{R}\{|\psi_s(O)_a|^2 - |\psi_s(O)_s|^2\} \tag{5.5}$$

δ is not absolute and for ^{57}Fe it is measured with respect to the centre of the iron metal spectrum. $\psi_s(O)$ is a function of the wave functions of p- and d-electrons as well as s-electrons. For example:

$$Fe^{2+}\ 3d^6\delta \sim 1.2\ \mathrm{mm\,s^{-1}}$$
$$Fe^{3+}\ 3d^5\delta \sim 0.5\ \mathrm{mm\,s^{-1}}$$

It is seen that there are difference shifts within the same s-electron configuration. In ^{57}Fe, $\Delta R/R$ is negative so a positive change in δ corresponds to a decrease in the s-electron density at the nucleus. The extra 3d electron in Fe^{2+} increases the screening of the nuclear charge from the 3s electrons and thus the 3s orbitals expand and the electron density is lowered.

5.2.2 *Quadrupole splitting (ΔE)*

Nuclear levels with $I > \frac{1}{2}$ have quadrupole moments, eQ. These interact with an electric field gradient (efg). This is a tensor quantity but it becomes diagonal in a suitable coordinate system and then has only two components, $V_{zz}(=eq)$ is the principal component $\partial^2V/\partial z^2$ and $\eta(=V_{xx} - V_{yy}/V_{zz})$ the asymmetry parameter.

The interaction Hamiltonian is:

$$H = \frac{eq}{2I[(2I-1)]}(V_{zz}I_z^2 + V_{yy}I_y^2 + V_{xx}I_x^2) \tag{5.6}$$

This has eigenvalues:

$$E_Q = \frac{e^2qQ}{4I(2I-1)}[3I_z^2 - I(I+1)]\left(1 + \frac{\eta^2}{3}\right)^{1/2} \tag{5.7}$$

For ^{57}Fe this gives two levels, $\pm\frac{3}{2}$, $\pm\frac{1}{2}$ split by:

$$\Delta E_Q = \frac{e^2qQ}{2}\left(1 + \frac{\eta^2}{3}\right)^{1/2} \tag{5.8}$$

This is commonly called the quadrupole splitting. It is normal to relate this to either the geometry (diamagnetic compounds) or to the electronic levels (paramagnetic compounds).

In general:

$$eq = V_{zz} = \frac{\int \rho(r) 3z^2 - r^2\, dr}{r^5} \tag{5.9}$$

There are two main contributions:

$$q_{\text{lattice}} = -\sum_i \frac{Z_i 3\cos^2\theta_i - 1}{r_i^3} \tag{5.10}$$

r_i and θ_i are the coordinates of the ith ion, $Z_i e$ is its charge. The other main contribution is from the valence electrons:

$$q_{\text{valence}} = -\sum_i \frac{\langle \psi_i | 3\cos^2\theta - 1 | \psi_i \rangle}{r^3} \tag{5.11}$$

where ψ_i are wavefunctions of the valence electrons. The core electrons, while not making direct contribution to the efg as they are in closed shells with spherical symmetry, modify q_{lattice} and q_{valence} by shielding and antishielding. These are reflected in the Sternheimer factors:

$$q = (1_\gamma - R)\, q_{\text{valence}} + (1_\gamma - \gamma_\alpha) q_{\text{lattice}} \tag{5.12}$$

All these are calculated by Hartree-Fock and similar methods. For high-spin iron(III), q_{valence} is zero and ΔE_Q is generally small and temperature independent. For high-spin iron(II), the extra electron over d^5 makes q_{valence} finite and dependent upon the orbital occupied. ΔE_Q is large and temperature dependent.

5.2.3 *Magnetic hyperfine splitting*

A nucleus with spin I interacts with a magnetic field B:

$$H_m = -g_n \beta_n B.I \tag{5.13}$$

g_n is the nuclear g-factor and β_n is the nuclear magneton.

The eigenvalues are:

$$E_m = g_n \beta_n B m_I \tag{5.14}$$

The Mössbauer measurement yields a value of B. It arises from a combination of external field, B_{app}, and a hyperfine field at the nucleus due to the atom's own electrons, this is B_{hf}. In general B_{hf} is much larger than B_{app}. The hyperfine field has three components:

$$B_{\text{hf}} = B_D + B_L + B_s \tag{5.15}$$

These are the dipolar, orbital, and Fermi contact contributions respectively.

$$B_D = 2\mu\beta\langle 3r(s.r)^{r-5} - sr^{-3}\rangle_{3d} \quad (5.16)$$

Thus, the dipolar field arises from the spin angular momentum of electrons.

$$B_L = 2\mu\beta\langle r^{-3}\rangle\langle L\rangle_{3d} \quad (5.17)$$

L is the orbital angular momentum. The well-known quenching mechanisms operate in iron compounds resulting in just a small orbital contribution arising from spin-orbit coupling. B_s, the Fermi contact term, arises from the difference between spin-up and spin-down s-electron density at the nucleus arising from core polarization.

$$B_s = -\frac{8\pi}{3}\mu\beta\Sigma\{|\psi_{s\uparrow}(O)|^2 - |\psi_{s\downarrow}(O)|^2\} \quad (5.18)$$

In iron compounds, this term is large and negative. All three terms relate to the ground state and low-lying levels of the unpaired electrons. The interactions between these fields and the nuclear magnetic moment depends upon their relative orientation and is best described using a tensor:

$$H_m = -S'.A.I \quad (5.19)$$

S' is the effective spin and A is the magnetic hyperfine interaction tensor. If A is diagonal and isotropic $-AS^1 \equiv -g_n\beta_n B_{hf}$ and $B_{hf} = AS^1/g_n\beta_n$. This is the effective field approximation. For further information see Greenwood and Gibb (1971) and Gibb (1976).

5.3 Interpretation of ^{57}Fe Mössbauer spectra

The interpretation of spectra and their correlation with structural features and other spectral parameters is dependent upon the type of compound under investigation. In the case of iron(II) compounds with weak ligands typified by hexa-aquo species such as $FeSO_4.7H_2O$, it is the normal practice to interpret the Mössbauer spectra of these high-spin complexes within the framework of crystal-field theory. Their low-spin analogues, being diamagnetic, are best treated by molecular-orbital methods as are the carbonyl and cyclopentadienyl compounds. Mössbauer spectroscopy has made significant contributions in iron–sulphur bond chemistry especially in the ferridoxin and rubredoxin field. That in turn exerted a stimulus on coordination compounds having sulphur donor ligands as model compounds were needed to confirm the very unusual shifts and splittings recorded for the naturally occurring substances.

5.3.1 Fe(II) compounds

These have electron configurations of the type $(t_g)^4$ $(e_g)^2$. The single extra electron over d^5 being a valence electron and close to the nucleus generates a strong electric field gradient at the nucleus as a consequence of which the majority of these compounds exhibit large quadrupole splittings, generally in the region of 2–4 mms^{-1}. The presence of a splitting requires that there be a low-symmetry component of the crystal field to raise the degeneracy of the t_{2g} orbitals. This seems to be true of all such compounds excepting the hexamine $[Fe(NH_3)_6]^{2+}$ for which ΔE is zero.

The magnitude of the low symmetry crystal field component is generally comparable with thermal energies with the result that the quadrupole splitting in these systems is temperature dependent and increases as the temperature is lowered. A detailed analysis of this temperature dependence gives the magnitude and sign of the small component of the crystal field splitting. The chemical shifts cluster around 1.0 mm s^{-1} but they depend on the electronic character of the six nearest neighbours, being highest for the most electronegative ligands. This observation suggests that the crystal field model is not entirely satisfactory but it is not easy to see how a more universal model might be developed.

If strong ligands are present in the coordination sphere then low-spin compounds result. These have the $(t_{zy})^6$ configuration which, itself, gives rise to no electric field gradient and no quadrupole splitting can arise from the d-electrons. However, these are strongly covalent compounds such as $[Fe(CN)_6]^{4-}$, $[Fe\ phen_3]^{2+}$, etc., so any inequalities in the covalent bonding arising from symmetry lowering can generate field gradients. In some special cases, e.g. $[Fe(CN)_5NO]^{2-}$, (1.71 mm s^{-1}) the splitting can be quite large: values less than 1.0 mm s^{-1} are normal as in $[Fe\ phen_2\ (CN)_2]$ and $[Fe(CN)_2(CNR)_4]$. In this last case, geometrical isomerism is possible: an additive model of quadrupole splitting strength has been found useful.

Thus high-spin iron(II) octahedrel complexes show appreciable quadrupole splittings while their diamagnetic low-spin analogues show little splitting. In complexes in which the crystal field is close to the critical value at which the change of high- to low-spin behaviour takes place, the phenomenon of spin-crossover is sometimes observed. Here, the complex is high-spin at room temperature but changes over to low-spin on cooling to 77 K or below. Both the effective magnetic moment and the Mössbauer spectrum are strongly temperature dependent, the former changing abruptly from *c.* 5 to 0.6 with the latter showing a marked decrease in quadrupole splitting. In the hope of developing useful optical storage devices, this aspect of iron(II) chemistry has been studied very closely and a satisfactory overall picture has emerged.

Tetrahedral iron(II) species show similar behaviour overall: their shifts are smaller in line with lower coordination, those of the $[FeS_4]$ type being particularly low, signalling the onset of considerable covalency. The quadrupole splittings are strongly temperature dependent and increase as the temperature is lowered. A crystal field model has been used to correlate the data but the increased covalency makes this a dubious procedure.

5.3.2 *Fe(III) compounds*

Iron(III) compounds are based on the $3d^5$ configuration. Removal of one electron increases the s-electron density at the nucleus through a deshielding mechanism. This causes the chemical shift to fall from 1.0 mm s^{-1} in Fe(II) compounds to values around 0.5 mm s^{-1}. This allows a clear distinction between iron(II) and iron(III) compounds to be made. Moreover, the $3d^5$ configuration is spherically symmetrical and cannot generate any electric field gradient. Iron(III) compounds show lower shifts and lower splittings than their iron(II) analogues. It is to be stressed that it is essential to compare like with like in making these generalizations: the archetype iron(III) complex could be taken to be $[FeCl_6]^{3-}$. Strong field ligands lead to low-spin compounds such as $[Fe(CN)_6]^{3-}$, $[Fe\ phen_3]^{3+}$, etc. These strongly covalent species often show shifts similar to their iron(II) analogues pointing to a similarity in their d^n configurations. The low-spin compounds are conveniently viewed as having a hole in a $(t_{2g})^6$ or $(t_2)^6$ configuration. The appreciable quadrupole splitting often displayed by these compounds is thus readily understood. We see here that the high-spin-low-spin pair is just the opposite of the iron(II) family in quadrupole splitting. If ligands of the critical strength are present, cooling the complex brings on a change from high- to low-spin. This change is heralded by a marked increase in quadrupole splitting and the Bohr magneton number falls from 6 to 2. The critical strength ligand field is generated by *N*,*N*-dialkyldithiocarbanate $R_2NCS_2^-$ or monothioaceytilacetonate $RCOCHCSR^-$.

5.3.3 *Covalent iron compounds*

Mössbauer spectroscopy has made significant contributions to the study of these compounds. It is true to say that the major impact now lies in the past at a time when X-ray techniques were not so advanced as today and when there was still considerable doubt about the identity and purity of certain reaction products.

These compounds are, for the most part, diamagnetic and this imposes quite a severe limit on its applicability here. Most of the compounds of interest have iron atoms at relatively low-symmetry sites, e.g. $Fe(CO)_5$ $[Fe_3(CO)_{12}]$, $[Fe(\eta^5\text{-}C_5H_5)_2]$, $[Fe(CO)_2(\eta^5\text{-}C_5H_5)]_2$, and $[Fe(\eta^6\text{-}C_6H_6)(\eta^5\text{-}C_5H_5)_2]^+$. These all show appreciable quadrupole splitting which is sensitive to substituent, oxidation state, isomerism, etc. Highlights include the reopening of the question of the molecular structure of $[Fe_3(CO)_{12}]$ and the support provided for the then molecular orbital theory treatment of ferrocene.

5.3.4 *^{57}Fe nuclear magnetic resonance*

Iron-57 with its 2.1% natural abundance has a detection sensitivity, relative to ^{13}C, of 0.004. Lacking a quadrupolar mechanism for relaxation, the upper spin state for ^{57}Fe will tend to saturate. This makes iron an unlikely element for NMR study (Harris and Mann, 1978). A pulse technique involving four irradiation frequencies (quadrupole Fourier transform) has been applied to $Fe(CO)_5$ and a signal to noise ratio of > 100 obtained. Using samples enriched to 82% in ^{57}Fe, Koridize *et al.* (1975) have employed a triple resonance technique involving observation of ^{13}C with successive decoupling of ^{1}H and then ^{57}Fe to extract chemical shifts for substituted ferrocenes. There does not seem to be much future for ^{57}Fe NMR.

5.3.5 *Electron paramagnetic resonance spectroscopy*

This branch of spectroscopy is one which has an enthusiastic following of scientists who are interested in the subject in its own right rather than in its applications to problems in iron chemistry. This is not to deny that their activity has provided a wealth of information about the electronic structures of iron complexes. Of the two major oxidation states of iron, EPR applies well to iron(III) systems with their odd number of d-electrons. Even-electron systems do not give EPR spectra. For Fe^{3+} the ground state is ^6S and there are no other sextet terms of higher energy, so EPR spectra are readily observed in any crystal field symmetry. The absence of spin-orbit coupling means that g-values are isotropic and close to 2.0023 (Goodman and Raynor, 1970).

References

Mössbauer, R.L. (1958) Nuclear resonance absorption of gamma rays in ^{191}Ir. *Naturwissenschaften* **45**, 58.

Mössbauer, R.L. (1959) Nuclear resonance absorption of gamma rays in ^{191}Ir. *Naturforsch.* **14a**, 211.

Wood, R.W. (1954) *Physical Optics*, Macmillan, New York.

Greenwood, N.N. and Gibb, T.C. (1971) *Mössbauer Spectroscopy*, Chapman and Hall, London.

Gibb, T.C. (1976) *Principles of Mössbauer Spectroscopy*, Chapman and Hall, London.

Sams, J.R. (1972) MTP International Review of Science: Magnetic Resonance. *Physical Chemistry Series One* **4**, 85.

Harris, R.K. and Mann, B.E. (1978) *NMR and the Periodic Table*, Academic Press, London.

Koridze, A.A., Petrovskii, P.S., Gubin, S.P. and Fedin, E.I. (1975) Iron-57 nuclear magnetic resonance spectra of ferrocene derivatives. *J. Organomet. Chem.* **93**, C26.

Goodman, B.A. and Raynor, J.B. (1970) *Advances in Inorganic Chemistry and Radiochemistry* **13**, 135.

Wertheim, G.K. (1964) *Mössbauer Effect: Principles and Applications*, Academic Press, New York.

May, L. (ed.) (1971) *An Introduction to Mössbauer Spectroscopy*, Plenum Press, New York.

6 Biological iron

J.G. LEIGH, G.R. MOORE and M.T. WILSON

6.1 Control of intracellular levels of iron

6.1.1 Introduction

The intracellular concentration of free iron, i.e. iron not bound to organic ligands, is tightly controlled in animals, plants, and microbes. This is partly a result of the poor solubility of iron under aqueous and aerobic conditions: the solubility product of $Fe(OH)_3$ is 4×10^{-38}, and even for $Fe(OH)_2$ it is only 8×10^{-16} (Neilands *et al.*, 1987; da Silva and Williams, 1991). Thus within the cell, free iron will generally precipitate as polymeric hydroxides. Another reason the level of free iron is under tight control is to minimise the production of harmful free radicals, such as the hydroxyl radical. This is formed by the Haber–Weiss reaction.

$$Fe^{2+} + H_2O_2 \rightarrow Fe^{3+} + OH^- + OH^{\cdot} \quad (6.1)$$

Free iron, whether in solution or precipitated as a polymeric hydroxide, can participate in this reaction, as can some iron–ligand complexes (Crichton, 1991).

The biochemical mechanisms that control intracellular iron concentrations occur at a number of levels: these include uptake into the organism, and transport and reversible storage within it. These processes involve a range of proteins, many of which have a requirement for reaction partners. This introduces various possibilities for biochemical control, in addition to control emanating from variation in the concentrations of key proteins as a result of events occurring at the levels of transcription and translation. Here the biochemical mechanisms and genetic controls for the uptake, transport and storage of iron by animals, plants and microbes are described.

6.1.2 Uptake of iron

Many bacteria acquire iron via a mechanism known as siderophore-mediated iron uptake (Crichton, 1991; Bullen and Griffiths, 1987; Hider, 1984; Winkelmann *et al.*, 1987; Neilands, 1990; and references therein). Small molecules with the ability to chelate iron tightly are secreted by the bacteria. These bind iron and the resulting iron–siderophore complexes are then absorbed by the bacteria via specific receptors in their outer membranes. Structural formulae and physicochemical properties of two important bacterial siderophores are given in Figure 6.1. To date several hundred siderophores have been characterised (see compilations in Bullen and Griffiths, 1987; Hider, 1984; Winkelmann *et al.*, 1987). Once absorbed the iron-siderophore complexes are broken down. Most seem to release their iron via its reduction to Fe^{2+}, which binds much less tightly to the siderophores. However, some appear to release their iron following enzymatic degradation of the siderophore itself.

Other mechanisms exist by which bacteria acquire iron (Griffiths, 1987; Weinberg, 1989). For example, some bacteria pathogenic to animals acquire iron by absorbing haem, while others may do so by a transferrin receptor mediated route which involves the bacteria absorbing the transferrin of the host.

Many fungi and plants also acquire iron via a siderophore mediated route (Winkelmann *et al.*, 1987; Crichton, 1991). In addition to this route, a membranous Fe^{3+} reductase system operates for many plants and fungi. This is

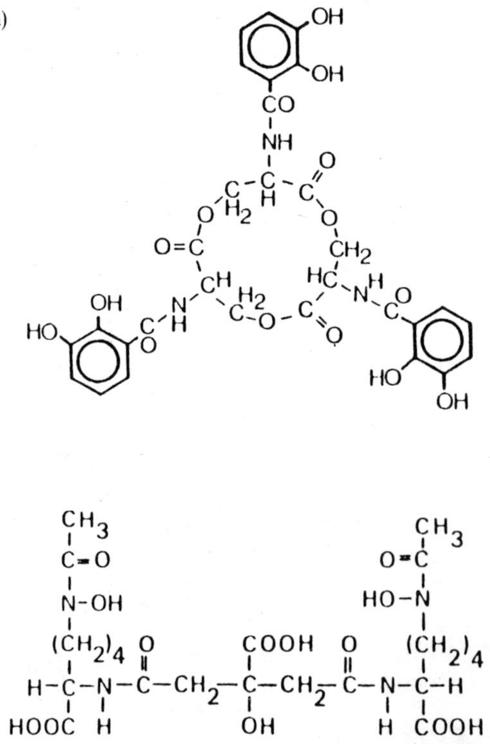

Figure 6.1 Structural formulae of (a) enterobactin; (b) aerobactin.

induced by low intracellular levels of iron which act as repressor signals for the gene that encodes for the enzyme.

The iron absorption pathways for animals are less well understood than those of microbes and plants. It appears that there are at least two routes by which iron can be transferred over the brush border membranes of the mucosal cells lining the lumen of the gut (Reddy *et al.*, 1987; Theil and Aisen, 1987; Crichton, 1991). Non-haem iron is absorbed by one of these routes, and haem iron by the second. Both of them may be receptor mediated (Crichton, 1991 and references therein).

The non-haem iron route requires a carrier, which may be ascorbate, citrate or other small molecules, or a specific carrier secreted into the lumen. Transferrin has been suggested to be such a carrier but there is no firm evidence indicating it is present in the gut lumen.

6.1.3 Transport of iron

Iron transport is best understood for animals. A family of iron-transport proteins, the transferrins, bind Fe^{3+} tightly and transport it to the tissues where it is required (Aisen and Listowsky, 1980; Brock, 1985; Theil and Aisen, 1987). The iron is carried across cell membranes via a process known as receptor-mediated endocytosis (Figure 6.2). This involves the iron–transferrin complex binding to a specific receptor on the cell surface and then being taken

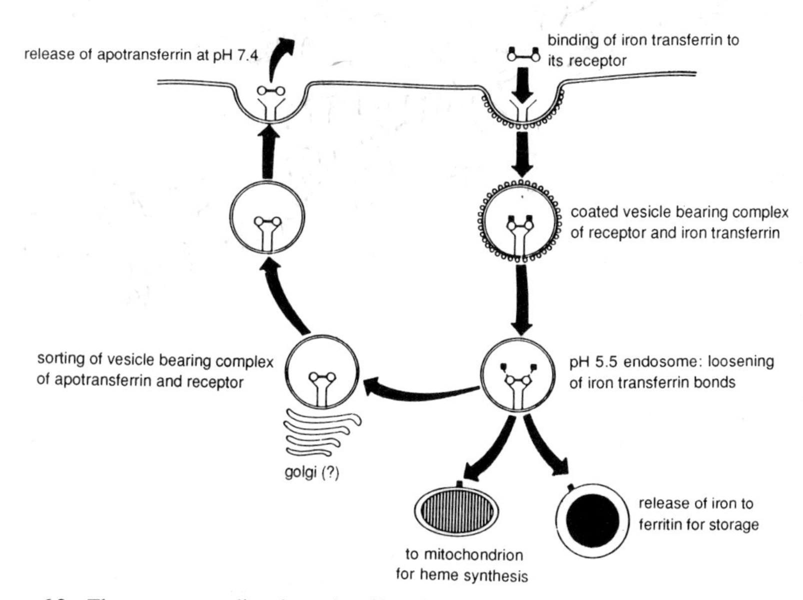

Figure 6.2 The receptor mediated uptake of iron into cells. (Reproduced with permission from Theil and Aisen, 1987.)

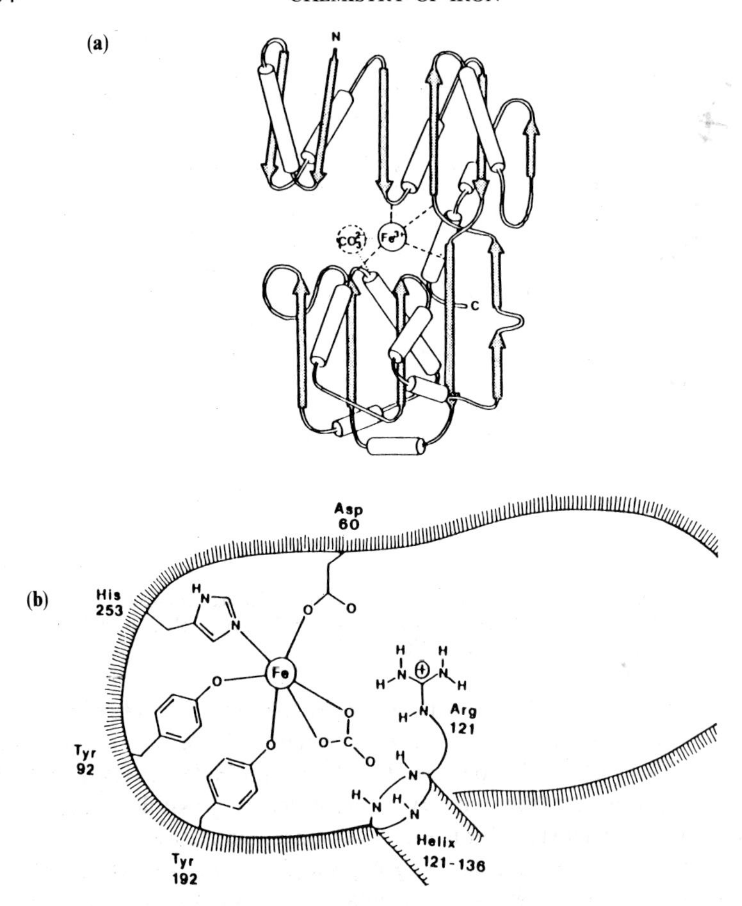

Figure 6.3 X-ray structure of lactoferrin. (Reproduced with permission from Baker *et al.*, 1987; Anderson *et al.*, 1990.) (a) The polypeptide fold of the N-terminal lobe of lactoferrin; (b) schematic representation of the iron and anion binding sites in the N-lobe of lactoferrin. The same arrangement of ligands is found in the C-lobe.

into the cell, within a vesicle, where the iron is released, probably following a reduction in pH.

The X-ray structures of lactoferrin and transferrin have been determined (Baker *et al.*, 1987; Anderson *et al.*, 1990; Bailey *et al.*, 1988). They are bi-lobed proteins with each lobe possessing a binding site for a single iron atom (Figure 6.3). Under physiological conditions the iron binds in the Fe^{3+} oxidation state with an extremely high affinity (Brock, 1985; Aisen *et al.*, 1978). However, for this high affinity binding to occur a carbonate anion must be bound with each Fe^{3+}. The anion locks the Fe^{3+} into the binding site (Figure 6.3).

Iron released from transferrin is incorporated into a cytosolic chelatable pool where it is probably bound to low molecular weight ligands (Jacobs,

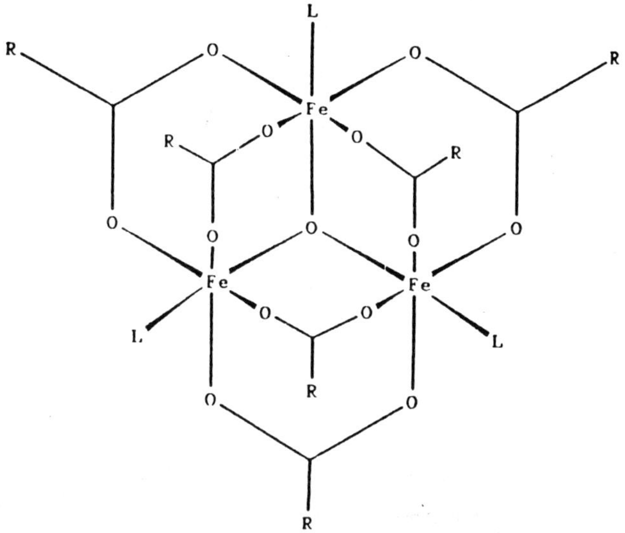

Figure 6.4 Structural formula of a μ_3-oxotriiron(III) cluster. In the intracellular pool of low molecular weight iron R may be glutamate or aspartate and L may be H_2O or OH- (Deighton and Hider, 1989).

1977; and see Crichton, 1991 and referençes therein). The identity of the ligands, and the nature of their iron complexes, have not been firmly established. Glutamate and aspartate (Deighton and Hider, 1989), and ATP (Weaver and Pollack, 1989), have been identified as possible ligands, and a μ_3-oxytriiron(III) cluster (Figure 6.4) has been suggested to be a key component of the low molecular weight pool (Deighton and Hider, 1989). The complexes forming this low molecular weight pool of iron may have only a short lifetime, but during that time they function as intracellular iron transporters, and may be involved directly in iron uptake and release by ferritin (see section 6.1.4).

A eukaryotic intracellular iron transport protein has not been described in detail but recent work indicates that one does exist (Cox and Kakepoto, 1991).

Efficient iron transport systems must exist in plants since iron has to be transported from the soil via the roots to the growing tips. However, the molecular details of such a transport system have not been described. Iron transport in the stems and trunks of plants occurs by passage of iron through the xylem and phloem, which both contain a variety of low molecular weight ligands such as amino acids (Marschner, 1986). Thus the iron may be transported in the form of low molecular weight complexes.

Microbial cytoplasmic iron transport systems have not been described, however, periplasmic Fe^{3+}-siderophore binding proteins have (e.g. see Pierce and Earhart, 1986; Pressler *et al.*, 1988). The latter transport specific

Fe(III)-siderophore complexes across the periplasm to the inner membrane where the complexes are then translocated across the membrane, or broken down (see Crichton, 1991 and references therein).

6.1.4 *Storage of iron*

The principal iron store in animals is the protein ferritin (Crichton, 1991; Ford *et al.*, 1984; Harrison *et al.*, 1991; and references therein). This has a plant analogue, phytoferritin (PFR), and a bacterial analogue, bacterioferritin (BFR). All of these proteins consist of a protein shell composed of 24 subunits, each of M_r 18–26 kDa, surrounding a central cavity of ~80 Å diameter (Figure 6.5). The iron is stored in this cavity as a non-haem-iron mineral.

BFR contains haem as isolated (Stiefel and Watt, 1979), though there are differences in the nature of the BFR haem binding in that for some BFRs (e.g. *P. aeruginosa*) the maximum haem loading is 24 per molecule (Kadir and Moore, 1990a), while for others (e.g. *E. coli*, *A. vinelandii*) maximum haem loading is only 12 per molecule (Stiefel and Watt, 1979; Smith *et al.*, 1989). However, all BFRs appear to have a common, and so far unique, *bis*-methionine ligation to haem (Cheesman *et al.*, 1990). A major difference between ferritin, PFR and BFR is that the former two proteins are isolated in a haem-free state. However, horse spleen ferritin does bind haem. The maximum loading is 16 per molecule (Kadir and Moore, 1990b; Kadir *et al.*, 1992) with *bis*-histidine axial ligation (Moore *et al.*, 1992a). The haem–ferritin complex has been named haemoferritin (Kadir *et al.*, 1992) and in both BFR and haemoferritin the haem is active in promoting iron release in *in vitro* reductive iron release assays (Kadir *et al.*, 1992; Moore *et al.*, 1992b).

Iron uptake and release occurs with the protein shell intact. The iron enters, or leaves, the protein by burrowing through the shell, probably by

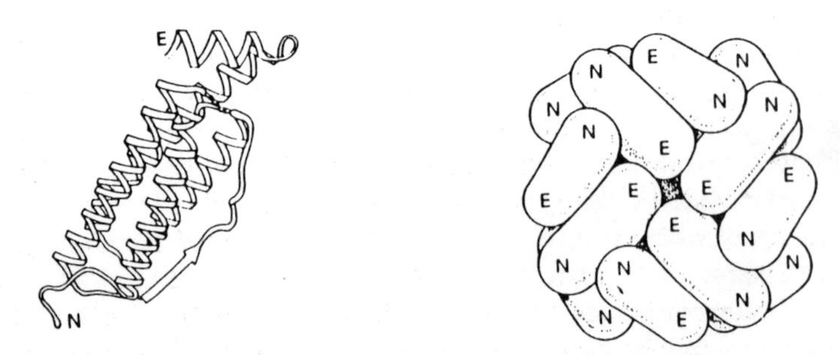

Figure 6.5 Schematic diagram of the structure of ferritin. Each subunit is composed of a 4-α-helical bundle with a short helix (labelled E) close to the C-terminus. N indicates the N-terminus. Channels form at the interfaces between subunits (right); a 4-fold channel and 3-fold channels can be seen in the figure. The protein shell is ~20 Å thick. (Reproduced with permission from Ford *et al.*, 1984.)

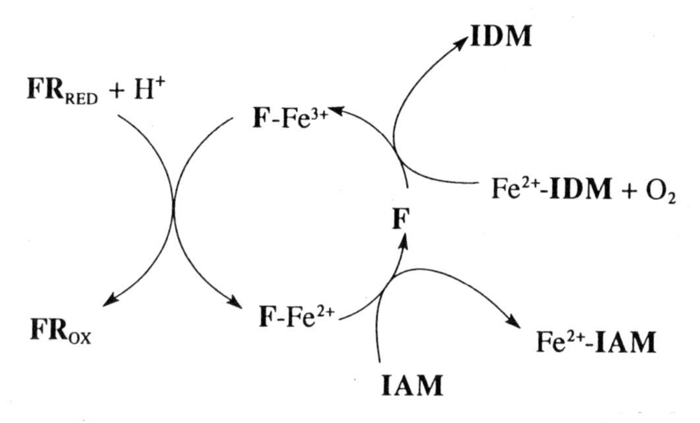

Figure 6.6 Iron enters ferritin (F) as Fe^{2+} and is oxidised in the protein with O_2 being the terminal electron acceptor. The Fe^{2+} is acquired from an Fe^{2+}-ligand complex, here termed the 'iron donor molecule' (IDM). For the iron to be released it must be in the Fe^{2+} state. Therefore an electron donor to ferritin, ferritin reductase (FR) is required. Once reduced, the Fe^{2+} is released to an 'iron acceptor molecule' (IAM). The identity of IAM, IDM and FR are not known (see text for further details). (Reproduced with permission from Moore *et al.*, 1992b.)

travelling through one of the inter-subunit channels. *In vitro*, rapid iron release only occurs if the iron is in the Fe^{2+} state (Clegg *et al.*, 1980; Theil, 1983; Fatemi *et al.*, 1991). A plausible mechanism for Fe release under physiological conditions is presented in Figure 6.6.

The key features of this model are that since the core iron needs to be Fe^{2+} to be released a ferritin (or BFR or PFR) reductase is required, and that this reductase does not need to have a lower redox potential than the core iron because an iron acceptor molecule is required to bind the released Fe^{2+}. This may well displace the equilibrium of the reduction step in an analogous fashion to chelators in reductive *in vitro* assays. Nevertheless, the redox potential of ferritin reductase may well influence the rate of electron transfer and thus iron release.

This model assumes that *in vivo* the core iron is in the Fe^{3+} state, an assumption supported by ^{57}Fe Mössbauer spectra of intact *P. aeruginosa* cells (Kadir *et al.*, 1991a) and from the low redox potentials of the core iron of ferritin (Watt *et al.*, 1985) and *A. vinelandii* BFR (Watt *et al.*, 1986). It also assumes that Fe^{2+} remains within the protein unless it is removed by chelators. This assumption is supported by a variety of chromatographic and spectroscopic data (Watt *et al.*, 1985, 1986, 1988; Kadir *et al.*, 1991b). However, Bauminger *et al.* (1991) have shown recently that Fe^{3+} and thus probably Fe^{2+}, can be exchanged by ferritin molecules. However, the rate of exchange is likely to be rather slow.

There have been many *in vitro* studies of the reductive release of Fe^{2+} from ferritin (reviewed by Crichton, 1991), with many workers showing that flavoproteins are able to achieve this. The earliest work was that of

Green and Mazur (1957), who showed that xanthine oxidase could release Fe^{2+}. Thus the search for a ferritin reductase has been underway for some time. There has been relatively little discussion in the literature concerning the identity of the iron acceptor molecule. However, Cox and Kakepoto (1991) describe the purification of a novel iron binding protein from human reticulocytes that participates in the intracellular transport of iron. It may be that this, and/or a low molecular weight pool of iron (see section 6.1.3), is the iron acceptor molecule (IAM) of Figure 6.6, and possibly the iron donor molecule (IDM) also. Deighton & Hider (1989) have specifically proposed that the IDM is a μ-oxotriiron(III) cluster compound (Figure 6.4).

The mechanism for iron uptake in Figure 6.6 is significantly different from that for iron release. Apart from the IDM and O_2 no other component may be needed. This is because ferritin has a ferroxidase activity that leads to the oxidation of Fe^{2+} in the presence of O_2 (see review by Crichton, 1991). The simplest reaction for Fe^{2+} uptake into ferritin would then be:

$$4Fe^{2+} + O_2 + 6H_2O \rightarrow 4FeOOH + 8H^+$$

This is consistent with the observed proton-electron stoichiometry required for reducing ferritin (Watt *et al.*, 1985), and with measurements of the Fe^{2+}/O_2 ratio at relatively high levels of ferritin iron loading. At low levels of iron loading (<240 iron atoms per molecule) the Fe^{2+}/O_2 ratio is ~ 2. Thus there is a biphasic reaction, but at no stage of core formation are additional oxidants required (Melino *et al.*, 1978; Treffry *et al.*, 1978; Xu and Chasteen, 1991). Hence, if the stoichiometries noted above occur *in vivo* the haem of BFR (and haemoferritin) may not have a role *in vivo* in iron uptake.

Numerous physico-chemical studies of the non-haem-iron mineral phase of ferritin have been carried out and it is clear that in some cases it is highly regular and crystalline, and in others it is amorphous. This can be demonstrated directly by high-resolution electron microscopy (EM) (St. Pierre *et al.*, 1989) and inferred from spectroscopic methods such as Mössbauer spectroscopy (St. Pierre *et al.*, 1986, 1989), EXAFS (Rohrer *et al.*, 1990) and EPR (Weir *et al.*, 1984; Deighton *et al.*, 1991; Cheesman *et al.*, 1992). One of the chemical features that appear to correlate with the amorphous nature of iron cores is the phosphate content: cores with a high level of phosphate tend to be amorphous (St. Pierre *et al.*, 1989). The comparison of *P. aeruginosa* BFR with horse spleen ferritin demonstrates this (Moore *et al.*, 1986). EM studies show that the cores have approximately the same size distribution even though the iron content of BFR is only 8.7% compared to that of horse spleen ferritin, in these comparative studies, of 29.0%. The Fe:P ratio for BFR was 1.7:1 whereas it was 21:1 for horse spleen ferritin. EXAFS measurements confirm that phosphate is coordinated to the iron of BFR (Rohrer *et al.*, 1990). Presumably the difference in core phosphate levels reflects the differences in the intracellular compositions of bacteria and animals, and it raises the question of whether BFR might also have a physiological role as a phosphate store (Fatemi *et al.*, 1991).

A striking feature of crystalline cores of ferritin is that the cores themselves generally consist of one large crystal with, perhaps, a few smaller crystals. This can be accounted for by a mechanism of core formation that comprises a nucleation stage and a crystal growth stage (Clegg *et al.*, 1980). The nucleation stage can occur many times in an individual ferritin molecule but once one of the nucleating centres has attained a particular size its further growth is catalytic and it rapidly expands to fill the central cavity.

The passage of iron to the nucleating site and growing crystal, and the location and nature of the ferroxidase centre, are currently being investigated by spectroscopic procedures (e.g. Chasteen *et al.*, 1985; Bauminger *et al.*, 1989; Hanna *et al.*, 1991; Cheesman *et al.*, 1992), and by site-directed mutagenesis (Levi *et al.*, 1988; Lawson *et al.*, 1991). The articles by Crichton (1991) and Harrison *et al.*, (1991) should be consulted for further details.

Ferritin can be degraded to form haemosiderin, an insoluble iron-rich material whose organic component includes the protein shell of ferritin. It appears that haemosiderin is produced by the lysosomal degradation of ferritin though the morphology of the iron containing phase of haemosiderin does not always resemble that of ferritin. Haemosiderin is the principle non-haem iron containing component present under conditions of severe iron-overload (see Weir *et al.*, 1984; Andrews *et al.*, 1988; Dickson *et al.*, 1988 for further details).

6.1.5 *Translational and transcriptional control of iron levels*

The concentrations of many of the iron uptake, transport and storage proteins are under tight genetic control. For example, in animals, large amounts of ferritin are produced only under conditions of plentiful iron, and the transferrin receptor protein levels increase when intracellular levels of iron are low (Theil, 1990 and references therein). Table 6.1 summarises the key effects resulting from iron limitation and excess for animals. The mechanism by which the iron level produces these effects is control of mRNA translation (Zähringer *et al.*, 1976; Casey *et al.*, 1988; Müllner *et al.*, 1989;

Table 6.1 Effect of iron on iron uptake and storage by eukaryotic cells, and on the synthesis of ferritin and transferrin receptors (see Theil, 1990, and references therein, for experimental data)

	Decrease in iron	Increase in iron
Iron uptake	Increases	Decreases
Transferrin receptor synthesis	Increases	Decreases
Transferrin receptor mRNA concentration	Increases	Decreases
Iron storage	Decreases	Increases
Ferritin synthesis	Decreases	Increases
Ferritin mRNA concentration	No change	No change

Koeller *et al.*, 1989). The mRNA for ferritin and the transferrin receptor protein contain non-translated regions known as 'iron-responsive elements' (IRE). Binding of the IRE-binding protein (IRE-BP) to the IRE of ferritin mRNA prevents mRNA translation but IRE-BP binding to the transferrin receptor protein mRNA prevents the mRNA from being degraded while also allowing it to be translated. Thus when IRE-BP is bound to mRNA the transferrin receptor protein is produced in relatively large amounts and ferritin is not. This occurs under conditions of low intracellular iron. With increasing iron the IRE-BP dissociates from the mRNA leading to an increase in ferritin synthesis and a decrease in the synthesis of the transferrin receptor protein. It seems likely that the IRE-BP itself is an Fe/S protein since it has a marked sequence similarity to aconitase (Rouault *et al.*, 1991), an Fe/S protein that catalyses the conversion of citrate to iso-citrate. Aconitase exists in an inactive 3Fe/4S form and an active 4Fe-4S form with Fe^{2+} binding causing the activation (Beinert and Kennedy, 1989), and it is an attractive hypothesis to suggest that the iron signal for the IRE-BP is similar.

IREs influence other mRNAs involved in iron metabolism in animals. For example, apart from the proteins mentioned above, the levels of δ-amino-laevolinic acid synthetase, a key enzyme in haem synthesis, are controlled by IREs (Cox *et al.*, 1991).

Regulation of proteins involved in iron metabolism in bacteria occur at the DNA level. It is well established that the synthesis of many proteins responsible for iron uptake into bacteria is repressed by the presence of iron (see Crichton, 1991 and references therein) and through a combination of genetic and biochemical experiments the molecular basis of this repression in *E. coli* is becoming clear (Hantke, 1981; Schäffer *et al.*, 1985; Bagg and Neilands, 1987a; Neilands *et al.*, 1990). A DNA binding protein, the Fe uptake regulatory protein, FUR, binds at a region of DNA known as an 'iron box'. The iron box forms part of the regulatory site for the various operons; e.g. for the aerobactin operon of *E. coli* which comprises genes coding for four enzymes responsible for biosynthesis of aerobactin and at least one gene responsible for the uptake of the Fe^{3+}-aerobactin complex (Bagg and Neilands, 1987b). FUR only binds to the iron box in the presence of Fe^{2+} and once it is bound transcription of the gene is repressed. FUR also represses other genes in the presence of Fe^{2+}, and perhaps Mn^{2+} also; e.g. the genes coding for the Mn and Fe containing superoxide dismutases and the gene for Shiga-like toxin (Hennecke, 1990; Fee, 1991). In other bacteria, e.g., *Vibrio anguillarum*, the transcriptional control is more complex and appears to involve iron regulated transcriptional activators (see Ralston and O'Halloran, 1990 and references therein).

The control of BFR concentrations in bacteria has not been established but in *Synechocystis* PCC6803 it is clear that BFR is not iron regulated (Laulhère *et al.*, 1991). This raises an intriguing question: if FUR only binds to DNA in the presence of Fe^{2+}, and there is a large iron storage capacity available in the form of BFR, what is the nature of the Fe^{2+} binding to FUR that allows FUR to compete effectively with BFR for iron?

6.1.6 *Iron as a major control element*

The previous section shows that the level of iron in a cell controls directly the intracellular concentration of free iron. However, the level of iron may have an even wider control over the activities within a cell. Many enzymes require iron for activity and in some cases the iron is weakly bound, as, for example, with aconitase (Beinert and Kennedy, 1989). da Silva and Williams (1991) have reviewed this area and noted that many enzymes involved in secondary metabolism require iron for activity (see their Table 12.12). Since some of these enzymes are involved in the production or removal of potentially toxic chemicals, including hormones, da Silva and Williams conclude that, at least in prokaryotes, there is a connection between the levels of free Fe^{2+} and the whole of the activity of the cell up to the point of initiation of cell division.

6.1.7 *Iron and human health*

There are three main areas where the amount of bioavailable iron affects human health (see Emery, 1991; Crichton, 1991 and references therein). The one that is most important numerically is iron deficiency, which usually leads to low haemoglobin levels and anaemia. Women, because of loss by menstruation, and children, because of an enhanced requirement for growth, are particularly susceptible to this. A large proportion of the world's population are iron deficient and in most cases iron sufficiency could be achieved relatively easily through a greater dietary intake of bioavailable iron (see chapter 8 for details).

Too much iron, or iron overload, is not so common but nor is it so readily treated (see Crichton, 1991 and references therein). Primary haemochromatosis results from a genetic defect leading to enhanced iron uptake. Secondary haemochromatosis occurs because excess iron has accumulated in the body as a result of the extensive blood transfusions required to ameliorate the effect of other genetic defects; for example, those which lead to thalassemia. Often, in such cases, the only treatment available is removal of excess non-haem iron by chelation therapy, or the removal of both haem and non-haem iron by bleeding (see chapter 8).

The third area is the role of iron in infection (Bullen and Griffiths, 1987; Crichton, 1991). Because iron is essential for the growth of microbes, pathogenic microbes growing in plants and animals usually do so under iron-limited conditions. Consequently, many microbes have evolved complex mechanisms to acquire their iron. These include the processes described in section 6.2, but they are often complemented by the secretion of other molecules, e.g. proteases, which damage transferrin and release its iron.

6.2 Iron-sulphur proteins

6.2.1 *Introduction*

This section considers a group of iron-sulphur enzymes. It should be remembered, however, that an enzyme is, strictly, a biological catalyst which works upon a substrate. Many interesting biological molecules are classified as enzymes, though neither their functions nor their substrates have been clearly defined. This is true of several of the iron-sulphur compounds discussed here, which are best described as a class of iron-sulphur proteins.

6.2.2 *Types of iron-sulphur protein*

The common characteristic of all these materials is that they contain one or more iron atoms ligated by sulphur, either sulphido-sulphur or thiolato-sulphur or both. They are distributed, albeit in small quantities, throughout plants, animals, and bacteria. Their principal but by no means only function would appear to be in electron transfer. They were recognised as a class of compound only some 30 years ago, possibly because their UV spectra are rather featureless and thus they avoided detection. Nevertheless, they possess very characteristic e.p.r. and electrochemical properties which make them relatively easy to identify in the right conditions (Thomson, 1985).

At the same time, and quite independently, model compounds representative of the known ferredoxin (multi-iron-sulphur) and rubredoxin (mono-iron-sulphur) species have been characterised (Harris, 1989). The interplay of bioinorganic and inorganic coordination chemistry has been particularly rewarding. Currently, models and/or proteins containing the following iron-sulphur cores are known: $[Fe(SR)_4]^{2,3-}$, $\{Fe_2S_2\}$, $\{Fe_3S_4\}$, $\{Fe_4S_4\}$, and $\{Fe_6S_6\}$ (Figure 6.7). There are also claims in natural systems for clusters which may contain multiples of these basic units, such as Fe_8S_8 ($= 2 \times Fe_4S_4$), Fe_7S_8 ($= Fe_3S_4 + Fe_4S_4$), and even $2 \times Fe$. These will be mentioned in the appropriate context.

6.2.3 *Single-iron systems*

Single-iron systems which contain iron(III) bound tetrahedrally by four sulphur (thiolato-) atoms occur naturally as rubredoxins. They may contain one or, less usually, two iron atoms and are found in both aerobic and anaerobic bacteria. They function as electron carriers. The rubredoxins are relatively low molecular weight proteins (up to *c.* 20 000 Dalton), and several structures have been determined. These show a recurring motif in the region of the protein backbone which binds the iron, namely two cysteinyl residues linked by two spacer amino acid residues designated Cys-X-X-Cys (Thomson, 1985). The cysteinyl sulphur atoms are ligands to the iron (Figure 6.8) in

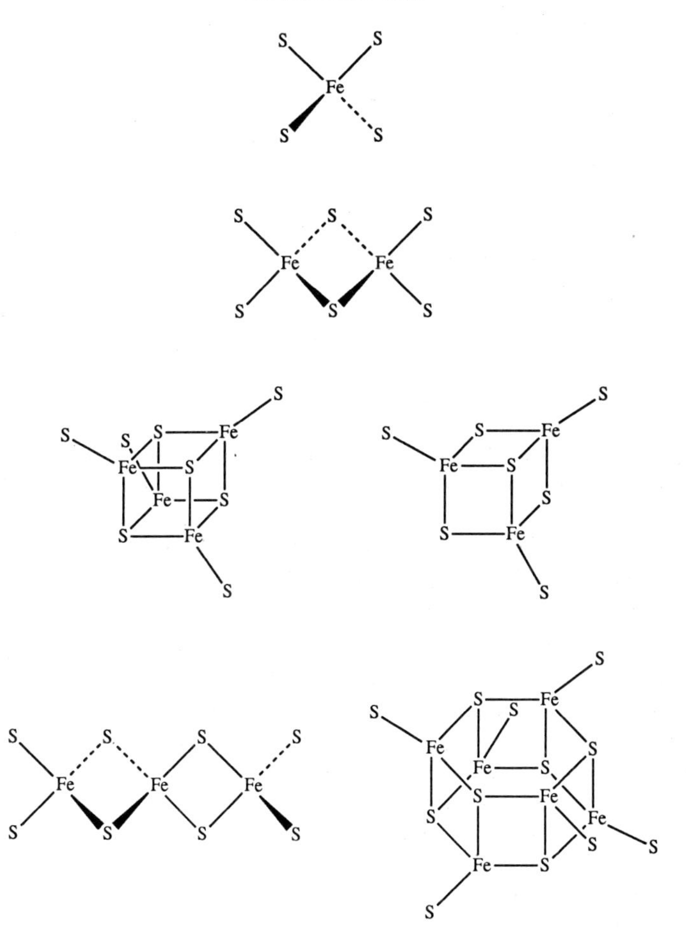

Figure 6.7 Principal iron-sulphur cores represented in model systems and in proteins.

```
(a) Clostridium pasteurianum     F-Met-Lys-Lys-Tyr-Thr-[Cys]-Thr-Val-[Cys]-Gly-Tyr-Ile-Tyr-Asp-Pro-
(b) Peptostreptococcus elsdenii    Met-Asp-Lys-Tyr-Glu-[Cys]-Ser-Ile-[Cys]-Gly-Tyr-Ile-Tyr-Asp-Glu-
(c) Micrococcus aerogenes          Met-Gln-Lys-Phe-Glu-[Cys]-Thr-Leu-[Cys]-Gly-Tyr-Ile-Tyr-Asp-Pro-
(d) Desulfovibrio vulgaris         Met-Lys-Lys-Tyr-Val-[Cys]-Thr-Val-[Cys]-Gly-Tyr-Glu-Tyr-Asp-Pro-

(a) Glu-Asp-Gly-Asp-Pro-Asp-Asp-Gly-Val-Asn-Pro-Gly-Thr-Asp-Phe-Lys-Asp-Ile-Pro-Asp-
(b) Ala-Glu-Gly-Asp-—-Asp-Gly-Asn-Val-Ala-Ala-Gly-Thr-Lys-Phe-Ala-Asp-Leu-Pro-Ala-
(c) Ala-Leu-Val-Gly-Pro-Asp-Thr-Pro-Asn-Gln-Asn-Gly-—-Ala-Phe-Glu-Asp-Val-Ser-Glu-
(d) Ala-Glu-Gly-Asp-Pro-Thr-Asn-Gly-Val-Lys-Pro-Gly-Thr-Ser-Phe-Asp-Asp-Leu-Pro-Ala-

(a) Asp-Trp-Val-[Cys]-Pro-Leu-[Cys]-Gly-Val-Gly-Lys-Asp-Glu-Phe-Glu-Glu-Val-Glu-Glu
(b) Asp-Trp-Val-[Cys]-Pro-Thr-[Cys]-Gly-Ala-Asp-Lys-Asp-Ala-Phe-Val-Lys-Met-Asp
(c) Asp-Trp-Val-[Cys]-Pro-Leu-[Cys]-Gly-Ala-Gly-Lys-Glu-Asp-Phe-Glu-Val-Tyr-Glu-Asp
(d) Asp-Trp-Val-[Cys]-Pro-Val-[Cys]-Gly-Ala-Pro-Lys-Ser-Glu-Phe-Glu-Ala-Ala
```

Figure 6.8 Representation of the protein chain in various rubredoxins, the conserved cysteines reponsible for binding the iron being shown in the boxes (Nakamura and Ueyama, 1989). Reprinted by permission from *Adv. Inorg. Chem.*, vol. 33, p. 46. Copyright 1989 Academic Press.

approximately tetrahedral fashion. The redox potentials of the rubredoxins are *c.* −50 mV, and it would be thought that models would be very easy to develop, but in fact, this is not a trivial problem.

Species such as $[Fe\{(SCH_2)_2C_6H_4\}_2]^-$ and $[Fe(SPh)_4]^{2-}$ were obtained relatively early (Lane *et al.*, 1977; Hagen *et al.*, 1981), but the isolation of a stable iron(III) analogue with monodentate ligands, $[Fe(SC_6HMe_4\text{-}2,3,5,6)_4]^-$, took rather longer (Millar *et al.*, 1982). These compounds undergo reversible Fe^{II}/Fe^{III} conversions, but at redox potentials considerably more negative than those of the natural rubredoxins. The couple $[Fe\{(SCH_2)_2C_6H_4\}_2]^{1-,2-}$ exhibits a potential *c.* 0.7 V more negative than that of the *Clostridium pasteurianum* rubredoxin. The couple $[Fe(SC_6HMe_4\text{-}2,3,5,6)_4]^{1-,2-}$ is about 0.6 V more negative (Christou *et al.*, 1977). Model complexes have been prepared in which synthetic peptides containing the motif Cys-X-X-Cys bind to a single iron atom. These have more positive redox potentials, indicative of some influence of the amino acid residues which is not yet understood, though some theories have been advanced (Nakamura and Ueyama, 1989). In addition, the conversion of rubredoxin iron atoms to four-iron four-sulphur ferredoxin-type clusters has also been reported (Christou *et al.*, 1979). This reflects on the phenomenon of so-called 'spontaneous self-assembly' (see section 6.2.4).

6.2.4 *Two-iron systems*

These are the smallest true clusters, and contain two iron atoms bridged by two sulphido-sulphur atoms, with additional thiolato-sulphur atoms co-ordinated to each iron so that it is approximately tetrahedral. They constitute the smallest group of ferredoxins, and are found in plant chloroplasts and are often termed plant ferredoxins. However, they are also found in photosynthetic bacteria and archaebacteria, so this is clearly a misnomer. These ferredoxins function in several ways, including as electron carriers in the photoreduction of $NADP^+$ in chloroplasts, in glutamate synthesis, in reduction processes of dinitrogen, nitrate, nitrite, and sulphite, and in hydroxylations. Others are found in the adrenal cortex (adrenodoxin), in liver (hepatoredoxin), and in kidneys (renoredoxin) and in other mammalian organs. These ferredoxins tend to differ in structure from the plant ferredoxins.

The structure of the 'plant ferredoxin' from *Spirulina platensis* (Figure 6.9) shows that the iron atoms are roughly tetrahedrally coordinated, using cysteinyl residues from the protein backbone, but although the usual Cys-X-X-Cys motif is present, it is involved in quite a different fashion to that involved in the rubredoxins, and more similar to the bridging mode found in the $\{Fe_4S_4\}$ ferredoxins. Cysteines 41 and 46 bind to one iron atom and cysteines 49 and 79 to the other. This may have significance for the assembly of four-iron systems from two-iron systems (Fukuyama *et al.*, 1980).

The two-iron ferredoxins have mid-point potentials generally within the

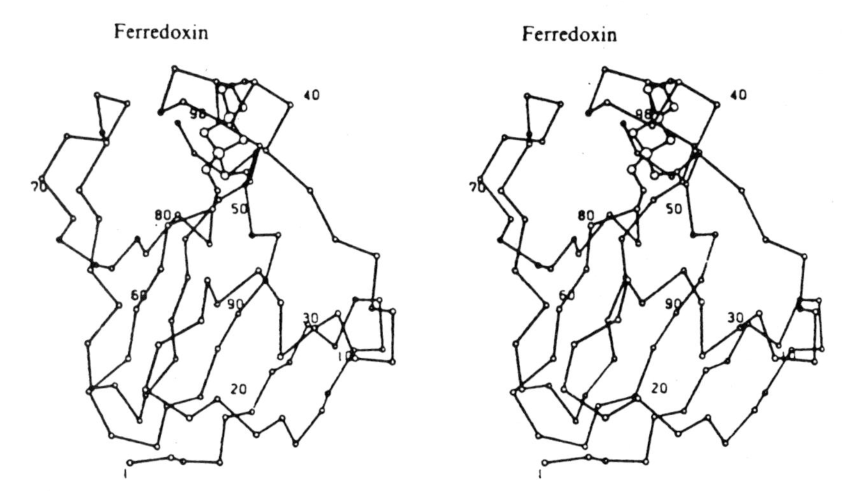

Figure 6.9 Representation of the structure of the ferredoxin of *S. platensis* at 2.5 Å resolution (Fukuyama *et al.*, 1980). (Reprinted by permission from *Nature*, vol. 286, p. 522. Copyright © 1980 Macmillan Magazines Ltd.)

range −350 to −450 mV for the plant types, and as low as *c.* −200 mV for the rest (Ueyama and Nakamura, 1988). The non-plant ferredoxins have few sequence relationships to the plant-type ferredoxins and, indeed, may also have quite different terminal ligands on the cluster. For example, the Rieske proteins, found in a range of bacterial types, probably have not more than two of the four terminal ligands which are cysteinyl residues, and the other two ligands may be nitrogen donors (Fee *et al.*, 1984). Their redox potentials are positive (+150 to +300 mV). This difference from the plant ferredoxins may be related to a difference in function (Ueyama and Nakamura, 1988).

Model compounds have been known since 1973, when $[\{Fe(SCH_2)_2C_6H_4\text{-}1,2\}_2(S)_2]^{2-}$ was reported (Mayerle *et al.*, 1973). Compounds with monodentate thiols such as $[Fe_2(SR)_4(S)_2]^{2-}$ have also been synthesised. These compounds contain formal iron(III), and they can be synthesised by reactions involving $[Fe(SR)_4]^-$ and hydrogensulphide, or $[Fe(SR)_4]^{2-,-}$ with elemental sulphur (Han *et al.*, 1986). The details of these reactions are fairly clear, and they establish a clear link between the rubredoxin centres and the two-iron ferredoxin centres (for more details, see chapter 7).

Other terminal ligands are found on this system. The chloro-complex $[Cl_2FeS_2FeCl_2]^{2-}$ (Wong *et al.*, 1978) reacts with anions such as pyrrolate and cresolate to yield compounds with oxygen- or nitrogen-donors in the terminal positions (Coucouvanis *et al.*, 1984). In fact, poly(sulphur) ligands such as $(S_5)^{2-}$ (Coucouvanis *et al.*, 1979) and organometallic ligands such as $(C_5Me_5)^-$ (Inomata *et al.*, 1991) are also observed. The two iron(III) atoms in these models are probably high-spin, and antiferromagnetically coupled. They undergo irreversible reductions at about −1.50 (vs. SCE). This differs considerably from the Rieske-type proteins, where spectroscopic properties (in particular Mössbauer parameters) and dimensions of the Fe_2S_2 core do not seem to vary very much with terminal ligands. Presumably terminal coordination and ligand type characteristic of peptide and protein binding have yet to be achieved in model compounds. It should also be noted that modification of the bridging $\{Fe_2S_2\}$ grouping, to form S–S bonds, or indeed to produce two independent $(S_2)^{n-}$ bridges, or even more complex systems have been detected (Inomata *et al.*, 1991; Brunner *et al.*, 1988; Tremel *et al.*, 1989).

6.2.5 *Three-iron systems*

These systems have been thoroughly characterised in Nature relatively recently. The model compounds are, however, better established. They readily

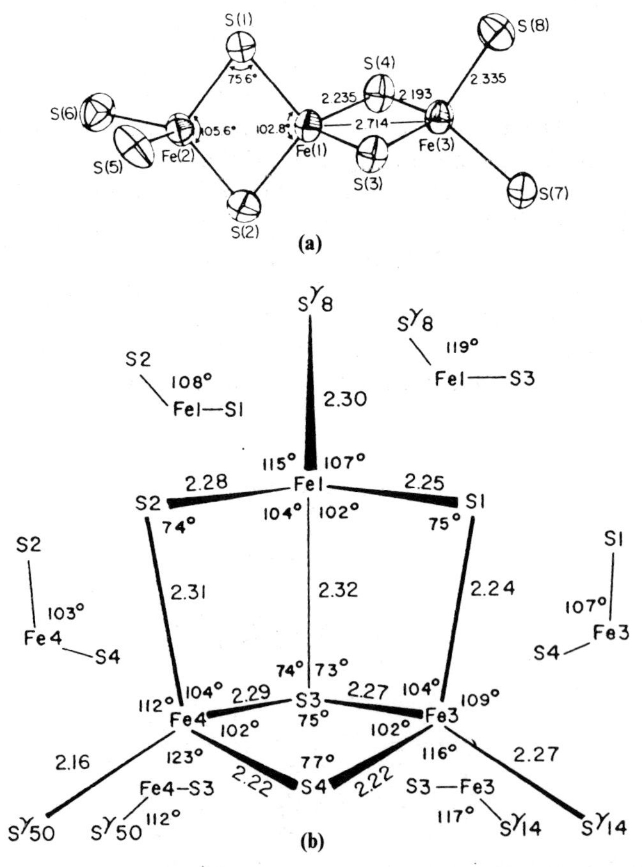

Figure 6.10 Representation of (a) the structure of $[Fe_3S_4(SPh)_4]^{3-}$, omitting the terminal phenyl groups (Hagen *et al.*, 1983) and (b) the three-iron cluster from *D. gigas* Ferredoxin II (Kissinger *et al.*, 1988). ((a) Reprinted by permission from *J. Am. Chem. Soc.*, vol. 105, p. 3909. Copyright 1983 ACS. (b) Reprinted by permission from *J. Am. Chem. Soc.*, vol. 110, p. 8722. Copyright 1988 ACS.)

take up a heterometal atom, M, to become four-metal clusters, and, in the particular case of M = Fe, the well known $\{Fe_4S_4\}$-type core (Ciurli *et al.*, 1990). However, the relationship of the three-iron models to the natural three-iron systems has been far from clear.

The first model systems containing $\{Fe_3S_4\}$ were of stoichiometry $[Fe_3S_4(SR)_4]^{3-}$ (R = Ph or Et) (Hagen *et al.*, 1983). They have three roughly tetrahedral iron atoms in a linear type of structure, as shown in Figure 6.10.

This is evidently not a $\{Fe_4S_4\}$ cubane structure (see below), though it retains certain features such as the Fe–Fe separation of *c.* 2.7 Å and the mixture of sulphido-and thiolato-sulphur atoms. The compounds are made by variations on the method applicable for most of the species discussed in this section, namely reaction of an iron(II) compound with thiolate and sulphur, and magnetic and Mössbauer measurements suggest that they contain high-spin, antiferromagnetically coupled iron(III) (Girerd *et al.*, 1984). The compounds react with more iron(II) in the form of $FeCl_2$ to yield $[Fe_4S_4(SR)_4]^{2-}$ (Hagen *et al.*, 1983). In the absence of sulphur, the preparative reaction can give rise to another three-iron species, $[\{FeCl_2\}_3(SPh)_3]^{3-}$ (Figure 6.11), which contains a six-membered ring of alternating iron and sulphur atoms (see below). The absence of elemental sulphur clearly excludes any redox reactions involving sulphur (Hagen and Holm, 1982).

Three-iron clusters are involved in a considerable number of ferredoxins, currently more than twenty. Some of these are more correctly described as seven-iron systems (Fe_3S_4 plus Fe_4S_4) whereas others are simple three-iron systems. Conversion to $\{Fe_4S_4\}$, and indeed to a range of $\{Fe_3MS_4\}$ (M = Zn, Co, Ni, Cd, etc.) clusters, within the ferredoxins or enzyme such as the ferredoxins of *Azotobacter vinelandii, Desulfovibrio africanus, Desulfavibrio gigas,* and *Pyrococcus furiosus* and in aconitase have been reported (see for

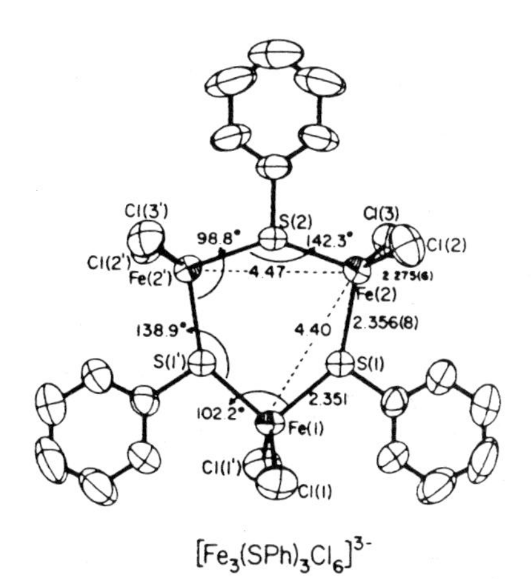

Figure 6.11 Representation of the structure of $[\{FeCl_2\}_3(SPh)_3]$ (Hagen and Holm, 1982). (Reprinted by permission from *J. Am. Chem. Soc.*, vol. 104, p. 5497. Copyright 1982 ACS.)

example Butt *et al.*, 1991; Emptage 1991; Conover *et al.*, 1991 and Macedo *et al.*, 1991). It was recognised early that the characteristic e.p.r. parameters of the three-iron centre in the model systems ($g = 4.3$, no signal near $g = 2.0$, $S = 1/2$ ground state) does not support a structural similarity with the three-iron cluster supported by proteins ($g = 2.0$, $S = 5/2$ ground state). Initial structural studies on the ferredoxin from *A. vinelandii* were interpreted in terms of a cyclic roughly planar structure for an $\{Fe_3S_3\}$ core (Ghosh *et al.*, 1981). Subsequent studies, on *A. vinelandii* (Stout, 1988, 1989; Jensen *et al.*, 1991; Stout *et al.*, 1988), on *D. gigas* (Kissinger *et al.*, 1989); Ferredoxin II, and on aconitase (Robbins and Stout, 1985, 1989) and a host of analytical results, suggest that the tri-iron cluster is always $\{Fe_3S_4\}$ and that it always takes the form of a 'voided cubane', that is, an Fe_4S_4 cluster from which one iron(II) corner has been removed, with relatively little change in the geometry of the residual atoms.

In many of these $\{Fe_3S_4\}$ systems, such as in aconitase from beef heart and *D. gigas* Ferredoxin II, there is evidence that a reversible iron extraction may be carried out, and that the iron is not removed randomly from the cubane corners. The three-iron systems appear often to be artefacts of the preparative procedure. The iron is

$$Fe^{2+} + \{Fe_3S_4\}^0 \leftrightharpoons \{Fe_4S_4\}^{2+}$$

lost from a specific corner, which must be defined by the ligation of the protein (Thomson, 1985; Holm *et al.*, 1990). The structure of the three-iron product from *D. gigas* Ferredoxin II has been determined by X-ray crystallography (Kissinger *et al.*, 1988). The ligands to these clusters (cf. Rieske proteins) are not all invariably cysteinyl residues. Indeed, where the cluster acts upon a substrate rather than simply in electron-transfer, invariable cysteinyl ligation might not be expected. However, the three-iron clusters do undergo a reversible redox process, $\{Fe_3S_4\}^+ \leftrightharpoons \{Fe_3S_4\}^0$ (Thomson *et al.*, 1990). Whether any three-iron system is active *in vivo* appears unlikely. The 'voided cubane' structure is presumably imposed upon the system by the protein, rather than by cluster thermodynamics. In a model system in which an $\{Fe_4S_4\}$ cluster is coordinated by a tridentate ligand, the complex has been shown to behave magnetically as a low-spin iron(II) subsite, together with a spin-isolated $\{Fe_3S_4\}^0$ fragment with the same ground state electronic structure as the singly reduced three-iron cluster of the proteins (Weigel *et al.*, 1990). It has been suggested that this fragment can be regarded as a ligand in its own right (Ciurli and Holm, 1991; Excoffen *et al.*, 1991). Other isomers for the model systems and for the protein clusters have been discussed, searched for, and even tentatively identified. There are no conclusive reports of any yet to hand.

6.2.6 *Four-iron systems*

Four-iron clusters were reported, both in Nature and in synthetic chemistry, in 1972. They all contain the $\{Fe_4S_4\}$ cubane, which has been structurally

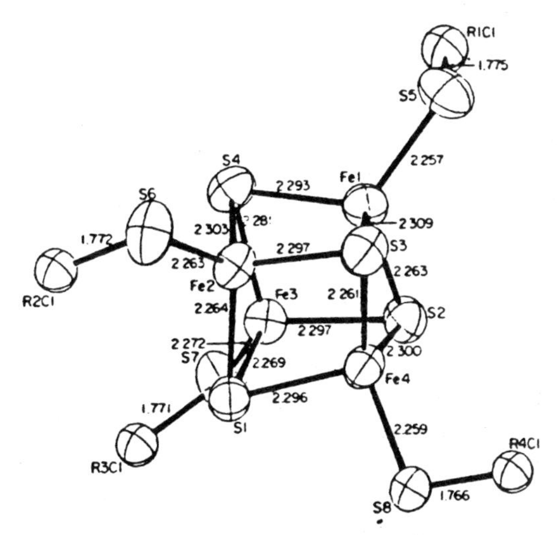

Figure 6.12 Representation of the structure of $[Fe_4S_4(SPh)_4]^{2-}$, showing characteristic distortion from perfect cubic shape (Que *et al.*, 1974). (Reprinted by permission from *J. Am. Chem. Soc.*, vol. 96, p. 4177. Copyright 1974 ACS.)

characterised in a wide range of proteins and models. The structures are always very similar (though not invariant), regardless of ligand-type and oxidation level. The characteristic distortion from the idealised cube involves a moving together of the pairs of iron atoms in each of the faces of the cube (Figure 6.12).

Model systems are known with a variety of anionic ligands bonded to the iron atoms, such as carbon donors (including η^5-C_5H_5), halides, oxygen donors (including phenoxide and latterly even methoxide), and sulphur donors (primarily thiolates, but including dithiocarbamates). In contrast, in the proteins the binding is primarily by cysteinyl sulphur, though exceptions to this are now beginning to accumulate (see below). More recently, as in $[Fe_4S_4I_2(SPPh_3)_2]$, clusters with neutral ligands have been characterised (Saak and Pohl, 1988; Pohl and Bierback, 1991).

Chemists have recently synthesised another four-iron cluster: this is the structurally characterised $[Fe_4S_6(SEt)_4]^{4-}$ (Figure 6.13) (Al-Ahmad *et al.*, 1991). It can be regarded as a higher homologue of $[Fe_2S_2(SEt)_4]^{2-}$ which is a model for the two-iron ferredoxins, the four terminal thiolate ligands being replaced by two bidentate $\{Fe(SEt)_2(S)_2\}^{3-}$ units. One can envisage a chain of $\{FeS_2\}$ units (cf. $KFeS_2$), but (with one exception) larger species have yet to be synthesised, and there are no examples of this $\{Fe_4S_6\}^0$ unit in proteins.

The synthetic route to Fe_4S_4 clusters has been discussed in some detail. The general procedure involves reaction of an iron(II) or iron(III) starting material with thiolate and elemental sulphur. There seems to be a system of complex equilibria, and the course of reaction in any specific circumstances apparently depends upon the proportions of reactants, the solvent, the pH, etc. Figure 6.14 shows one specific example (Stevens and Kurtz, 1985; Hagen *et al.*, 1981). Note that the Figure includes yet another four-iron cluster, which has also been

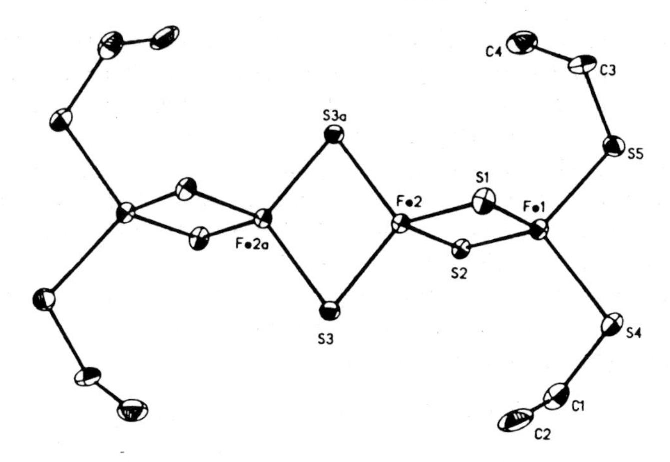

Figure 6.13 Representation of the structure of $[Fe_4S_6(SEt)_4]^{4-}$ (Al-Ahmad *et al.*, 1991). (Reprinted by permission from *Inorg. Chem.*, vol. 30, p. 1164. Copyright 1991 ACS.)

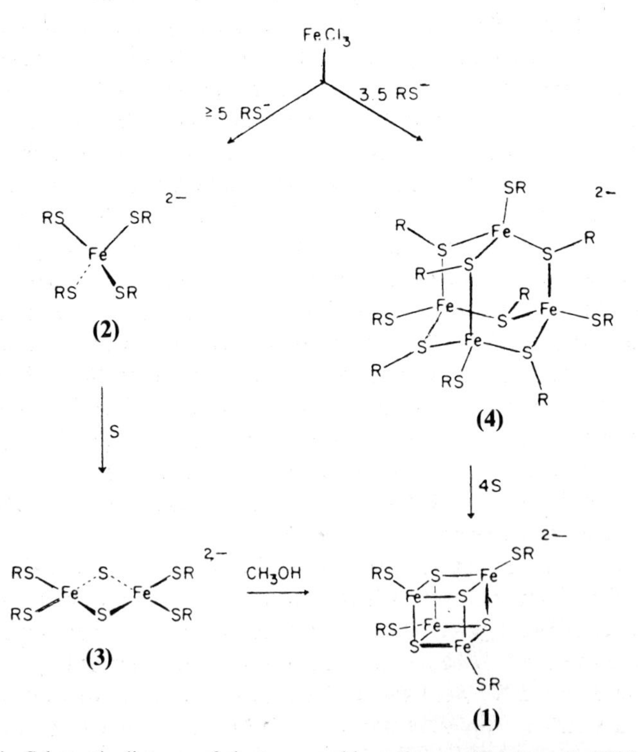

Figure 6.14 Schematic diagram of cluster assembly pathways (Hagen *et al.*, 1981). (Reprinted by permission from *J. Am. Chem. Soc.*, vol. 103, p. 4055. Copyright 1981 ACS.)

structurally characterised, $[Fe_4(SPh)_{10}]^{2-}$ (Stevens and Kurtz, 1985; Hagen *et al.*, 1980, 1981). The assembly of the cubanes is thermodynamically driven, hence the term 'spontaneous self-assembly', and this may indeed explain why they were available for adoption by biological systems during the evolutionary process. However, it has been shown that the *apo*-polypeptide from *C. pasteurianum* ferredoxin causes reconstitution of the ferredoxin itself 1–2 orders of magnitude faster than the self-assembly of $[Fe_4S_4(SR)_4]^{2-}$ under comparable conditions (Bonomi *et al.*, 1985).

The cubane model clusters can exist in four oxidation levels which, under appropriate conditions, are reversibly interchangeable.

Omitting ligands, this can be represented as:

$$\{Fe_4S_4\}^0 \rightleftharpoons \{Fe_4S_4\}^{1+} \rightleftharpoons \{Fe_4S_4\}^{2+} \rightleftharpoons \{Fe_4S_4\}^{3+}$$

If ligands are included, the formal charges are, of course, different.

$$[Fe_4S_4(SR)_4]^{4-} \rightleftharpoons [Fe_4S_4(SR)_4]^{3-} \rightleftharpoons [Fe_4S_4(SR)_4]^{2-} \rightleftharpoons [Fe_4S_4(SR)_4]^{1-}$$

The (4−) level of the cluster requires too strong a reducing agent for it to be accessible in biological systems, though it has been unequivocally demonstrated for R = Ph (Pickett, 1985) and isolated in $[Fe_4S_4(CO)_{12}]$ (Nelson *et al.*, 1982). Natural systems seem to occupy the oxidation levels (1+), (2+) and (3+). The four-iron ferredoxins have redox potentials in the range −250 to −650 mV, and cycle between the (2+) and (1+) levels (Thomson, 1985; Nakamoto *et al.*, 1988). There is another class of four-iron proteins, the so-called high-potential iron proteins (HIPIPs), which have redox potentials in the range +50 to +450 mV. These cycle between the (2+) and (3+) levels, and cannot be reduced to the (1+) level under normal physiological conditions (Thompson, 1985). Likewise, the ferredoxins cannot be oxidised to the (3+) level without destruction of the iron-sulphur core. The reason for these differences lies in the detailed ligation of the individual clusters (see below). The redox potentials of the model systems cover a wide range, consonant with the wide range of ligand types and solvents available (Blank *et al.*, 1991; Harris, 1989), but it needs to be explained why compounds such as $[Fe_4S_4(SC_6H_2Pr^i_3\text{-}2,4,6)_4]^{2-}$ have 1−/2− and 2−/3− redox potentials of −120 and −1200 mV, respectively (O'Sullivan and Millar, 1985). These values lie well outside the ranges assigned to ferredoxins and HIPIPs.

The almost invariable arrangement of amino acid residues in the peptide chain of the four-iron ferredoxins (and doubly so in the eight-iron systems) is the sequence Cys-X-X-Cys-X-X-Cys. The cysteinyl residues are bound to three iron atoms of the cube, whereas the fourth iron atom is ligated by a cysteine much further away in the protein backbone (Ueyama and Nakamura, 1988). Studies with synthetic peptides confirm the facility with which three such cysteinyl residues bind to an $\{Fe_4S_4\}$ core (Que *et al.*, 1974; Ueyama *et al.*, 1985). There is a high degree of amino acid conservation amongst the ferredoxins which have molecular weights in the region 6–12 kDalton (Meyer, 1988).

Structural evidence for the HIPIPs is much less plentiful, but it suggests they are quite different. Although there are four cysteinyl ligands bound to the cluster in *Chromatium vinosum* HIPIP (Shendan *et al.*, 1974: Carter *et al.*, 1974), there is no identifiable homology in the amino acid sequences with the ferredoxins (Backes *et al.*, 1991). The cluster is buried in the centre of a relatively spherical protein, whereas the two clusters in *Pseudomonas aerogenes* ferredoxin are situated at the ends of an elongated protein and closer to the protein surface. These differences must relate to the functions (unidentified as yet for HIPIP) and the redox behaviour.

It is argued persuasively that the detailed hydrogen-bonding in the cluster environment is a very significant feature of the structures of these proteins (Backes *et al.*, 1991; Sola *et al.*, 1991). The higher oxidation state attainable by HIPIPs may be a consequence of the more hydrophobic environment in which the clusters find themselves. Hydrogen bonding ($NH \cdots S_{bridge}$ and $NH\text{-}S_{cysteine}$) in several ferredoxins and HIPIPs has been evaluated by careful structural analysis and by resonance Raman spectroscopy, there being perhaps three times as many hydrogen bonds in the former as in the latter. Such bonding, it is argued, by removing charge from the cluster core, stabilises the (4−) level of the ferredoxins. Thus, hydrogen-bonding promotes ferredoxin redox properties, whereas bulky hydrophobic ligands promote HIPIP redox properties. The iron-sulphur-carbon(α)-carbon(β) dihedral angles at the cysteine ligands are also close to 180° in HIPIPs, but not in ferredoxins. This also influences redox potentials (Sola *et al.*, 1991).

Work with synthetic peptides is consistent with this analysis. Thus comparison of the redox properties of $\{Fe_4S_4\}$ complexed with Z-Cys-Gly-Ala-Cys-OMe or Z-Cys-lle-Ala-Cys-OMe (Z = benzoylcarbonyl) shows a positive shift for the former in dichloromethane (but not necessarily in protic solvents), ascribed to hydrogen-bonding between the Ala NH and the cysteinyl sulphur. This cannot occur with facility in the latter case. The potentials (as in other models) are solvent-dependent (Ueyama *et al.*, 1985; Ohno *et al.*, 1990).

Whatever the oxidation level of the cluster, these clusters contain linked iron atoms often formally in different oxidation states and they have provided a fertile field of study of magnetic and spectroscopic properties of all kinds. They will not be discussed here, and the reader is referred to the many reviews, and to the original literature.

All the four-iron systems discussed so far have pure cysteinyl-sulphur ligation. This is not, however, always the case. There are now known at least five ferredoxins which contain less than complete cysteinyl ligation. Four are eight-iron ferredoxins and one, from *Pyrococcus furiosus*, is a four-iron ferredoxin. This incomplete cysteinyl ligation appears to allow easy conversion $\{Fe_4S_4\} \rightleftharpoons \{Fe_3S_4\}$ and the incorporation of heterometals into $\{Fe_3S_4\}$, something that has been achieved in model systems starting from the isomeric

$\{Fe_3S_4\}$ (see above), and normally not from the cubane three-iron residue, which is not stable unless specific ligands forcing the cubane geometry are present. The $\{Fe_3S_4\}$ (voided cubane) group can be regarded as a distinct species, even when combined into a four-iron cluster. This concept has been very fruitful for preparative (Weigel and Holm, 1991; Weigel *et al.*, 1990) and theoretical studies (Noodleman *et al.*, 1988; Noodleman, 1991).

The ferredoxin from *P. furiosus* has an atypical arrangement of cysteinyl residues: X_{10}–Cys(11)–X_2–Asp(14)–X_2–Cys(17)–X_3–Pro–X_{25}–Cys–X_7–Cys–Pro–X_9. The cysteines at positions 11 and 17, as well as at 56, are 'normal', but the expected Cys at 14 is replaced by an aspartate. A similar atypical coordination has been postulated for the nitrogenase P-clusters and for iron hydrogenase H-clusters. It is believed that the aspartate and/or hydroxide bind at the atypical iron. Oxidation of the *P. furiosus* ferredoxin by ferricyanide leads to the formation of a $\{Fe_3S_4\}$ cluster which has a redox chemistry indicating interconvertible (0) and (1+) levels (models apparently have access to 3 oxidation levels, (1−), 0 and (1+)). This ferredoxin can also be induced to bind exogenous ligands such as cyanide, presumably to the atypical iron (Conover *et al.*, 1990, 1991).

As stated earlier, these $\{Fe_3S_4\}$ systems at least in oxidation level (0) can also incorporate other metals, to regenerate cubanes. For example, this has been achieved with Ni^{II} in *P. furiosus* (Conover *et al.*, 1990), and it may be that the nickel-iron cluster of *Rhodospirillum rubrum* carbon monoxide dehydrogenase is also an $NiFe_3S_4$ system (Stephens *et al.*, 1989).

It is not altogether clear whether the three-iron clusters identified in these systems (other examples are the ferredoxins of *A. chroococcum* and *D. africanus*) are always artefacts of the isolation procedure or not. They may be a consequence of the atypical ligation to one iron atom (perhaps connected with substrate binding) or, equally, the ready loss may be an aspect of regulation. Active aconitase from beef heart contains a four-iron cluster which can also lose an atypical iron atom (probably ligated by hydroxide), and it is believed that during its function a citrate chelates to this iron via a carboxylate and hydroxyl group, and the citrate then loses the elements of water to form aconitate, which then gives rise to *iso*-citrate (Emptage, 1988).

Model systems have yet to show such chemistry, but it is now clear that any amino acid carrying functional groups is capable of binding to $\{Fe_4S_4\}$ clusters (Evans and Leigh, 1991).

Even this does not exhaust all the possibilities. *E. coli* sulphite reductase contains a haem, an $\{Fe_4S_4\}$ cluster and a siroheme (Christner *et al.*, 1981; McKee *et al.*, 1986). It is likely that one of the four cluster cysteinyl residues is shared through the sulphur with the siroheme iron.

6.2.7 *Six-iron systems*

Model cluster containing $\{Fe_6S_6\}$ cores have been appearing in the literature since 1984 (Saak *et al.*, 1984; Coucouvanis *et al.*, 1984; Kanatzidis *et al.*, 1984).

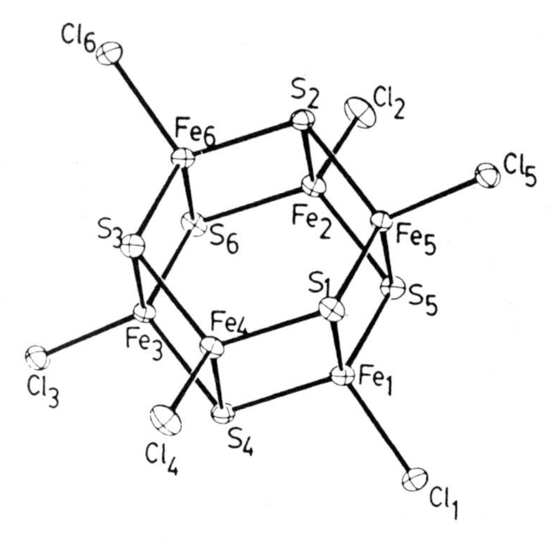

Figure 6.15 Representation of the structure of $[Fe_6S_6Cl_6]^{3-}$ (Coucouvanis *et al.*, 1984). (Reprinted by permission from *J. Am. Chem. Soc.*, vol. 106, p. 7998. Copyright 1984 ACS.)

The compound $[Fe_6S_6Cl_6]^{3-}$ was initially obtained from the reaction of $FeCl_2$, KSPh, S, and $[Et_4N]Cl{\cdot}H_2O$ in MeCN. A similar reaction, but with $[Ph_4P]Cl$ in place of $[Et_4N]Cl$, yields $[Fe_4S_4Cl_4]^{2-}$. The structure of the 'prismane' is shown in Figure 6.15. The bond lengths are very similar to those of the cubane, and, in fact, the prismane changes to the cubane upon heating in MeCN solution. Conversion of cubane to prismane occurs upon mild oxidation. The ion $[Fe_6S_6Cl_6]^{3-}$ can be reversibly reduced to the (4−) level at −0.82 V (S.C.E., MeCN) (Saak *et al.*, 1984; Coucouvanis *et al.*, 1984; Kanatzidis *et al.*, 1984).

Since then, prismanes with a range of ligands, Cl, Br I, SR, OR, etc., have been prepared (Kanatzidis *et al.*, 1985, 1986). They all undergo quasi-reversible one-electron reduction and reversible one-electron oxidation. The prismane structure is apparently favoured by OPh ligands as compared to SPh ligands, but the dependence of the redox potentials upon ligand parallels that of the cubanes. The prismanes can act as ligands themselves through the bridging sulphido-sulphur atoms to yield species such as $[Fe_6S_6Cl_6\{Mo(CO)_3\}_2]^{3-}$, some of which have been structurally characterised by X-ray crystallography (Coucouvanis *et al.*, 1988; Al-Ahmad *et al.*, 1990). They have been considered structural models for the molybdenum-iron cluster of the molybdenum nitrogenases (see below).

By careful control of reagents, reagent proportions, and reaction conditions it is possible to isolate further species with six-iron clusters. Thus $Fe(SR)_3$ and Na_2S_2 in methanol can lead to the isolation of $[Fe_6S_9(SCH_2Ph_2)]^{4-}$, or $[Fe_6S_9(SMe)_2Na_2]^{6-}$. The basic structure is shown in Figure 6.16 (Snasdeit *et al.*, 1984; Hagen *et al.*, 1983; Christou *et al.*, 1982). They contain Fe_6S_9 fragments made up of Fe_2S_2 rhombs with dimensions not very different from

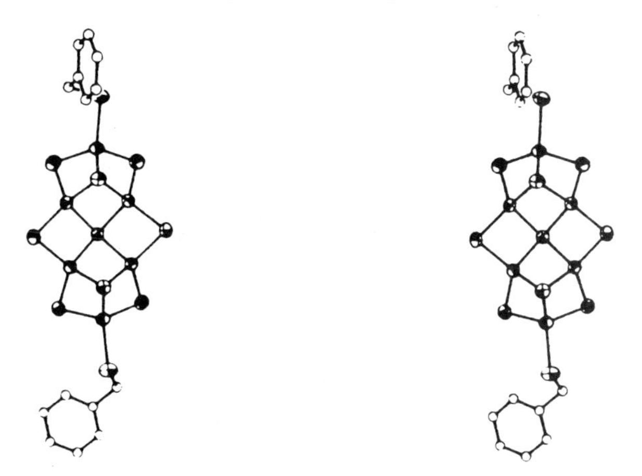

Figure 6.16 Basic structures of species containing the $[Fe_6S_9(SR)_2]^{4-}$ units (Snasdeit *et al.*, 1984). (Reprinted by permission from *Inorg. Chem.*, vol. 23, p. 1818. Copyright 1984 ACS.)

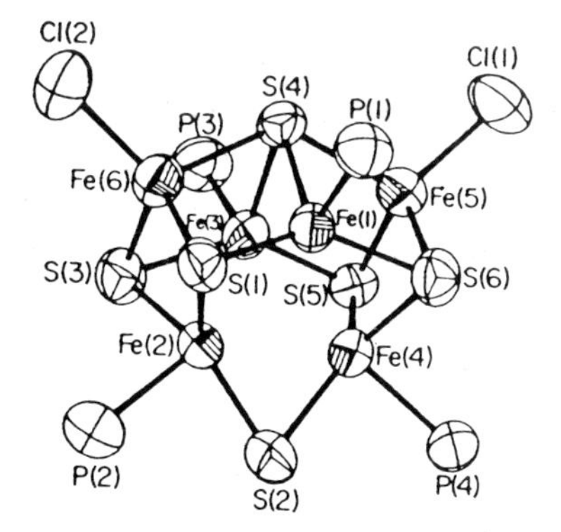

Figure 6.17 Representation of the structure of $[Fe_6S_6(PBu^n{}_3)_4Cl_2]$ (Snyder *et al.*, 1988). (Reprinted by permission from *Inorg. Chem.*, vol. 27, p. 586. Copyright 1988 ACS.)

those found in other complexes, but the overall structure is without parallel in nature.

A further variant may be obtained by the reaction of $[Fe_6S_6X_6]^{3-}$ with $[FeX_2(PEt_3)_2]$ (X = Cl, Br, or I) (Snyder *et al.*, 1988, 1991; Snyder and Holm, 1988). The structure (Figure 6.17) may be regarded as being made up of non-planar $\{Fe_2S_2\}$ rhombs to give an $\{Fe_6(\mu\text{-}S)(\mu_3\text{-}S)_4(\mu_4\text{-}S)\}^{2+}$ core. In $[Fe_6S_6(PBu^n{}_3)_4Cl_2]$ it has been described as a basket, with the phosphine-bearing iron atoms carrying the single μ-S as the handle (Snyder and Holm, 1988; Snyder *et al.*, 1988, 1991). In addition to these, the variant $[Fe_6S_8(PEt_3)_6]^-$ has been described (Snyder and Holm 1988; Snyder *et al.*, 1988, 1991). It has a closely related structure and the structures of all the Fe_mS_n species have been discussed (Reynolds and Holm, 1988). The general structures of all the species

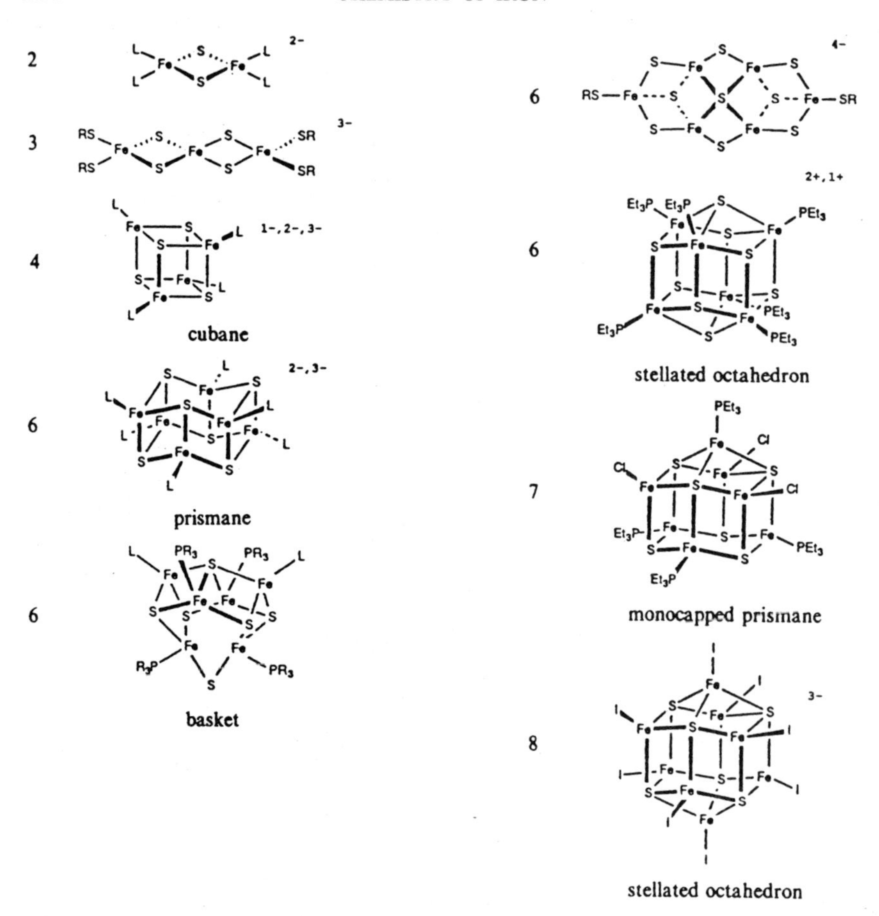

Figure 6.18 Schematic structures of iron–sulphur clusters (Reynolds and Holm, 1988). (Reprinted by permission from *Inorg. Chem.*, vol. 27, p. 4498. Copyright 1988 ACS.)

discussed so far are shown in Figure 6.18. Note that $\{Fe_8S_6\}^{3+}$ is an analogue of the molybdenum capped prismane described above, but that it has yet to be observed, in nature or in models. The core $\{Fe_7S_6\}^{3+}$ is, however, known. It has been characterised in $[Fe_7S_6(PEt_3)_4Cl_3]$ (Figure 6.19) (Noda *et al.*, 1986).

A further variant, $[Fe_6S_8(PEt_3)_6]^{+/2+}$, which has been isolated from $[Fe(H_2O)_6]^{2+}$, H_2S, and PEt_3 (Cecconi *et al.*, 1987; Agresti *et al.*, 1985). They form part of a series of reversible one-electron redox reactions between $\{Fe_6S_8\}^{4+}$ and $\{Fe_6S_8\}^0$. The structure (Figure 6.20) may be regarded as a octahedron of iron atoms, each face of which is capped by a sulphido-sulphur atom. A structure which is essentially of the $[Fe_6S_6(PR_3)_4Cl_2]$ type, but in which the 'handle' sulphide is replaced by an SPh^- has also been reported as has another variant of formula $[Fe_6S_6(SPh)(PBu^n{}_3)_4(SPh)_2]$ (Chen *et al.*,

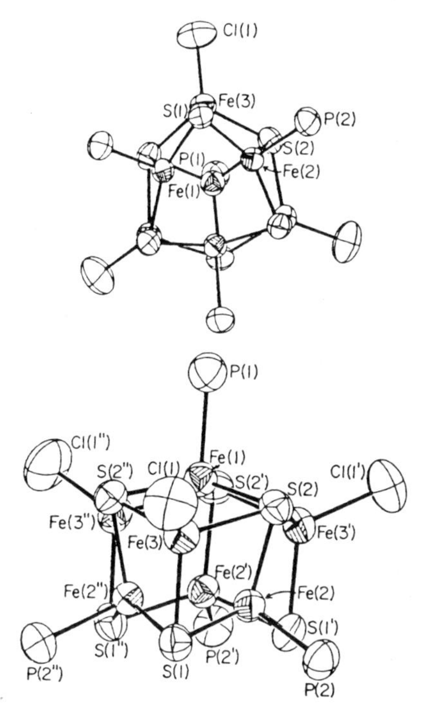

Figure 6.19 Representation of the structure of [$Fe_7S_6(PEt_3)_4Cl_3$] (Noda *et al.*, 1986). (Reprinted by permission from *Inorg. Chem.*, vol. 25, p. 3852. Copyright 1986 ACS.)

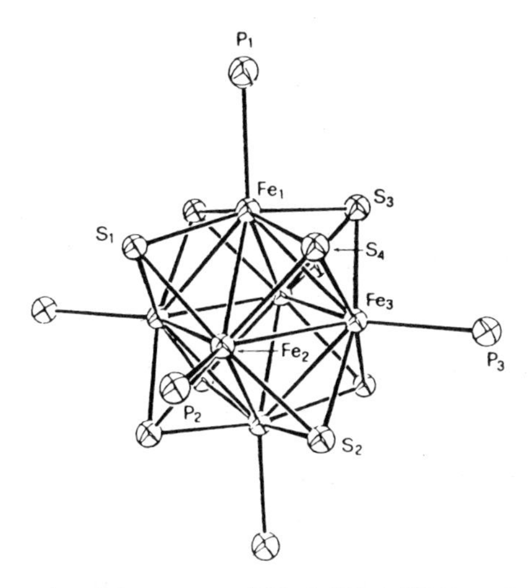

Figure 6.20 Representation of the structure of [$Fe_6S_8(PEt_3)_6$]$^{2+}$ (Agresti *et al.*, 1985). (Reprinted by permission from *Inorg. Chem.*, vol. 24, p. 690. Copyright 1985 ACS.)

1990). The electronic properties of baskets and related species have been discussed (Snyder and Holm, 1990).

Clearly, iron and sulphur can produce a vast range of structures. An eighteen-iron system, essentially a ring of fourteen FeS_2 units, with an interior bridge of $2Fe_2S_4$ units and two sodium cations yields an essentially planar $[Na_2Fe_{18}S_{30}]^{8-}$ (You *et al.*, 1990).

To date, none of these structures has been unequivocally identified in a natural system. However, the e.p.r. spectrum of a six-iron system isolated from *Desulfovibrio vulgaris* has been reported, and it is believed to contain a prismane, $\{Fe_6S_6\}$ (Hagen *et al.*, 1989, 1991). More recently, analogous clusters have been suggested to occur under certain conditions in *Desulfovibrio vulgaris* sulphite reductase (Hagen *et al.*, 1989, 1991), in *Methanotrix soehngenii* carbon monoxide dehydrogenase (Hagen *et al.*, 1989, 1991), in *A. vinelandii* nitrogenase (Hagen *et al.*, 1989, 1991), and in *Desulfovibrio desulfuricans* (Ravi *et al.*, 1991). These are all characterised by an e.p.r. spectrum characteristic of a species with spin 9/2. *Mycobacterium smegmatis* may also contain such a protein (Imai *et al.*, 1991).

6.2.8 *Heterometal iron-sulphido systems*

Mention has already been made of the incorporation of heterometal ions into formal $\{Fe_3S_4\}^0$ clusters, or at faces of prismanes. A specific class of heteroatom cluster is worthy of mention because of its relationship to nitrogenases, even if this is only one of an almost limitless set of metal-sulphur clusters, which, in range, may be comparable to metal-oxygen clusters (metal oxides). Nitrogenases consist of two proteins, one an electron-donor to the other. The electron donor contains an $\{Fe_4S_4\}$ cluster, and the electron acceptor is believed to contain the dinitrogen reducing site. In the molybdenum nitrogenases, this active site is postulated to be a molybdenum atom, which is apparently part of a metal-sulphur cluster of approximate stoichiometry $MoFe_6S_4$ (Eady, 1991). There has been a tremendous effort to prepare clusters approaching this stoichiometry and with nitrogenase activity, though with little success with regard to the latter property. Some of these clusters are described here.

Complexes with three iron atoms and one molybdenum or tungsten atom at the corners of a tetrasulphur cubane were first synthesised by 'spontaneous self-assembly', in the presence of tetrathiomolybdate or tetrathiotungstate to supply the hetero-metal (Holm, 1981; Wolff *et al.*, 1980; Christou and Garner, 1980). Later a similar procedure was developed to prepare vanadium-substituted clusters from $[VS_4]^{3-}$ (Ciurli and Holm, 1989). Typical examples of the first kinds of material reported are $[Mo_2Fe_6S_8(SEt)_9]^{3-}$ and $[Mo_2Fe_7S_8(SEt)_{12}]^{3-}$, with structures as shown in Figure 6.21 (Armstrong *et al.*, 1982). The compounds contain $\{MoFe_3S_4\}$ clusters, with bridges composed of $(SEt)_3$ or $Fe(SEt)_6$, respectively. The clusters undergo ligand substitution reactions

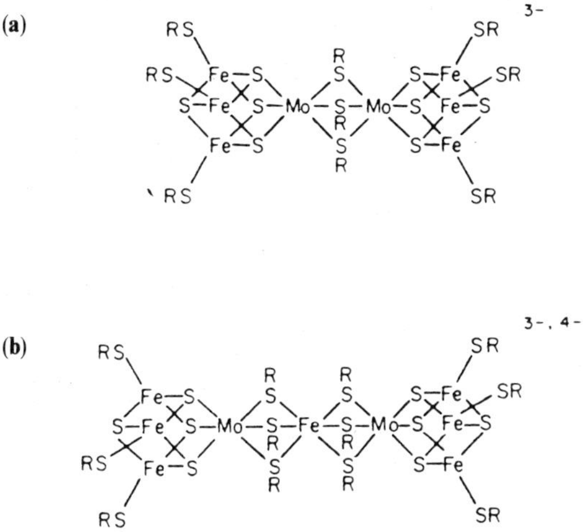

(a)

(b)

Figure 6.21 Representations of the structures of (a) $[Mo_2Fe_6S_8(SEt)_9]^{3-}$ and (b) $[Mo_2Fe_7S_8(SEt)_{12}]^{3-}$ (Armstrong *et al.*, 1982). (Reprinted by permission from *J. Am. Chem. Soc.*, vol. 104, p. 4373. Copyright 1982 ACS.)

(cf. $\{Fe_4S_4\}$), but not at the bridges, only at the terminal ligands (Palermo *et al.*, 1982). All the clusters show at least two reversible redox changes. The bridges can also contain alkoxide groups, and addition of catechols results in complete bridge cleavage yielding monocubanes, via double cubanes of a novel type, as shown in Figure 6.22 and the reaction (Mascharak *et al.*, 1983; Zhang *et al.*, 1987):

$$[Mo_2Fe_7S_8(SEt)_{12}]^{3-} + C_6H_2Pr_2(OH)_2\text{-}3,6$$

$$\xrightarrow{\text{solvent}} [Mo_2Fe_6S_8(SEt)_6(C_6H_2Pr_2O_2)_2]$$

$$\xrightarrow[L]{} [MoFe_3S_4(SEt)_3L(C_6H_2Pr_2O_2)]^{2-}$$

This single cubane type has been isolated and characterised, structurally, chemically and magnetically (Mascharak *et al.*, 1983), and there are significant parallels with the properties of the iron-molybdenum cluster in nitrogenases (Palermo *et al.*, 1984), especially as judged by EXAFS and XANES (Conradson *et al.*, 1985).

More recently, the 'spontaneous self-assembly' of a vanadium cluster $[VFe_3S_4Cl_3(dmf)_3]^-$, a structural model for the core of vanadium nitrogenases, has been reported (Carney *et al.*, 1987; Kovacs and Holm, 1986). This again has EXAFS properties similar to those of the corresponding nitrogenase, and it undergoes a variety of substitution reactions (Figure 6.23) (Kovacs and Holm, 1987). However, despite all the sophisticated analysis, no dinitrogen chemistry has yet to be reported for these clusters, although nitrogenase substrates (including hydrazine) have been persuaded to bridge between

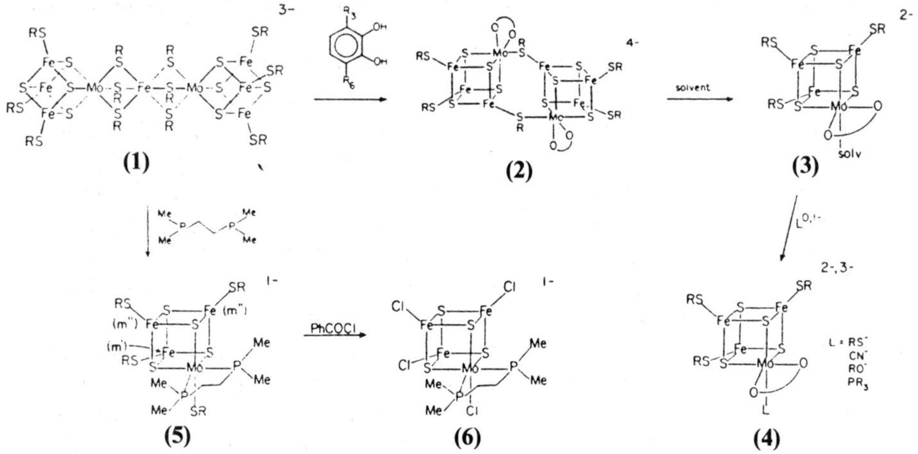

Figure 6.22 Formation of monocubanes by cleavage of heterometalladicubanes (Kovacs and Holm, 1987). (Reprinted by permission from *Inorg. Chem.*, vol. 26, p. 695. Copyright 1987 ACS.)

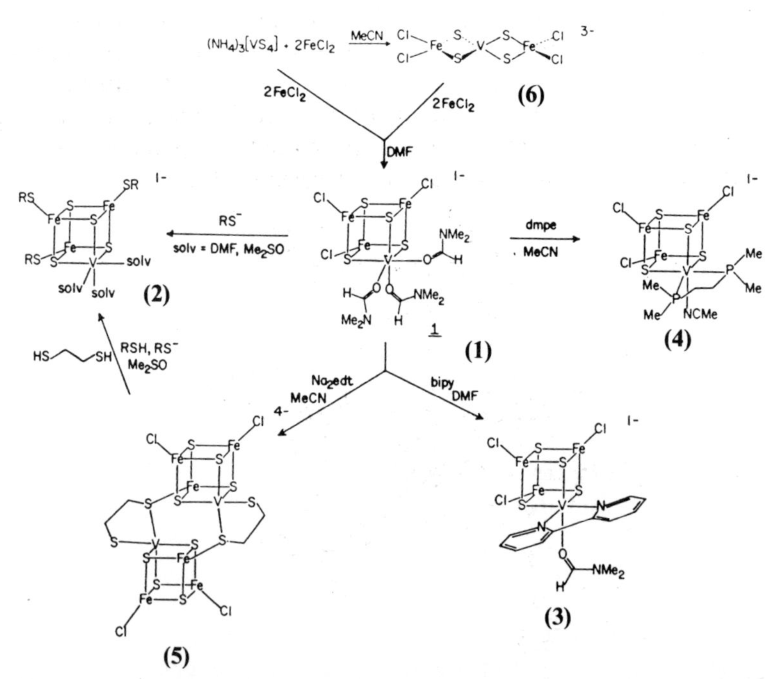

Figure 6.23 Some reactions of iron–vanadium–sulphur clusters (Kovacs and Holm, 1987). (Reprinted by permission from *Inorg. Chem.*, vol. 26, p. 705. Copyright 1987 ACS.)

cubanes, as in $[\{MoFe_3S_4Cl_2(C_6Cl_4O_2)\}_2(\mu\text{-}S)(\mu\text{-}N_2H_4)]^{4-}$ (Challen *et al.*, 1990).

These are not the only structural 'nitrogenase models' in the literature. Mention has already been made of the adducts to the $\{Fe_6S_6\}$ core. This work has been pursued with some vigour, as with the species $[Fe_6S_6(OC_6H_4R)_6\{M(CO)_3\}_2]^{n-}$, where R = Me, OMe, NMe_2, or H, M = Mo or W, and n may be 3 or 4 (Coucouvanis *et al.*, 1988; Al-Ahmad *et al.*, 1990; Kanatzidis *et al.*, 1986). These have quite distinct electronic and structural properties which resemble those of nitrogenases 'only qualitatively'. The heteromonocubanes $\{VFe_3S_4\}^{2+}$ and $\{MoFe_3S_4\}^{3+}$ have been combined into the core cavitand designed to ligate three corners of a cubane (Ciurli and Holm, 1989). This leaves the hetero-atom available for chemistry. To date, little nitrogenase chemistry has been uncovered. Finally, other clusters approaching nitrogenase stoichiometry have been reported. An example is the group $[MoFe_6S_6(CO)_{16}]^{2-}$, $[MoFe_4S_3(CO)_{13}(PEt_3)]^{2-}$, and $[Mo_2Fe_2S_2(CO)_{12}]^{2-}$ (Eldridge *et al.*, 1991). The first has the structure shown in Figure 6.24. It is a very asymmetric structure, and may be considered as being formed from the coordination of three $\{FeS_2Fe\}$ units to a Mo. It does not appear to approach the geometry of molybdenum nitrogenases.

There will doubtless be more structural models. The factors which determine the structures of these complex clusters when they form from a given set of reagents are not understood though basic structural units such as Fe_2S_2 rhombs are often evident. The protein can often impose a structure on the heterometal cluster (as in the three-iron ferredoxin-derived species) which is not thermodynamically favoured in its absence. Consequently, despite the wealth of chemistry accumulating from structural model studies, understanding of the structure of the nitrogenase clusters and of their special properties may become clear only from studies of the nitrogenases themselves.

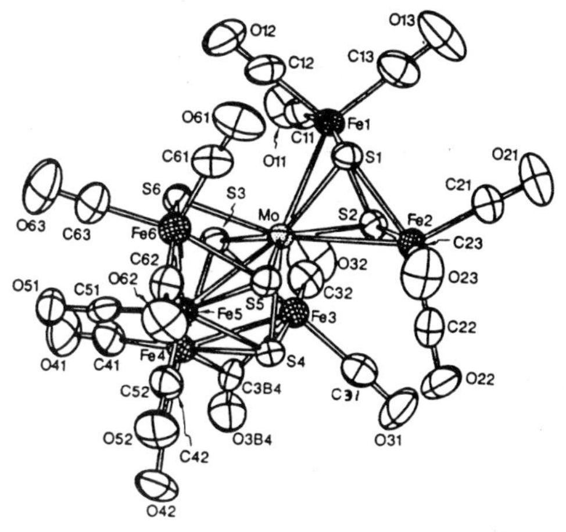

Figure 6.24 Representation of the structure of $[MoFe_6S_6(CO)_{16}]^{2-}$ (Eldridge *et al.*, 1991). (Reprinted by permission from *Inorg. Chem.*, vol. 30, p. 2369. Copyright 1991 ACS.)

6.3 Haem proteins

6.3.1 Introduction

Apart from some interesting but numerically insignificant exceptions, such as organisms living near geothermal outwellings on the floor of the ocean, all life on earth is sustained by energy from the sun. This free (in both senses) energy drives an elaborate redox cycle involving the whole biosphere. In essence the energy of light quanta are used by photosynthetic organisms to oxidise water, releasing oxygen and separating protons from electrons; the latter eventually recombine reducing carbon dioxide and thus fixing it into carbohydrate. In non-photosynthetic organisms (e.g. the animal kingdom) which use oxygen in respiration, this electron acceptor is reduced by electrons donated from ingested carbohydrate, or equivalent, releasing carbon dioxide to the atmosphere and forming water. Each half of this cycle is exergonic and supplies free energy for vital processes such as biosynthesis, movement, etc. The whole redox cycle is, therefore, a transducer permitting efficient conversion of electromagnetic energy into chemical-free energy (Figure 6.25).

To accomplish these redox processes, which involve electron transfer and ligand binding reactions, the living cell has made use of transition metals (most abundantly, but not exclusively, Fe and Cu) coordinated into molecules designed to donate specific ligands to and confer specific geometries upon the central metal atom (see for example Fransto da Silva and Williams, 1991). The most widespread of these transition metal complexes is the haem group comprising a porphyrin moiety and a central iron atom. The porphyrin consists of four linked pyrrole rings, each of which provides a nitrogen ligand to the iron atom which is positioned at the centre of the highly conjugated

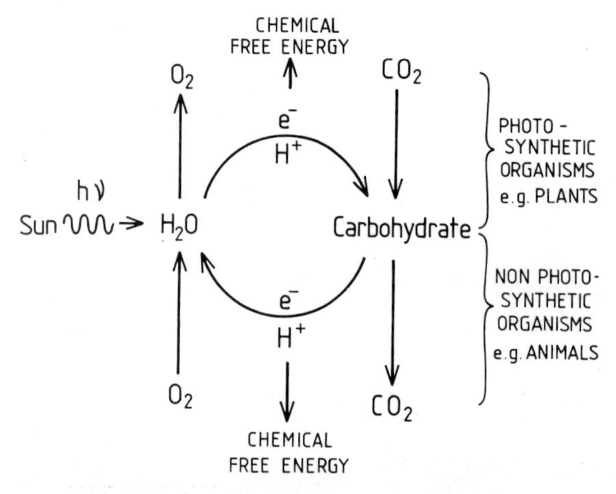

Figure 6.25 The redox cycle of the biosphere.

quasi-planar macrocycle (Dolphin, 1978). This haem group, being hydrophobic, is incorporated into the interior of the protein which provides one and often two further ligands to fill the fifth and sixth coordination positions of the iron atom.

The complex of a haem group and a globular protein, the haemprotein, is wonderfully varied in its properties; properties which, through evolutionary processes, have been tuned to fit the protein to the specific role it is to play in the redox cycle outlined above or to related ligand binding and electron transfer functions. The flexibility and utility of these properties accounts for the ubiquity of haemproteins in nature, being found in all branches of the plant and animal kingdoms.

Selection of the properties of the haem group in association with its protein is achieved in a number of ways, some of the most important of which may be outlined as follows.

6.3.1.1 The chemical structure of the haem group. The structures of some naturally occurring haem groups are given in Figure 6.26. The most widespread of these is termed protoporphyrin IX.Fe or haem b. This is the non-covalently bound prosthetic group of, amongst others, myoglobins, haemoglobins, erythrocruorins (large extracellular respiratory pigments of annelids), the b type cytochromes and peroxidases. In the c type cytochromes the vinyl groups of haem b form thioether bonds with cysteine residues donated by the protein, thus covalently linking the haem group to the protein. This change in ring substituents modifies the affinity of the central iron to ligands, e.g. binding constant of the ferrous form for sulphur ligands is enhanced (Schejter and Plotkin, 1988). Thus c type ferrous haems bind very tightly their *in vivo* methionine ligands. In cytochrome c oxides (cytochrome aa_3) two haem a groups (see Figure 6.26), each bearing a formyl group and a farnesyl side chain,

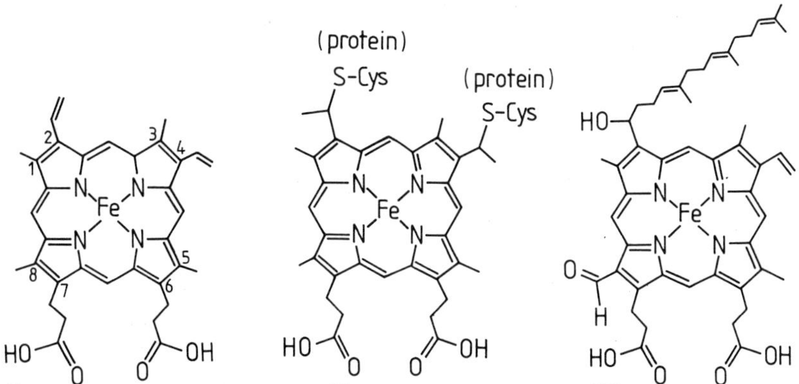

I proto haem IX, haem b   II haem c   III haem a

Figure 6.26 The structures of some commonly occurring haem groups.

are found. Several other substituted or modified haem groups (e.g. haem d of nitrite reductase) are also found in nature.

6.3.1.2 The ligands donated by the protein. The availability of the haem to bind extrinsic ligands such as oxygen and the spin state of the iron are controlled by the ligand set provided by the protein. The redox potential of the haemprotein is also influenced by these ligands. All dioxygen binding haemproteins, which function as oxygen carriers, provide one histidine residue, the N(ε) which acts as the fifth nitrogen ligand to the central ferrous high spin iron, the remaining coordination site being vacant or filled with a weak field ligand. Similarly, other haemproteins which bind oxygen or hydrogen peroxidase as substrate at the iron provide one ligand; examples include a histidine nitrogen in the peroxidases, a tyrosine oxygen in catalase or a cysteine sulphur in cytochrome P_{450}. In this latter case the electron donating properties of the ligands play an important role in oxygen activation. Electron transfer proteins (e.g. the cytochromes) similarly donate one histidine nitrogen but, in general, also provide a sixth, strong field ligand rendering the iron low spin and inaccessible to extrinsic ligands. This sixth ligand may be, for example, a histidine nitrogen as in most b type cytochromes, a methionine sulphur in the c type cytochromes or an amino group as suggested for cytochrome f (see Pettigrew and Moore, 1987 for review).

6.3.1.3 Access to the haem group. Haem proteins control access to the prosthetic group buried within their hydrophobic cores. Thus the rate at which ligands such as dioxygen bind to or dissociate from the central iron atom depends critically upon protein dynamics around the haem and the location of specific amino acid residues; e.g. a specific histidine, the distal histidine, in haemoglobins interacts with iron-bound oxygen strongly influencing its rate of dissociation and hence the molecule's affinity for this ligand (see for example Perutz, 1990a). Similarly, in electron-transfer proteins such as cytochrome c, only the haem edge is exposed to the solvent (Dickerson *et al.*, 1971), allowing facile outer sphere electron transfer to its *in vivo* partners. Access of the bulk phase water to the haem group has, through its dipole interactions, a strong influence on the redox potential (Moore and Rogers, 1985) of the haemprotein and is also controlled by the protein composition and structure.

This chapter briefly examines representative examples of the important classes of haemproteins and the way in which the factors outlined above are important to their function. Apart from the important structural information provided by crystallographic studies much of what is known has come from spectroscopic studies. Because of their iron-containing prosthetic group, haem proteins have a wealth of spectroscopic signatures. All, for example, have characteristic optical spectra, with bands in the 'Soret' region, γ-band ($\sim$390–450 nm) with high extinction coefficients ($\sim 10^5\ M^{-1}cm^{-1}$) and others in the 'visible' (500–600 nm) region, which in low spin ferrous proteins are well resolved into α- and β-bands (Adar, 1979). These absorption bands are sensitive to the redox and ligation state of the iron and allow spectroscopic, ligand binding, and kinetic studies to be performed at low ($\sim 10^{-6}$ M) concentrations (see for example Antonini and Brunori, 1971). In addition ESR, Mössbauer, NMR and MCD spectroscopic methods have all provided extensive and valuable information on the ligand sets and structure of haemproteins together with a knowledge of their spin state and magnetic properties.

6.3.2 *Oxygen carriers: myoglobin and haemoglobin*

Myoglobin comprises a single polypeptide chain containing approximately 150 amino acid residues, depending on species, and an associated haem b. The first X-ray structure of sperm whale myoglobin shows the 153 amino acids of the protein form eight α-helical sections (designated A to H) with short non-helical regions between them (termed AB, CD, etc.) (see Dickson and Geis, 1983, for extensive review of both myoglobin and haemoglobin structures with references). The helices pack together to form essentially a box (44A*44A*25A), hydrophilic externally with a hydrophobic interior into which the haem group fits leaving one edge, bearing the propionate groups, exposed to solvent. The protein donates an imidazole nitrogen of histidine (F8), termed the proximal histidine, as the fifth ligand, the sixth position in the ferrous form being reserved for dioxygen (Figure 6.27a).

In muscle, myoglobin acts as an intermediate store of oxygen to facilitate the diffusion of this ligand to the mitochondria where it is reduced to water in cellular respiration. As such the haem iron is maintained in the ferrous state, ferric iron being unable to bind dioxygen reversibly. Within the hydrophobic pocket surrounding the haem and close to the oxygen binding site is placed the N(ε) of a second histidine residue, the distal histidine, which forms a hydrogen bond with bound oxygen and constrains the Fe–O–O angle to 121°. This binding geometry and the relatively hydrophobic environment, limiting the access of anions (e.g. Cl^-), ensures the iron is not oxidised to the ferric form and the protein rendered functionally useless.

In the unliganded state the haem group is domed, the iron displaced out of the haem plane 0.55 Å towards the proximal histidine and the Fe–N bond makes an angle of 11° with a perpendicular to the haem. On binding dioxygen the iron undergoes a high to low spin transition and moves towards the haem plane.

Although in solution, the combination of oxygen with myoglobin is very fast ($k \sim 10^7\ M^{-1}\ s^{-1}$, Antonini and Brunori, 1971); it appears from the crystal structure that the route to the iron from the aqueous phase is occluded and that the approach of oxygen to its binding site is highly sterically hindered. The resolution of this apparent paradox is to recognise that the structure of the protein in solution is freed from the restraints imposed by intermolecular interactions within the crystal and is continually perturbed from the average structure by thermal motions. Fluorescence quenching studies, initially by Weber (1972), have shown that dynamic fluctuations in protein structure occur on the picosecond (and faster) time scale and this 'breathing' of the molecule, as it is sometimes termed, allows rapid passage of oxygen into the interior of myoglobin.

Myoglobin binds oxygen with relatively high affinity in a process which conforms to a single one-step mechanism, yielding a hyperbolic binding curve (Figure 6.27b) ($K_d = 10^{-6}$ M or P_{50}, partial pressure of O_2 required for half saturation 1 mmHg). With the advent of site-directed mutagenesis it is now

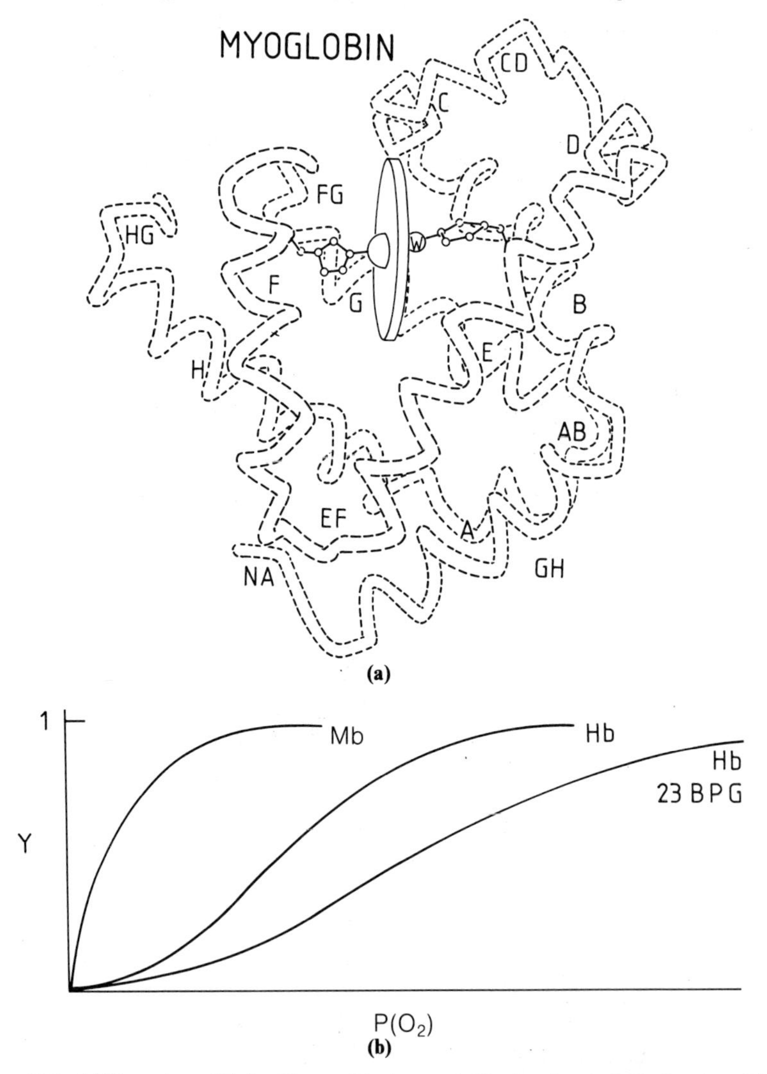

Figure 6.27 (a) The polypeptide backbone of ferric myoglobin showing the helical regions folded to form a hydrophobic cone into which the haem b fits. The histidine F8 provides the fifth ligand. The sixth coordination position is filled by a water molecule in the ferric protein and is empty in the ferrous form, available for O_2 binding. The 'distal' histidine is also shown. Other amino acid side chains (not shown here) completely surround the haem, which is thus hidden from incoming ligands (see text). (b) Schematic representation of the oxygen binding curves for myoglobin (Mb), haemoglobin (Hb) and haemoglobin in the presence of 2.3 DPG. Y is the fractional saturation. (c) Oxygen binding to the R and T state of haemoglobin. (Adapted from Perutz, 1990.) (d) Rotation of the α,β-dimers on transition from the R to T state. (Adapted from Perutz, 1990.)

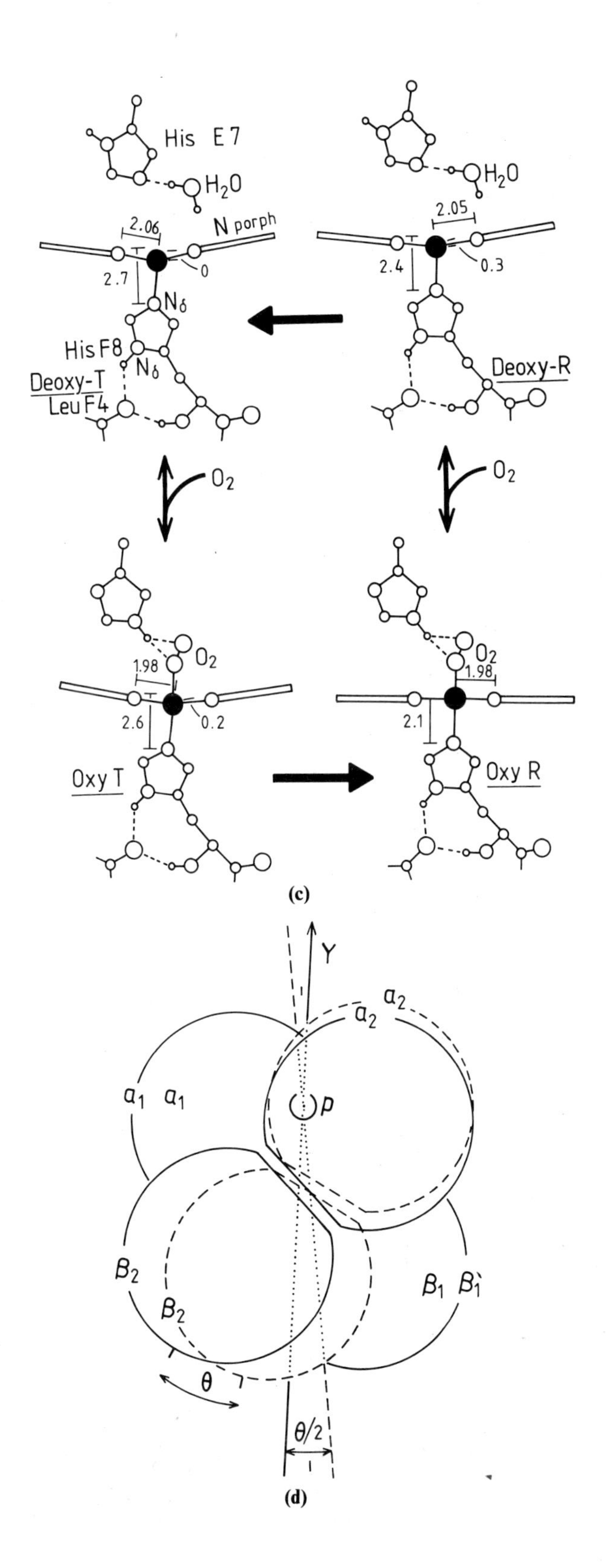

His E7
H_2O
N porph
2.06
2.7
0
N_δ
His F8
N_δ
Deoxy-T
Leu F4
2.05
2.4
0.3
Deoxy-R
O_2
O_2
O_2
1.98
2.6
0.2
Oxy T
O_2
1.98
2.1
Oxy R
(c)
Y
α_2
α_2
α_1 α_1
P
β_2
β_2
β_1 β_1
θ
θ/2
(d)

clear that this affinity (and for haemoglobin also) is highly affected by the nature of a 'distal' amino acid residue which, through its interactions with bound oxygen, controls the oxygen dissociation rate constant. Generally, this residue is a histidine and the O_2 dissociation rate constant is approx. 20 s^{-1} (Antoni and Brunori, 1971). When this is substituted by residues unable to form hydrogen bonds with oxygen, this rate may be as high as 1600 s^{-1} (Olson *et al.*, 1988). It is interesting to note that glutamine may also hydrogen bond to dioxygen and this is the distal residue in elephant myoglobin.

Haemoglobins form a large class of proteins, the function of which is, in general but not exclusively, to carry oxygen in the blood stream from the external environment by way of lung or gill, to the respiring tissues. Other functions include regulation of the gas content of fish swim bladders (Brunori *et al.*, 1973), buoyancy organs, and warming brain cells (in sword fish) through the enthalpy of O_2 dissociation (Gibson and Carey, 1982). Human, and most other vertebrate, haemoglobin is a tetrameric molecule made up of four proteins, 2 α-chains (each 141 amino acids) and 2 β-chains (each 146 amino acids). In fact, it is often better to think of haemoglobin as a dimer of $\alpha\beta$ units. Each chain bears a strong chemical and structural resemblance to myoglobin in which specific residues have been substituted so that intersubunit interactions stabilise the tetramer.

The essential features of oxygen binding are similar to those described for myoglobin except, and most importantly, the movement of the iron and hence proximal histidine which occurs on oxygen binding (see Figure 6.27c) is coupled to the tertiary and quaternary structure of the molecule. It is this coupling which leads to a lower oxygen affinity for oxygen than myoglobin and cooperative oxygen binding manifest in the classic sigmoidal (cooperative) oxygen binding curve exhibited by haemoglobin (Figure 6.27). In solution haemoglobin tetramers exist in an equilibrium between two conformations, the R (relaxed) and T (tense) states; the former has the higher oxygen affinity and is the state almost fully populated in the oxy form (i.e. $Hb(O_2)_4$) while the T-state has lower oxygen affinity and constitutes the form overwhelmingly populated in deoxy haemoglobin (Hb). This lower affinity of the T-state is related to the haem remaining domed on oxygenation due to constraints imposed by the tightly packed side chains of the globin around the haem (Perutz, 1990a, b; Perutz *et al.*, 1987). To a very close approximation it appears that haemoglobin behaves in a 'concerted' manner, the molecule being either wholly in the R- or T-state, with no hybrid forms in which different chains are in different conformations (Perutz *et al.*, 1987). The T- to R-state transition consists of a rotation of the dimer $\alpha_1\beta_1$ relative to dimer $\alpha_2\beta_2$ by $\sim 15°$ and a transition of one dimer with respect to the other of 0.8 Å (Figure 6.27d). The current understanding of the exact stereochemical steps involved in the 'switch' from the T- to R-states on oxygenation have been described in detail elsewhere (Perutz, 1990b; Perutz *et al.*, 1987). Here it will suffice to note that cooperativity arises from the preferential binding of oxygen to, and stabilisation of, the R (high affinity) state. The equilibrium, favouring the T-state in the deoxy form, is thus perturbed on oxygenation in favour of the R-state (Figure 6.27c). Because the transition is cooperative, stabilisation of partially oxygenated tetramers in the R-state brings complete T-state tetramers through the equilibrium to the R-state, thus increasing the proportion of the high affinity sites available for oxygen binding. The consequent shift from low (T) to high (R) affinity through the course of oxygen titrations gives rise to the sigmoidal oxygen equilibrium curve (Figure 6.27b).

This cooperative behaviour ensures that a larger fraction of the binding curve is employed in the transfer of oxygen from the lung to the tissue than would be the case if a myoglobin-like non-cooperative haemoglobin was the oxygen carrier.

The rotation of the α,β-dimers (Figure 6.27d) leads to a relative shift in the two β-chains so that in the deoxy (T) state (but not the oxy R) a pocket is formed into which a molecule of 2-3,bisphosphoglyceric acid (2,3 BPG) may fit, the negative phosphate groups forming salt bridges with the positive α NH_3^+ groups of the β-chains. The binding of this organic phosphate stabilises the T-state relative to the R-state and thus lowers the affinity of the haemoglobin for oxygen, increasing the P_{50} value and thus 'right-shifting' the binding curve. BPG is an important allosteric effector of haemoglobin and its concentration within the red blood cell regulates the oxygen affinity of the erythrocyte. It is depletion of BPG on blood storage which largely makes 'aged' blood unsuitable for transfusion purposes. It is also the relative insensitivity of foetal haemoglobin ($(\alpha.\gamma)_2$) to BPG which ensures the one-way transport of oxygen from mother to foetus by ensuring the maternal blood ($(\alpha.\beta)_2$) has the lower oxygen affinity.

6.3.3 Cytochrome c

Cytochromes of the c type are generally relatively small (molecular weight of horse cyt. *c.* 12 360), water soluble proteins, and it is partly for this reason that they have been so widely studied (see Pettigrew and Moore, 1987, for extensive review). The eukaryote, mitochondrial forms, which are located in the space between the inner and outer mitochrondrial membranes, function to transfer electrons from the bc_l complex to cytochrome c oxidase, but also participate in electron transfer with cytochrome b_5 and sulphite oxidase, in animals, and cytochrome c peroxidase and cytochrome b_2 in yeasts. In all cases they consist of a single polypeptide chain folded to provide a hydrophobic pocket into which the haem group is inserted. Unlike other haemproteins the haem is covalently bound to the protein through thioether linkages formed between the porphyrin vinyls and cysteine sulphydryls close in the primary sequence (-Cys-X-X-Cys). The structures of several mitochondrial and bacterial cytochromes c have been determined by X-ray diffraction, some to high resolution (see Mathews, 1985) (Figure 6.28).

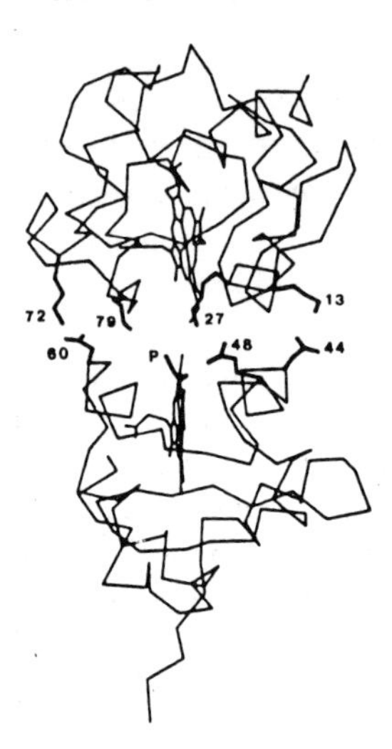

Figure 6.28 Complex between cytochrome c (top) and cytochrome b_5 (bottom). The heavy lines (numbered) denote the interaction residues.

All cytochromes c have histidine (N) and methionine (S) as the fifth and sixth ligands. In the ferric form the methionine may be displaced by ligands with high affinity for ferric iron, e.g. cyanide, but such reactions are slow and require relatively high extrinsic ligand concentrations to compete with the intrinsic methionine. The methione ligand may, however, be replaced by another protein ligand, a lysine, in a pH-dependent transition which in eukaryotes has a pK ~ 9. The function, if any, of this replacement is unclear but the mechanism may involve deprotonation of a specific water molecule buried within the hydrophobic core close to the methionine ligand and which forms a network of hydrogen bonds with a tyrosine and a threonine residue (see Wilson and Greenwood, in press, for review). In the ferrous form the haem iron is very tightly bound by the intrinsic ligands and under normal conditions it is impossible to make adducts of ferrous cytochrome c with added ligands. It is, for example, only at pH values above 12–13 that a CO complex can be formed.

Mammalian cytochromes c are highly basic (PI > 10) due to their high lysine content (horse 19 Lys residues). They have rather positive redox potentials (~ + 260 mV) and are able to participate in facile outer sphere electron transfer through the haem edge (Marcus and Sutin, 1985) which is surrounded by positively charged amino acids (lysines) by which cytochrome c orientates and 'docks' with its *in vivo* redox partners. There is little structural perturbation on changing redox state, the central iron remains low spin and electron transfer with its oxidase (Van Bauren *et al.*, 1974) can be accomplished, at low ionic strength, with rate constants approaching $\sim 10^8\ M^{-1} s^{-1}$. This rate constant is highly salt concentration-dependent reflecting the electrostatic interaction between the proteins which bring the haem c into juxtaposition with the primary electron acceptor(s) in cytochrome oxidase.

6.3.4 b-Type cytochromes

These form a large class of electron transfer proteins all of which have protohaem IX as the non-covalent bound prosthetic groups buried within the hydrophobic interior of their respective proteins.

The protein generally donates two histidine ligands to fill the fifth and sixth coordination positions of the iron but histidine/methionine ligation (cf. cytochrome c) occurs in *E. coli* cytochrome b_{562} (Moore *et al*., 1985) and the b haem site in cellobiose oxidase (Cox *et al*., 1992) from *P. chrysosporium*.

The haem is generally low spin and the optical spectra are characteristic having well defined α and β bands in the reduced form around 560 and 530 nm respectively.

Cytochrome b_5 is one of the most extensively studied of this class of proteins. It exists in a soluble form in erythrocytes where it functions to transfer electrons to haemoglobin keeping the latter in the ferrous state and thus capable of binding oxygen reversibly. An abundant form of cytochrome b_5 can be isolated from microsomes where it functions to transfer electrons to fatty acid desaturase and to the P_{450} enzyme system (see below). It is anchored to the membrane by a hydrophobic tail and may be isolated in a soluble form by tryptic cleavage of the globular haem containing domain from the tail. The haem b containing domain has been crystallised and its structure determined by X-ray diffraction (Argos and Mathews, 1975). Figure 6.28 shows the structure. The haem group is buried within a very hydrophobic pocket of low dielectric constant with one edge having limited exposure to the solvent. One of the haem propionates extends into solution while the other folds back into the protein and may form a weak ionic bond with the ferric ion partially neutralising its charge, thus playing a role in lowering the redox potential to ~ 0.02 V by stabilising the oxidised state.

The exposed haem edge is surrounded by acidic residues and these are thought to interact electrostatically with a reductase. In the case of the mitochrondrial outer membrane b_5 these interactions could be with the positive charges surrounding the haem pocket of cytochrome c (see above), thus bringing the haem groups into suitable relationship for facile outer electron transfer which with cytochrome c proceeds with a rate constant $> 10^7\ M^{-1} s^{-1}$ depending on ionic strength (Strittmatter, 1964; Dingh, 1991).

The b-type cytochromes also play an essential role in energy transduction in the inner mitochrondrial membrane where they act together with cytochrome c to transfer electrons from ubiquinol (UQ) to cytochrome c. In yeast the b,c complex (complex III) containing cytochrome b, cytochrome c and an iron-sulphur protein is composed of 9 subunits. The haem b containing protein has 8 helices which transverse the inner mitochondrial membrane. There are two haems b termed b_{562} (or b_h), a high potential form and b_{566} (or b_l) a low potential haem. These haems are located towards the inner and outer side of the membrane and form a transmembrane electronic circuit between the two redox sites of UQ within the b, c complex. The b_{566} is close to the water/lipid interface on the positive side of the membrane (Tron *et al.*, 1991). Similar electron transfer functions are carried out by the b, f complex (cytochrome b, cytochrome f) in the chloroplast membrane.

6.3.5 *Cytochrome P_{450}*

Cytochrome c P_{450} dependent monooxygenases belong to the class of oxygen activating enzymes which are known to transfer electrons to molecular oxygen thereby catalysing biological oxidations of the type

$$RH + O_2 + NADPH + H^+ \rightarrow ROH + H_2O + NADP^+ \qquad (6.2)$$

They function to hydroxylate a wide range of componds and thus act as an initial step in the detoxification of xenobiotics (e.g. drugs, hydrocarbons) as well as in the *in vivo* metabolism of compounds such as steroids. In mammals they are membrane-bound and their structure is unknown, but a water-soluble bacterial form has been crystallised and its structure determined by X-ray diffraction (Poulos *et al.*, 1988).

The activation of oxygen is required because dioxygen exists in its ground state as a triplet, $(\sigma_s)^2(\sigma_s^*)^2(\sigma_p)^2(\pi_p)^4(\pi_p^*)^1(\pi_p^*)^1$. Its reactions with organic molecules, which do not have unpaired electrons, are spin forbidden and have high activation energy (420–460 kJ/mol) (Malmström, 1982). In addition, single electron reduction of oxygen to superoxide (O_2^-) is an endergonic process ($\Delta G + 58$ kJ mol^{-1}). It is fortunate that these barriers exist, as the overall reduction of oxygen to water is highly favoured energetically, and without these restrictions on oxygen reduction organic matter would spontaneously combust. The enzymes discussed below all activate oxygen by oxygen bond scission and formation of ferryl (FeIV) intermediates.

The term P_{450}, which derives from the absorbance peak at 450 nm in the ferrous CO adduct, encompasses a number of enzymes. They have in common haem b as prosthetic group but have different specificities for substrates (RH in equation 6.2). During the catalytic cycle oxygen binds to ferrous haem b as in myoglobin, but unlike myoglobin and haemoglobin, the proximal ligand donated by the protein is not histidine but a thiolate group of a cysteine residue. It is this unusual ligation which is responsible for the shift of the Soret band in the absorption spectrum of the ferrous CO derivative 450 nm. The thiolate haem complex induces strong π-back donation from the iron to bond dioxygen and is a step towards reductive splitting of the oxygen molecule (Ruckpaul *et al.*, 1989).

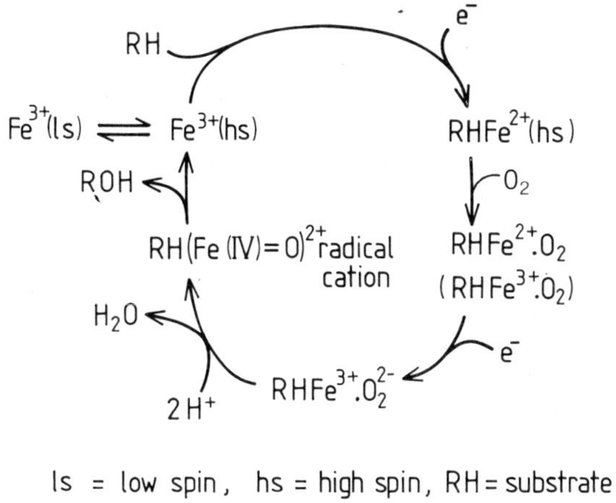

RH
e^-
Fe^{3+}(ls) ⇌ Fe^{3+}(hs)
$RHFe^{2+}$(hs)
ROH
O_2
$RH(Fe(IV)=O)^{2+}$ radical cation
$RHFe^{2+}.O_2$ ($RHFe^{3+}.O_2^-$)
H_2O
e^-
2 H^+
$RHFe^{3+}.O_2^{2-}$

ls = low spin, hs = high spin, RH = substrate

Scheme 6.1 Catalytic cycle P_{450}.

Ferric P_{450} in the absence of substrate exists in a low spin/high spin equilibrium ($K = 0.08$ favouring low spin) characterised by a high ΔS^0 value ($\sim 30\ \mathrm{J\,mol^{-1}\,K^{-1}}$). On binding substrate, at a site close to the haem but not coordinated to it, this equilibrium is shifted in favour of the high spin form. This shift in equilibrium is coupled to a change in the very negative redox potential of the haem to a more positive value (-360 mV to -296 mV, Ristau *et al.*, 1978). This high spin form is stabilised by reduction of the iron by donation of one electron from the second substrate, NADPH (via an associated iron sulphur protein). The ferrous high spin complex (cf. Mb, Hb) can now bind oxygen and after a second electron enters the haem the complex is formally equivalent to a ferric peroxide anion adduct. This step, preceding substrate conversion, is considered to be rate limiting. The oxygen atoms are further apart in this O_2^{2-} adduct than in O_2 and the bond between the oxygen atoms is subsequently split following protonation and water is released. The second oxygen atom, activated in a ferryl (Fe(IV) = O) cation radical complex, is protonated and hydroxylates the substrates. The ferric form of the enzyme is now available for the next cycle of catalysis (see Scheme 6.1).

The essential role played by the proximal sulphur ligand in the control of electron transfer to the haem iron has recently been emphasised by Baldwin *et al.* (1991) (see also Halliwell, 1991). These authors propose (Figure 6.29) that on binding reduced putidaredoxin to ferric P_{450}, the proximal ligand Cys 357 (putida) switches from coordination to the iron to form a covalent bond with tryptophan 106. In a subsequent step the ligand returns to the iron carrying with it the electron delivered to the tryptophan from putidaredoxin and reducing the iron which now binds oxygen. A second electron (rate limiting, see above) transferred by the same mechanism forms the peroxide intermediate and dioxygen bond scission follows. It is suggested that on

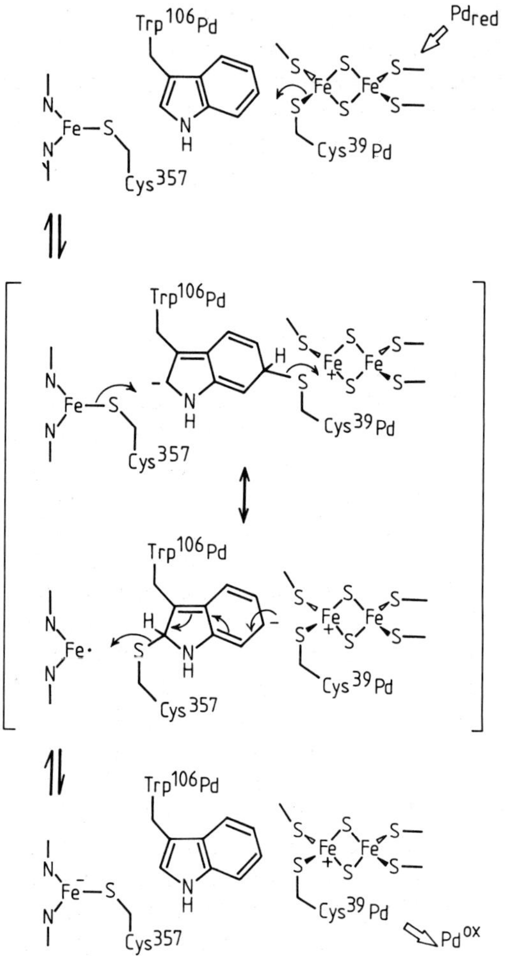

Figure 6.29 (Adapted from Baldwin *et al.* (1991) (see also Halliwell, 1991).)

formation of the activated oxy species (ferryl), Cys 357 dissociates thus protecting the sulphur from oxidation. Once hydroxylation of the substrate has taken place, Cys 357 returns to coordinate the ferric iron. This mechanism provides a switch which when 'on' allows electron flow to the haem and when 'off' insulates it.

The control of access to the haem by substrate binding and by distal haem pocket amino acid residues has recently (Shimizu *et al.*, 1991) been investigated by monitoring the rapid kinetics of CO binding to the ferrous haem of P_{450} from rat liver and genetically engineered mutants (Glu 318 Asp and Thr 322 Ala). Substrates may increase (7,8-benzoflavone) or decrease (phenanthrene) the K_D values for CO indicating they distort the Fe–CO bond or effect CO access. These effects are unaltered by mutation. The CO 'on' rates to the P_{450} substrate complexes are, however, markedly higher in the mutants indicating that in the wild type enzyme the route of CO to the haem is sterically hindered by amino acid residues within the distal pocket. The results of comparable experiments with oxygen as the incoming ligand will be of considerable interest.

6.3.6 *Peroxidases and catalase*

These enzymes use hydrogen peroxide to oxidase substrates, catalase being a special case in which the substrate is itself another H_2O_2 molecule. Peroxidases form a very large class of enzymes found in plants, animals and microorganisms. Horseradish peroxidase and the cytochrome c peroxidase of yeast, for which a high resolution structure is available (Poulos *et al.*, 1980), are perhaps the most widely studied. In mammals thyroid peroxidase catalyses the oxidation of iodide in thyroxin formation and myeloperoxidase is found in neutrophils where it catalyses the formation of oxidised halides, e.g. hypochlorite, which act as antibacterial agents. Peroxidases (lignin peroxidase and manganese dependent peroxidases) also play a central role in lignin (wood) degradation and hence are critically important to carbon flow in the biosphere. The function of catalase is to protect cells from the toxic effect of hydrogen peroxide formed as a consequence of an aerobic lifestyle.

Both peroxidases and catalase contain iron porphyrin IX, haem b. The proximal ligand is a histidine in the peroxidases and a negatively charged tyrosine oxygen in catalase. The classic work of Chance (1951) established that in the catalytic cycle the reaction between ferric peroxidase and hydrogen peroxide, or an organic peroxide, forms an intermediate, compound I, which contains two more oxidising equivalents than the native enzyme. The reducing substrate donates an electron to compound I forming compound II which in turn is reduced by a second substrate molecule reforming the ferric enzyme (Scheme 6.2).

In both compound I and compound II the haem is in the ferryl form [Fe(IV)O]. The extra oxidising equivalent in compound I forms a cation radical either on the porphyrin or on a nearby amino acid residue (e.g. a trytophan in cytochrome c peroxidase). The proximal histidine of peroxidases, unlike the sulphydryl of P_{450} is not strongly electron donating and may not aid oxygen activation. However, the properties of the histidine are apparently different to those of the distal histidine in myoglobin. These differences may be due to hydrogen bonding involving residues in the distal pocket.

The catalytic cycle is similar in catalase, the oxidising equivalent in compound I being located as a radical cation on the porphyrin ring. Reaction of compound I with a second H_2O_2 molecule returns catalase to its original ferri form. The overall reaction catalysed is $2H_2O_2 \rightarrow 2H_2O + O_2$.

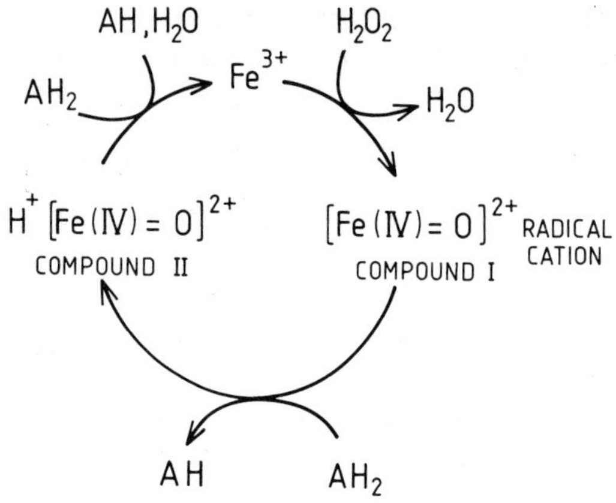

AH, H_2O — H_2O_2

AH_2 → Fe^{3+} → H_2O

H^+ $[Fe(IV)=O]^{2+}$ COMPOUND II — $[Fe(IV)=O]^{2+}$ RADICAL CATION COMPOUND I

AH — AH_2

Scheme 6.2 Catalytic cycle of peroxidases

6.3.7 *Guanylate cyclase*

It has come as a great surprise to most that nature has chosen nitric oxide as a potent 'hormone' involved in, amongst others, the relaxation of vascular smooth muscle (see Snyder and Bredt, 1992 for review). The NO is produced by a specific enzyme, NO synthase, from L-arginine and the receptor for NO is guanylate cyclase, a haem b containing enzyme which catalyses the reaction of GTP (guanidine triphosphate) to cyclic GMP (guanidine monophosphate). It has been suggested that the NO triggers its effect by binding the reduced enzyme to a ferrous haem coordinated, as are most haem proteins, to a histidine residue. Nitric oxide is known to greatly lower the affinity of ferrous haem for histidine and thus on binding NO it is suggested (Traylor and Sharma, 1992) that the proximal histidine dissociates.

This reaction has the following consequences:

(1) It provides a base (uncoordinated histidine) within the haem pocket for catalysis.
(2) It breaks the haem histidine bond, perhaps permitting conformational changes in the rest of the molecule.
(3) It allows the haem to change location within the protein.

Any one of these events could 'switch on' the GTP cyclase activity of the enzyme in response to NO.

6.3.8 *Cytochrome c oxidase*

This enzyme is the central enzyme in cellular respiration acting as the terminal electron acceptor of the mitochondrion, the inner membrane of which it spans. It functions to reduce oxygen to water and to transduce the free energy of this highly exergonic reaction into a form which may be used by the cell. To do this, the enzyme is responsible (with other complexes in the membrane) for building an electrochemical potential gradient of protons across the membrane.

The overall reaction catalysed by the enzyme is:

$$4Cytc^{2+} + O_2 + 8H^+_{in} \rightarrow 4Cytc^{3+} + 2H_2O + 4H^+_{out}$$

where the suffixes 'in' and 'out' refer to the inside and outside of the inner mitochondrial membrane.

The enzyme accomplishes the truly amazing multiple tasks of:

(1) coupling the single electron donor, cytochrome c, to the four-electron acceptor oxygen;
(2) activating oxygen for reduction (see above) but ensuring that none of the potentially harmful intermediates (e.g. O_2^-, H_2O_2, $OH^{\cdot}$) escapes the active site into solution;
(3) ensuring that the substrate (scalar) protons are taken up inside the mitochrondrion;
(4) pumping (on average) one proton from the inside to the outside of the mitochondrial membrane for each electron passed to oxygen.

The enzyme contains two identical, non covalently bound haem a groups (Figure 6.26) which, however, inhabit environments within the protein which confer distinct properties on them. One is low spin, hexacoordinate with two histidine nitrogen ligands provided by the protein and functions as an electron transfer complex termed cytochrome a. The other termed cytochrome a_3 is high spin, having only a single histidine nitrogen as ligand. The enzyme also contains copper atoms termed Cu_A and Cu_B. The former acts also as an electron transfer site which, like cytochrome a, passes electrons to the oxygen binding site (see Brunori *et al.*, 1988; Babcock and Wikström, 1992 for review). Cu_A until recently was considered to be a single copper atom coordinated, probably, to two SH, and two nitrogen (histidine) ligands. This may well be the case but a new interpretation of its EPR spectrum and comparison with nitrous oxide reductase has suggested that it may be a mixed valance, CuI/CuII, site (Kroneck *et al.*, 1988).

The other copper, Cu_B, together with the haem of cytochrome a_3 forms a magnetically coupled binuclear centre, displaying no EPR signal despite the metals being in their oxidised state, which acts to bind and reduce oxygen.

The mammalian enzyme comprises 13 protein subunits but the redox active metals are all located in two of the largest, hydrophobic, mitochrondrially coded, proteins. The bacterial enzymes have fewer subunits and there is strong homology between their metal bearing subunits and those from mammalian sources. The structure of the enzyme is not known, but a wide range of methods involving electronmiscroscopy, protein chemistry and spectroscopy together with site-directed mutagenesis have given a general picture of the metal sites and their distribution (Figure 6.30) (see Brunori *et al.*, 1988; Babcock and Wikström, 1992 for reviews).

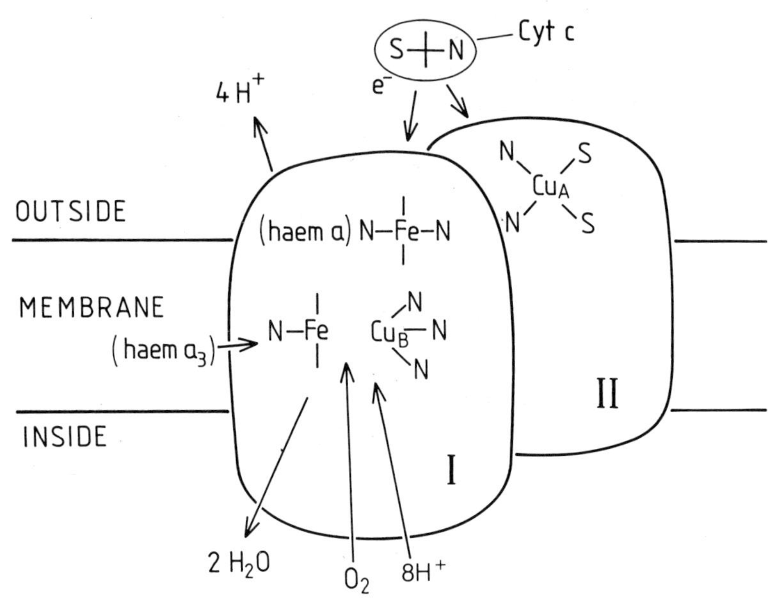

Figure 6.30 Diagrammatic representation of the metal centres within subunits I and II of mammalian cytochrome c oxidase.

Electrons enter the complex through cytochrome a and/or Cu_A with which it is in rapid electron transfer equilibrium and then pass to the binuclear centre in what is considered to be the rate limiting step (however, see Morgan and Wikström, 1991, for an alternative view). Oxygen binds to the binuclear centre only when both haem a_3 and Cu_B are reduced. This mechanistic device ensures that once oxygen binds to ferrous haem a_3 (cf. Mb), it rapidly receives two electrons, one from each metal, trapping it as a tightly bound peroxide and bypassing the thermodynamically unfavourable one electron addition which forms superoxide. Subsequent additions, of a third electron, and protonation lead to oxygen bond scission and formation of water (cf. P_{450}), probably retained bound to Cu_B. The remaining activated oxygen is bound to ferryl iron (Fe(IV) = 0). The haem a_3 is not oxidised to a radical cation, this role of electron addition having been taken by the nearby Cu_B atom. The transfer of a fourth electron, from either cytochrome a or Cu_A, to the binuclear centre and proton uptake forms the second molecule of water and returns the binuclear centre to its initial ferric/cupric state (Scheme 6.3).

Transfer of the third and fourth electrons to the peroxy and ferryl complexes are highly exergonic processes and recent data have provided evidence that each of these electron transfer reactions is coupled to vectoral proton translocation across the membrane (Wikström, 1989). The mechanism of this coupling is still to be elucidated, but it is this vectoral transport which builds a pH and electrical potential gradient across the membrane which is

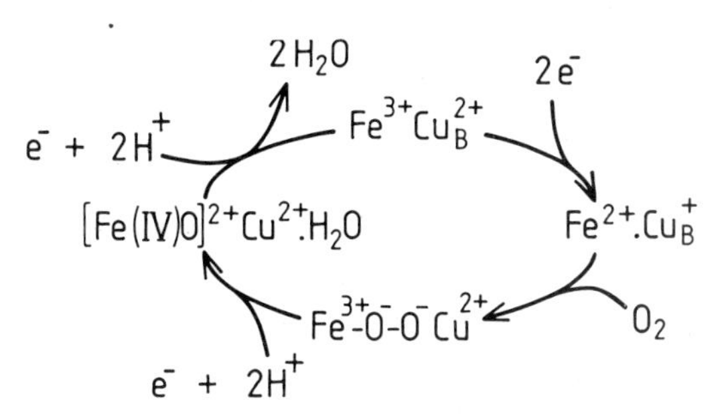

Scheme 6.3 Catalytic cycle of cytochrome c oxidase

subsequently used to form ATP and hence couple the ΔG rich process of oxygen reduction to free energy-requiring life processes.

6.4 Non-haem and non-Fe/S proteins

6.4.1 Introduction

Proteins containing haem iron or Fe/S centres represent two major classes of iron-containing proteins, but there are many others that do not fall into either of these categories. These are the subject of this section. They will not be discussed in great detail; rather the aim here is to briefly consider a limited number of proteins containing either monomeric or dimeric iron centres, and then to review the general chemistry of non-haem iron centres in proteins.

This is an extremely active area of current research, as indicated by the recent review of certain aspects of iron biochemistry given by da Silva and Williams (1991), and by the presentation of 55 posters dealing with non-haem and non-Fe/S proteins at the ICBIC 5 conference held in August 1991 (*J. Inorg. Biochem.*, **43**, 75–716).

6.4.2 Structurally characterised proteins containing monomeric Fe centres

Apart from the iron-transporting transferrins, which are considered in section 6.1.3, these are only two kinds of structurally characterised protein containing monomeric non-haem and non-Fe/S iron centres: superoxide dismutase and dioxygenases.

Superoxide dismutase catalyses the following reaction:

$$2O_2^{-} + 2H^+ \rightarrow H_2O_2 + O_2 \qquad (6.3)$$

The iron, which in the active form of the enzyme is high-spin Fe^{3+}, is bound to the protein by side chains of three histidines and one aspartate in a distorted tetrahedral coordination (Carlioz *et al.*, 1988; Stoddard *et al.*, 1990). The reaction cycle for the enzyme is probably relatively simple with the first O_2^- reducing the iron to Fe^{2+} and the second O_2^- re-oxidising it to Fe^{3+} (Fridovich, 1989).

The dioxygenases are more complex enzymes. These couple the cleavage of O_2 with the incorporation of both O atoms into organic substrates. The best characterised example is the protocatechuate-3,4-dioxygenase from *Pseudomonas aeruginosa*. The protein is a complex oligomer composed of six $\alpha_2\beta_2$ tetramers. Each β-subunit contains iron bound to two tyrosines, two histidines and one H_2O molecule (Ohlendorf *et al.*, 1988).

In the active enzyme the iron is high-spin Fe^{3+} (Whittaker and Lipscomb, 1984, and references therein). The mechanism of reaction is not clear (Que *et al.*, 1977; Whittaker and Lipscomb, 1984) but it proceeds by the substrate binding to the Fe^{3+}, probably with displacement of the coordinated H_2O. The Fe^{3+} is then proposed to act as a Lewis acid and to make the carbon at the 4-position susceptible to nucleophilic attack by O_2. The resulting peroxy group is bound to the Fe^{3+}. Cleavage of the peroxy O–O bond then occurs to generate a Fe^{3+}–OH which attacks the carbon at the 4-position. Note that throughout this proposed mechanism the redox state of the iron does not change.

6.4.3 *Proteins containing Fe–O–Fe units*

An Fe–O–Fe centre was first detected in the invertebrate O_2 transport protein haemerythrin. The protein provides seven amino acid residues that bind the unit (Figure 6.31) which, in its fully reduced state, is capable of binding O_2 to generate a diferric-peroxide complex (Figure 6.31) (Sheriff *et al.*, 1987; Lippard, 1988). One of the unanswered questions concerning this complex is: why does

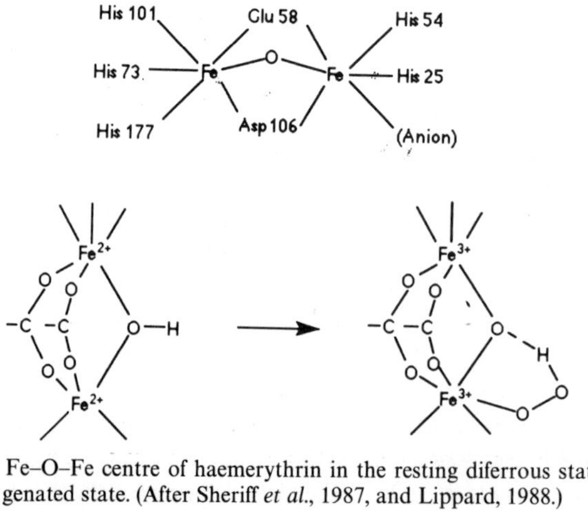

Figure 6.31 The Fe–O–Fe centre of haemerythrin in the resting diferrous state and the oxygenated state. (After Sheriff *et al.*, 1987, and Lippard, 1988.)

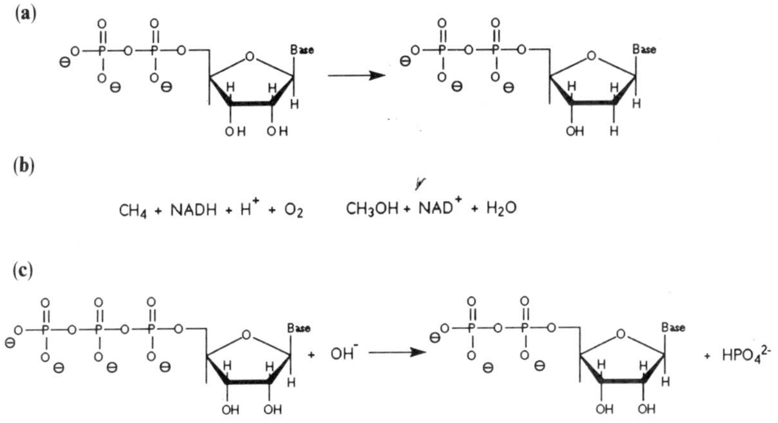

(a)

(b)

$CH_4 + NADH + H^+ + O_2$ $CH_3OH + NAD^+ + H_2O$

(c)

Figure 6.32 Reactions catalysed by: (a) ribonucleotide reductase; (b) methane monoxygenase; (c) purple acid phosphatases.

it regenerate the diferrous protein by releasing O_2 rather than oxidising the protein to its diferric state? In light of the fact that the Fe–O–Fe unit is used in a variety of enzymes, it will be of interest to see how the redox properties of haemerythrin compare with those of its related enzymes.

It is becoming clear that the Fe–O–Fe unit is a common and versatile catalytic centre. It is present in, among others, ribonucleotide reductase (Nordlund *et al.*, 1990), methane monooxygenase (Ericson *et al.*, 1988; Fox *et al.*, 1988), and certain purple acid phosphatases, such as uteroferrin (Averill *et al.*, 1987; True *et al.*, 1991). The reactions catalysed by these enzymes are given in Figure 6.32. Only one of them has been characterised in detail by X-ray crystallography–ribonucleotide reductase–and its Fe–O–Fe centre is illustrated in Figure 6.33. The similarity with haemerythrin is striking, though there are significant differences; for example, in the number of histidine ligands to each iron, and in the fact that there is only one, instead of two, bridging carboxylates. The ligands to the Fe–O–Fe unit of utoferrin are also different to those of haemerythrin. Spectroscopic data indicate that this protein contains tyrosinate ligands to at least one of the irons in addition to the histidine ligation to both irons (Averill *et al.*, 1987).

The mechanisms of reaction for these enzymes have not been defined. However, some progress has been made with ribonucleotide reductase. The *E. coli* enzyme is an $\alpha_2\beta_2$ tetramer: the α_2 unit is called protein B1 or R1, and the β_2 unit protein B2 or R2 (Reichard, 1988; Nordlund *et al.*, 1990). Protein B1 contains three groups of redox active cysteines that either interact with the reductants thioredoxin and glutaredoxin, or interact with substrate (Nordlund *et al.*, 1990). Protein B2 contains Fe–O–Fe units, one in each β subunit. Associated with each Fe–O–Fe unit is an extremely stable, and neutral,

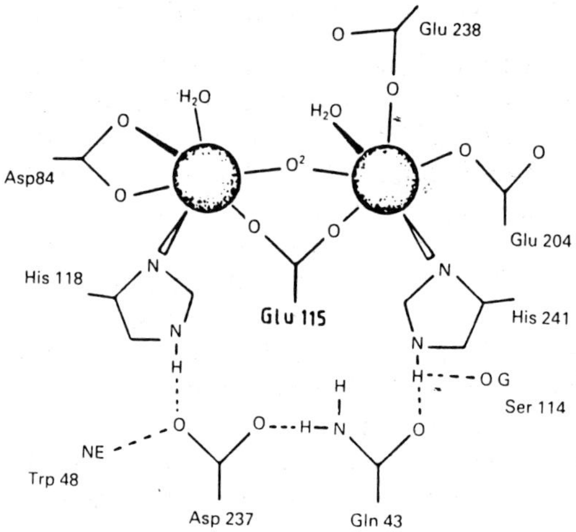

Figure 6.33 The Fe–O–Fe centre of ribonucleotide reductase. (Reproduced from Nordlund *et al.*, 1990.)

tyrosine free-radical. This is located at position 122, only ~5 Å from the nearest Fe^{3+}. The rolc of the diferric Fe–O–Fe centre appears to be to stabilise the tyrosine free-radical. This state of the enzyme is generated by the reaction of O_2 with the diferrous enzyme. The mechanism of this reaction is unclear (Stubbe, 1989; Que and True, 1990). A μ-peroxodiferric intermediate has been proposed, as have high-valent iron species. Nordlund *et al.* (1991) point out that the available X-ray structure is for the diferric form of the enzyme and they indicate that the diferrous form may be different by analogy with the dimanganous protein B2 which contains an additional carboxylate bridge between the Mn^{2+} ions in place of an oxo bridge.

Ashley and Stubbe (1985) have proposed that once the tyrosine radical is formed, it abstracts a hydrogen atom from the 3′-position of the ribonucleotide to generate a ribonucleotide radical. Protonation of the 2′-hydroxyl followed by elimination of H_2O leads to a ribonucleotide cation radical that, via a series of further radical species, is converted to the deoxyribonucleotide. This mechanism is consistent with chemical data (Ashley and Stubbe, 1985; Reichard, 1988).

A free-radical reaction pathway also appears likely for the key step in methane monooxygenation (Figure 6.32). The enzyme system catalysing this reaction consists of three proteins: a reductase; a hydroxylase, which is the component containing the Fe–O–Fe unit; and a regulatory protein (Colby *et al.*, 1977; Lund *et al.*, 1985). A plausible mechanism involves a $Fe^{4+}{=}O$ species abstracting a hydrogen atom from CH_4 to generate $CH_3^{\bullet}$, and a hydroxyl radical attached to the iron centre reacting with $CH_3^{\bullet}$ to form methanol.

6.4.4 *Structurally uncharacterised iron-containing proteins*

There are a large number of iron-containing proteins whose iron centres have not been structurally defined. These range from mononuclear iron proteins such as lipoxygenase (which catalyses the oxygenation of unsaturated fatty acid (Vliegenthart and Veldink, 1982), and whose iron may be coordinated by 4 histidines and 2 oxygen donors (Navaratnam *et al.*, 1989)), and the ferroquinone complex of photosystem II (which acts in electron transfer (Petrouleas *et al.*, 1991)), to proteins containing binuclear Fe/Zn reaction centres, such as red kidney bean purple acid phosphatase (Beck *et al.*, 1986; Körner *et al.*, 1991; Priggemeyer *et al.*, 1991). This latter protein catalyses the hydrolysis of activated phosphoric acid monoesters.

Probably when these and other non-haem and non-Fe/S iron-containing proteins have been structurally characterised it will be found that they exhibit a wider variety of structural types than do the entire family of Fe/S proteins. Partly because of limitations of space further consideration of structurally uncharacterised iron-containing proteins will be restricted to two groups of proteins that require Fe^{2+} for activity. Examples of the reactions catalysed by these groups of proteins are given in Figure 6.34.

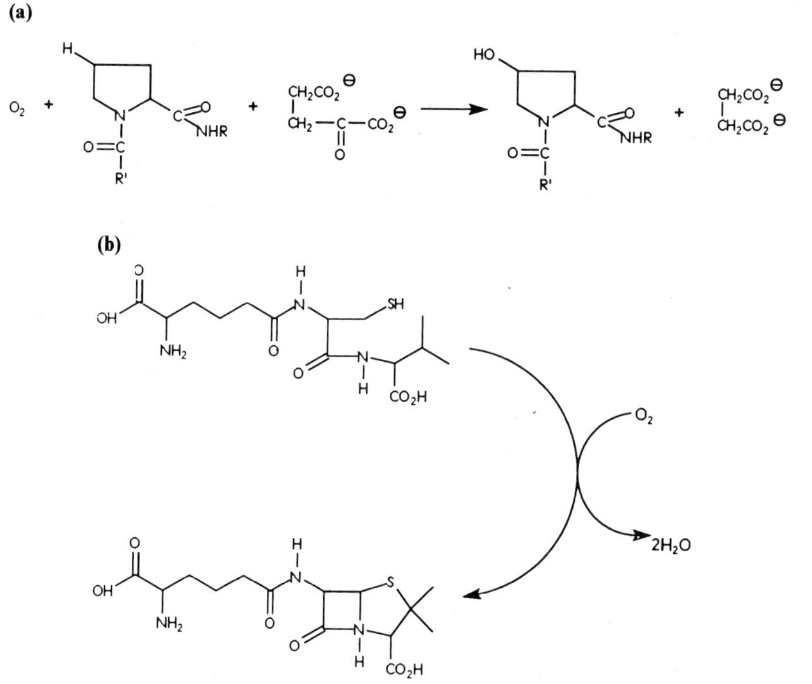

Figure 6.34 Reactions catalysed by: (a) prolyl-4-hydroxylase (R′ and R represent the remainder of the protein); (b) isopenicillin-N-synthetase.

The reaction of prolyl hydroxylase, as with that of the related lysyl hydroxylase, require α-ketoglutarate. This is stoichiometrically decarboxylated with one atom of the O_2 being incorporated into the resulting succinate and the other atom into the hydroxyl group of 4-hydroxyproline (or hydroxylysine). The reaction starts with the ferrous enzyme which reacts with α-ketoglutarate and O_2 to generate an $Fe^{4+}{=}O$ species which hydroxylates the substrate (Majamaa *et al.*, 1984, and references therein).

Isopenicillin N synthase (IPNS) catalyses a key step for the biosynthesis of penicillin. Its mononuclear iron atom, which spectroscopic data indicate may be coordinated to three histidines, one aspartate and one H_2O in the resting Fe^{2+} state (Chen *et al.*, 1989; Fujishima *et al.*, 1991; Ming *et al.*, 1990, 1991), coordinates the substrate via the sulphur, probably by displacing one of the amino acid ligands. Then reaction with O_2 leads to the double ring closure with the ferryl group being a key intermediate. The oxygen atoms end up in H_2O (Baldwin and Abraham, 1988).

6.4.5 General aspects of non-haem iron centres in proteins

There are three general aspects of the non-haem and non-Fe/S iron containing proteins that merit further consideration: their binding affinities for iron, their involvement in hydrolytic reactions, and the production of the ferryl ion.

Fe^{2+} tends to have rather low binding constants in comparison to those of Fe^{3+} (Williams, 1982), and though tightly bound Fe^{2+} requiring proteins do exist (e.g. lipoxygenase in some of its reactions (Vliegenthart and Veldink, 1982; Slappendel *et al.*, 1982)) many such proteins bind Fe^{2+} weakly (e.g. IPNS (Baldwin and Bradley, 1990)). These often lose their iron during isolation and thus for *in vitro* experiments need Fe^{2+} to be added. This weak binding of Fe^{2+} may be a control feature (Williams, 1982) (see section 6.1). Fe^{2+} present in mixed valent or diferrous Fe–O–Fe centres appears to be relatively tightly bound because Fe^{2+} does not readily dissociate from such enzymes.

Iron is used in some proteins as a Lewis acid catalyst; e.g. in uteroferrin, which has maximal activity in the mixed valent Fe^{2+}–O–Fe^{3+} form (Vincent *et al.*, 1992); in the Fe^{3+}/Zn^{2+} phosphatases; and in the dioxygenases described in section 6.2. Fe^{3+} is a good Lewis acid catalyst because it has both a strong polarising tendency and relatively rapid ligand exchange kinetics (Crumbliss and Garrison, 1988). However, Zn^{2+} and Mg^{2+} are the common Lewis acid catalysts in biology and da Silva and Williams (1991) speculate that Fe^{3+} is employed in acidic conditions when the dipositive cations will not be so effective.

The ferryl ion is a powerful oxidant. The Fe^{4+}/Fe^{3+} reduction potential is estimated to be ~1 volt by da Silva and Williams (1991). This is far higher than that of most oxidants in biology and thus the ferryl ion is not particularly common in proteins. Where it is proposed to occur it is produced by the action of H_2O_2 on ferrihaem proteins (e.g. with catalase and peroxidase – see section

6.3.6) or O_2 on non-haem Fe^{2+} proteins, such as IPNS and prolyl hydroxylase. In both cases the ferryl ion is proposed to be in combination with O^{2-} as $(Fe{=}O)^{2+}$. Why O_2 reacts with some Fe^{2+} proteins to create $(Fe{=}O)^{2+}$, and with others to give a dioxygen ferrous ion complex or a ferric protein plus H_2O or HO_2, is not known.

Acknowledgements

G.R. Moore wishes to thank the Wellcome Trust and the TRAMPS panel of the Science and Engineering Research Council for supporting his work on iron metabolism, and Professors A.J. Thomson (UEA, Norwich) and P.M. Harrison (Sheffield) and Dr A.G. McEwan (UEA, Norwich) for many helpful discussions.

References

Adar, F. (1979) in *The Porphyrins*, Vol. III, (ed. D. Dolphin), Academic Press, New York, Chapter 2.

Aisen, P., Leibman, A. and Zweier, J. (1978) Stoichiometric and site characteristics of the binding of iron to human transferrin. *J. Biol. Chem.*, **253**, 1930–1937

Aisen, P. and Listowsky, I. (1980) Iron transport and storage proteins. *Ann. Rev. Biochem.*, **49**, 357–393

Agresti, A., Bacci, M., Cecconi, F., Ghilardi, C.A. and Midollini, S. (1985) *Inorg. Chem.*, **24**, 689–695.

Al-Ahmad, S.A., Salifoglou, A., Kanatzidis, M.G., Dunham, W.R. and Coucouvanis, D. (1990) *Inorg. Chem.*, **29**, 927–938.

Al-Ahmad, S.A., Kampf, J.W., Dunham, R.W. and Coucouvanis, D. (1991) *Inorg. Chem.*, **30**, 1163–1164.

Anderson, B.F., Baker, H.M., Norris, G.E., Rice, D.W. and Baker, E.N. (1989) Structure of human lactoferrin: crystallographic structure analysis and refinement at 2.8 Å resolution. *J. Mol. Biol.*, **209**, 711–734.

Anderson, B.F., Baker, H.M., Norris, G.E., Rumball, S.V. and Baker, E.N. (1990) Apolactoferrin structure demonstrates ligand-induced conformational change in transferrins. *Nature*, **344**, 784–787.

Andrews, S.C., Brady, M.C., Treffry, A., Williams, J.M., Mann, S., Cleton, M.I., de Bruijn, W. and Harrison, P.M. (1988) Studies on haemosiderin and ferritin from iron-loaded rat liver. *Biol. Metals*, **1**, 33–42.

Antonini, E. and Brunori, M. (1971) *Hemoglobin and Myoglobin in their Reactions with Ligands*, North-Holland, Amsterdam.

Argos, P. and Mathews, F.S. (1975) *J. Biol. Chem.*, **250**, 747–751.

Armstrong, W.H., Mascharak, P.K. and Holm, R.H. (1982), *J. Am. Chem. Soc.*, **104**, 4373–4383.

Ashley, G.W. and Stubbe, J. (1985) Current ideas on the chemical mechanism of ribonucleotide reductases. *Pharmacol. Ther.* **30**, 301–329.

Averill, B.A., Davis, J.C., Burman, S., Zirino, T., Sanders-Loehr, J., Loehr, T.M., Sage, J.T. and Debrunner, P.G. (1987) Spectroscopic and magnetic studies of the purple acid phosphatase from bovine spleen. *J. Am. Chem. Soc.*, **109**, 3760–3767.

Babcock, H.T. and Wikström, M. (1992) *Nature*, **356**, 301–309.

Backes, G., Mino, Y., Loehr, T.M., Meyer, T.E., Cusanovich, M.A., Sweeney, W.V., Adman, E.T. and Sanders-Loehr, J., (1991) *J. Am. Chem. Soc.*, **113**, 2055–2064.

Bagg, A. and Neilands, J.B. (1987a) Ferric uptake regulation protein acts as a repressor, employing iron(II) as a cofactor to bind the operator of an iron transport operon in *Escherichia coli*. *Biochemistry*, **26**, 5471–5477.

Bagg, A. and Neilands, J.B. (1987b) Molecular mechanism of regulation of siderophore-mediated iron assimilation. *Microbiol. Rev.*, **51**, 509–518.

Bailey, S., Evans, R.W., Garratt, R.C., Gorinsky, B., Hasnain, S., Horsburgh, C., Jhoti, H., Lindley, P.F., Mydin, A., Sarra, R. and Watson, J.L. (1988) Molecular structure of serum transferrin at 3.3 Å resolution. *Biochemistry*, **27**, 5804–5812.

Baker, E.N., Rumball, S.V. and Anderson, B.F. (1987) Transferrins: insights into structure and function from studies on lactoferrin. *Trends Biochem. Sci.*, **12**, 350–353.

Baldwin, J.E. and Abraham, E.P. (1988) *Nat. Prod. Rep.*, **5**, 129–145.

Baldwin, J.E. and Bradley, M. (1990) Isopenicillin N synthase: mechanistic studies. *Chem. Revs.*, **90**, 1079–1088.

Baldwin, J.E., Morris, G.M. and Richards, W.G. (1991) *Proc. R. Soc. Lond.*, **B245**, 43–51.

Bauminger, E.R., Harrison, P.M., Nowik, I. and Treffry, A. (1989) Mössbauer spectroscopic study of the initial stages of iron-core formation in horse spleen apoferritin: evidence for both isolated Fe(III) atoms and oxo-bridged Fe(III) dimers as early intermediates. *Biochemistry*, **28**, 5486–5493.

Bauminger, E.R., Harrison, P.M., Hechel, D., Nowik, I. and Treffry, A. (1991) Iron(III) can be transferred between ferritin molecules. *Proc. R. Soc. Lond. B*, **244**, 211–217.

Beck, J.L., McConachie, L.A., Summors, A.C., Arnold, W.N., de Jersey, J. and Zerner, B. (1986) Properties of a purple phosphatase from red kidney bean: a zinc-iron metalloenzyme. *Biochim. Biophys. Acta*, **869**, 61–68.

Beinert, H. and Kennedy, M.C. (1989) Engineering of protein bound iron-sulfur clusters. *Eur. J. Biochem.*, **186**, 5–15.

Bell, S.H., Weir, M.P., Dickson, D.P.E., Gibson, J.F., Sharp, G.A. and Peters, T.P. (1984) Mössbauer spectroscopic studies of human haemosiderin and ferritin. *Biochim. Biophys. Acta*, **787**, 227–236.

Brock, J.H. (1985) Transferrins, in *Metalloproteins, Part 2.* (ed. P.M. Harrison), Verlag-Chimie, Weinheim, pp. 183–262.

Blank, H.L., Kierit, O., Roth, E.K.H., Jordanov, J., van der Linden, J.G.M. and Steggerda, J.J. (1991) *Inorg. Chem.*, **30**, 3231–3234.

Bonomi, F., Werth, M.T. and Kurtz, D.M., (1985) *Inorg. Chem.*, **24**, 4331–4335.

Brunner, H., Merz, A., Pfauntsch, J., Serhadle, O., Wachter, J. and Ziegler, M.L. (1988). *Inorg. Chem.*, **27**, 2055–2058.

Brunori, M., Bonaventura, J., Bonaventura, C., Giardina, B., Bossa, F. and Antonini, E. (1973) *Mol. and Cell. Biochem.*, **1**, 189–196.

Brunori, M., Antonio, G., Malatesta, F., Sarti, P. and Wilson, M.T. (1988) *Adv. Inorg. Biochem.*, 7, 93–153.

Bullen, J.J. and Griffiths, E. (1987) *Iron and Infection*, Wiley, New York.

Butt, J.N., Armstrong, F.A., Breton, J., George, S.J., Thomson, A.J. and Hatchikan, E.C. (1991) *J. Am. Chem. Soc.*, **113**, 6663–6670.

Carney, M.J., Kovacs, J.A., Zhang, Y-P., Papaefthymiou, G.C., Spartalian, K., Frankel, R.B. and Holm, R.H. (1987). *Inorg. Chem.*, **26**, 719–724.

Carlioz, A., Ludwig, M.L., Stallings, W.C., Fee, J.A., Steinman, H.M. and Touati, D. (1988) Iron superoxide dismutase: Nucleotide sequence of the gene from *Escherichia coli* K12 and correlation with crystal structures. *J. Biol. Chem.*, **263**, 1555–1562.

Carter, C.W., Krant, J., Freer, S.T. and Alden, R.A. (1974) *J. Biol. Chem*, **349**, 6339–6346.

Casey, J.L., Hentze, M.W., Koeller, D.M., Caughman, S.W., Rouault, T.A., Klausner, R.D. and Harford, J.B. (1988) Iron-responsive elements: regulatory RNA sequences that control mRNA levels and translation. *Science*, **24**, 924–928.

Cecconi, F., Ghilardi, C.A., Midollini, S., Orlandini, A. and Zanello, P. (1987). *J. Chem. Soc., Dalton Trans.*, 831–835.

Challen, P.R., Koo, S-M., Kim, C.G., Dunham, W.R. and Coucouvanis, D., (1990) *J. Am. Chem. Soc.*, **112**, 8606–8607.

Chance, B. (1951) *Adv. Enzymol.*, **12**, 153–190.

Chasteen, N.D., Antanaitis, B.C. and Aisen, P. (1985) Iron deposition in apoferritin. *J. Biol. Chem.*, **260**, 2926–2929.

Cheesman, M.R., Thomson, A.J., Greenwood, C., Moore, G.R. and Kadir, F.H.A. (1990) Bis-methionine axial ligation of haem in bacterioferritin from *Pseudomonas aeruginosa. Nature*, **346**, 771–773.

Cheesman, M.R., Kadir, F.H.A., Al-Basseet, J., Farrar, J., Greenwood, C., Thomson, A.J. and

Moore, G.R. (1992) EPR and MCD spectroscopic characterization of bacterioferitin from *Pseudomonas aeruginosa* and *Azotobacter vinelandii. Biochem. J.*, in press.
Chen, V.J., Orville, A.M., Harpel, M.R., Frolik, C.A., Surerus, K.K., Münck, E. and Lipscomb, J.D. (1989) Spectroscopic studies of isopenicillin N synthase. *J. Inorg. Biochem.*, **264**, 21677–21681.
Chen, C., Cai., J., Liu, Q., Wu, D., Lei, X., Zhao, K., Kang, B. and Lu, J. (1990) *Inorg. Chem.*, **29**, 4878–4881.
Christner, J.A., Münck, E., Janick, P.A. and Siegel, L.M. (1981) *J. Biol. Chem.*, **256**, 2098–2101.
Christou, G., Ridge, B. and Rydon, H.N. (1977) *J. Chem. Soc., Chem, Commun.*, 908–909.
Christou, G., Ridge, B. and Rydon, H.N. (1979) *J. Chem. Soc., Chem, Commun.*, 20–21.
Christou, G., Sabat, M., Ibers, J.A. and Holm, R.H. (1982) *Inorg. Chem.*, **21**, 3518–3526.
Christou, G. and Garner and C.D. (1980) *J. Chem. Soc., Dalton Trans.*, 2354–2362.
Ciurli, S. and Holm, R.H. (1989). *Inorg. Chem.*, **28**, 1685–1690.
Ciurli, S. and Holm, R.H. (1991) *Inorg. Chem.* **30**, 743–750.
Ciurli, S., Yu, S., Holm, R.H., Srivastava, K.P.P. and Münck, E. (1990) *J. Am. Chem. Soc.*, **112**, 8169–8171.
Clegg, G.A., Fitton, J.E., Harrison, P.M. and Treffry, A. (1980) Ferritin: molecular structure and iron-storage mechanisms. *Prog. Biophys. Mol. Biol.*, **36**, 56–86.
Colby, J., Stirling, D.I. and Dalton, H. (1977) The soluble methane mono-oxygenase of *Methylococcus capsulatus* (Bath). *Biochem. J.*, **165**, 395–402.
Conover, R.C., Park, J-B., Adams, M.W.W. and Johnson, M.K. (1990) *J. Am. Chem. Soc.*, **112**, 4562–4564.
Conover, R.C., Kowal, A.T., Fu, W., Park, J-B., Aino, S., Adams, M.W.W. and Johnson, M.K. (1990) *J. Boil. Chem.*, **265**, 8533–8541.
Conover, R.C., Park, J-B., Adams, M.W.W. and Johnson, M.K. (1991) *J. Am. Chem. Soc.*, **113**, 2799–2800.
Conover, R.C., Finnegan, M.G., Park, J-B., Adams, M.W.W. and Johnson, M.K. (1991) *J. Inorg. Biochem.*, 43, 245.
Conradson, S.C., Burgess, B.K., Newton, W.E., Hodgson, K.O., McDonald, J.W., Rubinson, J.F., Gheller, S.F., Mortenson, L.E., Adams, M.W.W., Mascharak, P.K., Armstrong, W.A. and Holm, R.H. (1985) *J. Am. Chem. Soc.*, **107**, 7935–7940.
Coucouvanis, D., Swenson, D., Stremple, P. and Baenziger, N.C. (1979) *J. Am. Chem. Soc.*, **101**, 3392–3394.
Coucouvanis, D., Salifoglou, A., Kanatzidis, M.G., Simopoulos, A. and Papaefthymiou, V. (1984a) *J. Am. Chem. Soc.*, **106**, 6081–6082.
Coucouvanis, D., Kanatzidis, M.G., Dunham, W.R. and Hagen, W.R. (1984b) *J. Am. Chem. Soc.*, **106**, 7998–7999.
Coucouvanis, D., Salifoglou, A., Kanatzidis, M.G., Dunham, W.R., Simopoulos, A. and Kostikas, A. (1988) *Inorg. Chem.*, **27**, 4066–4077.
Cox, T.M. and Kakepoto, G.N. (1991a) Isolation and purification of a novel iron transport protein from human reticulocytes. *Transactions of the 10th International Conference on Iron and Iron Proteins*, Oxford, p. P60.
Cox, T.C., Bawden, M.J., Bhasker, C.R., Bottomley, S.S., Kühn, L.C. and May, B.K. (1991b) Is iron/protoporphyrin homeostasis mediated through erythroid-5-aminolevulinate synthase. *Transactions of the 10th International Conference on Iron and Iron Proteins*, Oxford p. 043.
Cox, M.C., Rogers, M.S., Cheeseman, M., Jones, G., Thomson, A.J., Wilson, M.T. and Moore, G.R. (1992) Spectroscopic identification of the haem ligands of cellobiose oxidase, *FEBS, Lett.*, in press.
Crichton, R.R. (1991) *Inorganic Biochemistry of Iron Metabolism*, Ellis Horwood, New York, London.
Crumbliss, A.L. and Garrison, J.M. (1988) A comparison of some aspects of the aqueous coordination chemistry of aluminum(III) and iron(III). *Comments Inorg. Chem.*, **8**, 1–26.
da Silva, J.J.R.F. and Williams, R.J.P. (1991) *The Biological Chemistry of the Elements*, Clarendon Press, Oxford.
Deighton, N. and Hider, R.C. (1989) Intracellular low molecular weight iron. *Biochem. Soc. Trans.*, **17**, 490.
Deighton, N. Abu-Raqabah, A., Rowland, I.J., Symons, M.C.R., Peters, T.J. and Ward, R.W. (1991) Electron paramagnetic resonance studies of a range of ferritins and haemosiderins *J. Chem. Soc. Faraday Trans.*, **87**, 3193–3197.

Dickerson, R.E., Takano, T., Eisenberg, D., Kallai, O.B., Samson, L., Cooper, A. and Morgoliash, E. (1971) *J. Biol. Chem.*, **246**, 1511–1535.

Dickerson, R.E. and Geis, I. (1983) *Hemoglobin, Structure, Function, Evolution and Pathology*, The Benjamin/Cummings Publishing Co. Inc., Menlo Park, California.

Dickson, D.P.E., Reid, N.M.K., Mann, S., Wade, V.J., Ward, R.J. and Peters, T.J. (1988) Mössbauer spectroscopy, electron microscopy and electron diffraction studies of the iron cores in various human and animal haemosiderins. *Biochim. Biophys. Acta*, **957**, 81–90.

Dingh, H. (1991) *Investigation of Haem Orientational Disorder in Cytochrome* b_5, PhD Thesis, University of Essex.

Dolphin, D. (ed) (1978) *The Porphyrins* (Vol. I and II), Academic Press, New York, San Francisco, London.

Eady, R.R. (1991) *Adv. Inorg. Chem.*, **36**, 77–102.

Eldridge, P.A., Bose, K.S., Barber, D.E., Bryan, R.F., Sinn, E., Rheingold, A. and Averill, B.A. (1991) *Inorg. Chem.*, **30**, 2365–2375.

Emery, T.F. (1991) *Iron and Your Health: Facts and Fallacies*, CRC Press, Inc. Boca Raton.

Emptage, M.H. (1988), in *Metal Clusters in Proteins*, (ed. L. Que) ACS Symposium Series, **392**, Chapter 17.

Emptage, M.H. (1991) *J. Inorg. Biochem.*, **43**, 255.

Ericson, A., Hedman, B., Hodgson, K.O., Green, J., Dalton, H., Bentsen, J.G., Beer, R.H. and Lippard, S.J. (1988) Structural characterization by EXAFS spectroscopy of the binuclear iron in protein a of methane monooxygenase from *Methylococcus capsulatus* Bath. *J. Am. Chem. Soc.*, **110**, 2330–2332.

Evans, D.J. and Leigh, G.J. (1991) *J. Inorg. Biochem.*, **42**, 25–35.

Excoffen, P., Langier, J. and Lamotte, B. (1991). *Inorg. Chem.*, **30**, 3075–3081.

Fatemi, S.J.A., Kadir, F.H.A., Williamson, D.J. and Moore, G.R. (1991) The uptake, storage, and mobilization of iron and aluminium in biology. *Adv. Inorg. Chem.*, **36**, 409–448.

Fee, J.A., Findling, K.L., Yoshida, T., Hille, R., Tarr, G.E., Hearshen, D.O., Dunham, W.R., Day, E.P., Kent, T.A. and Münck, E. (1984) *J. Biol. Chem.*, **259**, 124–133.

Fee, J.A. (1991) Regulation of *sod* genes in *Escherichia coli*: relevance to superoxide dismutase function, *Molecular Microbiology*, **5**, 2599–2610.

Ford, G.C., Harrison, P.M., Rice. D.W., Smith, J.M.A., Treffry, A., White, J.L. and Yariv, J. (1984) Ferritin: design and formation of an iron-storage molecule. *Phil. Trans. Roy Soc. Lond.*, **304**, 551–565.

Fridovich, I. (1989) Superoxide dismutases: An adaptation to a paramagnetic gas. *J. Biol. Chem.*, **264**, 7761–7764.

Fox, B.G., Surems, K.K., Munck, E. and Lipscomb, J.D. (1988) Evidence for a μ-oxo-bridged binuclear iron cluster in the hydroxylase component of methane monooxygenase. *J. Biol. Chem.*, **263**, 10553–10556.

Fujishima, Y., Schofield, C.J., Baldwin, J.E., Charnock, J.M. and Garner, C.D. Recent physical and mechanistic studies on isopenicillin N synthase. (1991) *J. Inorg. Biochem.*, **43**, 564.

Fukuyama, K., Hase, T., Matsumoto, S., Tsukihara, T., Katsube, Y., Tanaka, N., Kakudo, M., Wada, K. and Matsubara, H. (1980) *Nature*, **286**, 522–524.

Ghosh, D., Furey, W., O'Donnell, S. and Stout, C.D. (1981) *J. Biol. Chem.*, **256**, 4185–4192.

Gibson, Q.H. and Carey, F.G. (1982) *Adv. Exp. Med. Biol.*, **148**, 49–65.

Girerd, J.J., Papaefthymiou, G.C., Watson, A.D., Gamp, E., Hagen, K.S., Edelstein, N., Frankel, R.B. and Holm, R.H. (1984) *J. Am. Chem. Soc.*, **106**, 5941–5947.

Green, S. and Mazur, A. (1957) Relation of uric acid metabolism to release of iron from hepatic ferritin. *J. Biol. Chem.*, **227**, 653–668.

Griffiths, E. (1987) The iron-uptake systems of pathogenic bacteria, in *Iron and Infection*, (eds. J.J. Bullen and E. Griffiths), Wiley, New York, pp. 69–137.

Hagen, K.S., Berg, J.M. and Holm, R.H. (1980). *Inorg. Chim. Acta*, **45**, L17–18.

Hagen, K.S., Reynolds, J.G. and Holm, R.H. (1981), *J. Am. Chem. Soc.*, **103**, 4054–5063.

Hagen, K.S. and Holm, R.H. (1982) *J. Am. Chem. Soc.*, **104**, 5496–5497.

Hagen, K.S., Watson, A.D. and Holm, R.H. (1983), *J. Am. Chem. Soc.*, **105**, 3905–3913.

Hagen, W.R., Pierick, A.J. and Veeger, C. (1989) *J. Chem. Soc., Faraday Trans. I*, **85**, 4083–4090.

Hagen, W.R., Pierick, A.J., Wolbert, R.B.G., Wassink, H., Haaker, H., Veeger, C., Jetten, M.K., Stams and A.J.M. and Zehnder, A.J.B., (1991) *J. Inorg. Biochem.*, **43**, 237.

Halliwell, B. (1991) *Nature*, **354**, 191–192.
Han, S., Czernuszewicz, R.S. and Spiro, T.G. (1986) *Inorg. Chem.*, **25**, 2276–2277.
Hanna, P.M., Chen, Y. and Chasteen, N.D. (1991) Initial iron oxidation in horse spleen apoferritin. *J. Biol. Chem.*, **266**, 886–893.
Hantke, K. (1981) Regulation of ferric iron transport in *Escherichia coli* K12: isolation of a constitutive mutant. *Mol. Gen. Genet.* **182**, 288–292.
Harris, S. (1989) *Polyhedron*, **8**, 2843–2882. (This is a general review of all cubane-like clusters).
Harrison, P.M., Andrews, S.C., Artymiuk, P.J., Ford, G.C., Guest, J.R., Hirzmann, J., Lawson, D.M., Livingstone, J.C., Smith, J.M.A., Treffry, A and Yewdall, S.J. (1991) Probing structure-function relations in ferritin and bacterioferritin. *Adv. Inorg. Chem.*, **36**, 449–486.
Hennecke, H. (1990) Regulation of bacterial gene expression by metal-protein complexes. *Molecular Microbiology*, **4**, 1621–1628.
Hider, R.C. (1984) Siderophores mediated absorption of iron. *Structure and Bonding*, **58**, 25–88.
Holm, R.H. (1981), *Chem. Soc. Rev.*, **10**, 455–490.
Holm R.H., Ciurli, S. and Weigel, J.A., (1990) Prog. Inorg. Chem., **38**, 1–74.
Imai, T., Sakamoto, Y., Saito, H., Urushiyama, A., Ohta, K., Tobari, J. and Ohmori, D. (1991) *J. Inorg. Biochem.*, **43**, 253.
Inomata, S., Tobita, H. and Ogino, H. (1991) *Inorg. Chem.*, **30**, 3039–3043.
Jacobs, A. (1977) Low molecular weight intracellular iron transport compounds. *Blood*, **50**, 433–439.
Jensen, G.M., Vasquez, A., Burgess, B.K. and Stephens, P.J. (1991) *J. Inorg. Biochem.*, **43**, 240.
Kadir, F.H.A. and Moore, G.R. (1990a) Bacterial ferritin contains 24 haem groups. *FEBS Lett.* **271**, 141–143.
Kadir, F.H.A. and Moore, G.R. (1990b) Haem binding to horse spleen ferritin. *FEBS Lett.*, **276**, 81–84.
Kadir, F.H.A., Read, N.M.K., Dickson, D.P.E., Greenwood, C., Thompson, A. and Moore G.R. (1991a) Mössbauer spectroscopic studies of iron in *Pseudomonas aeruginosa. J. Inorg. Biochem.*, **43**, 753–758.
Kadir, F.H.A., Al-Massad, F.K., Fatemi, S.J.A., Singh, H.K., Wilson, M.T. and Moore G.R. (1991b) Electron transfer between horse ferritin and ferrihaemoproteins. *Biochem. J.*, **278**, 817–820.
Kadir, F.H.A., Al-Massad, F.K. and Moore G.R. (1992) Haem binding to horse spleen ferritin and its effect on the rate of iron release *Biochem. J.*, **282**, 867–871.
Kanatzidis, M.G., Dunham, W.R., Hagen, W.R. and Coucouvanis, D. (1984) *J. Chem. Soc., Chem. Commun.*, 356–357.
Kanatzidis, M.G., Salifoglou, A. and Coucouvanis, D. (1986) *Inorg. Chem.*, **25**, 2460–2468.
Kanatzidis, M.G., Hagen, W.R., Dunham, W.R., Lester, R.K. and Coucouvanis, D. (1985) *J. Am. Chem. Soc.*, **102**, 953–961.
Kissinger, C.R., Adman, E.J., Sieker, L.R. and Jensen, L.H. (1988) *J. Am. Chem. Soc.*, **112**, 8721–8723.
Kissinger, C.R., Adman, E.T., Sieker, L.C., Jensen, L.H. and LeGall, J. (1989) *FEBS Lett.*, **244**, 447–450.
Koeller, D.M., Casey, J.L., Hentze, M.W., Gerhardt, E.M., Chan, L-N. L., Klausner, R.D. and Harford, J.B. (1989) *Proc. Natl. Acad. Sci. USA.*, **86**, 3574–3578.
Körner, M., Suerbaum, H., Witzel, H., Kappl, R., Hüttermann, J., Priggemeyer, S. and Krebs, B. (1991) The interaction of oxoanions with the binuclear Fe^{3+}/Zn^{2+}-center of the violet phosphatase from red kidney beans as investigated by UV/Vis, EPR and EXAFS spectroscopy. *J. Inorg. Biochem.*, **43**, 540.
Kovacs, J.A. and Holm, R.H. (1986) *J. Am. Chem. Soc.*, **108**, 340–341.
Kovacs, J.A. and Holm, R.H. (1987) *Inorg. Chem.*, **26**, 702–711; 712–718.
Kroneck, P.M.H., Anthroline, W.A., Riesterer, J. and Zumft, W.G. (1988) *FEBS Lett.*, **242**, 70–74.
Lane, R.W., Ibers, J.A., Frankel, R.B., Papaefthymiou, G.C. and Holm, R.H. (1977) *J. Am. Chem. Soc.*, **99**, 84–98.
Lawson, D.M., Artymiuk, P.J., Yewdall, S.J., Smith, J.M.A., Livingstone, J.C., Treffry, A., Luzzago, A., Levi, S., Arosio, P., Cesarini, G., Thomas, C.D., Shaw, W.V. and Harrison, P.M. (1991) Solving the structure of human H chain ferritin by genetically engineering intermolecular crystal contacts. *Nature*, **349**, 541–544.

Laulhère, J.P., Labouré, A.M., van Wuytswinkel, O. & Briat, J.F. (1991) Structure, function and synthesis of bacterioferritin from cyanobacteria *Synechocystis* PCC6803. *Transactions of the 10th International Conference on Iron and Iron Proteins*, Oxford p P9.

Levi, S., Luzzago, A., Cesareni, G., Cozzi, A., Franceschinelli, F., Albertini, A. and Arosio, P. (1988) Mechanism of ferritin iron uptake: activity of the H-chain and deletion mapping of the ferro-oxidase site. *J. Biol. Chem.*, **263**, 18086–18092.

Lippard, S.J. (1988) Oxo-bridged polyiron centres in biology and chemistry. *Angew. Chem. int. Ed. Engl.*, **27**, 344–361.

Lund, J., Woodland, M.P. and Dalton, H. (1985) Electron transfer reactions in the soluble methane monooxygenase of *Methylococcus capsulatus* (Bath). *Eur. J. Biochem.*, **147**, 297–305.

Macedo, A.L., Moreno, C., Moura, I., LeGall, J. and Moura, J.J.G. (1991) *J. Inorg. Biochem.*, **43**, 243.

Majamaa, K., Hanauske-Abel, M., Günzler, V. and Kivirikko, K.I. (1984) The 2-oxoglutarate binding site of prolyl 4-hydroxylase. *Eur. J. Biochem.*, **138**, 239–245.

Malmström, B.G. (1982) *Ann. Rev. Biochem.*, **51**, 21–59.

Marcus, R. and Sutin, N. (1985) *Biochim. Biophys. Acta*, **811**, 265–322.

Marschner, H. (1986) *Mineral Nutrition of Higher Plants*, Academic Press.

Mascharak, P.K., Armstrong, W.H., Mizobe, Y. and Holm, R.H. (1983a) *J. Am. Chem. Soc.*, **105**, 475–483.

Mascharak, P.K., Papaefthymiou, G.C., Armstrong, W.H., Foner, S., Frankel, R.B. and Holm, R.H. (1983b) *Inorg. Chem.*, **22**, 2851–2858.

Mathews, F.S. (1985) *Prog. Biophys. Molec. Biol.*, **45**, 1–56.

Mayerle, J.J., Frankel, R.B., Holm, R.H., Ibers, J.A., Phillips, W.D. and Weiher, J.F. (1973) *Proc. Natl. Acad. Sci., U.S.A.*, **70**, 2429–2433.

Mckee, D.E., Richardson, D.C., Richardson, J.S. and Siegel, L.M. (1986) *J. Biol. Chem.*, **261**, 10277–10281.

Melino, G., Stefanini, S., Chiancone, E., Antonini, E. and Finazzi-Agró, A. (1978) Stoichiometry of iron oxidation by apoferritin. *FEBS Lett.*, **86**, 136–138.

Meyer, J. (1988) *TREE*, **3**, 222–226.

Millar, M., Lee, J.F., Koch, S.A. and Fikar, R. (1982) *Inorg. Chem.*, **21**, 4105–4106.

Ming, L.-J., Que, L., Jr. Kriauciunas, A., Frolik, C. and Chen, V.J. (1990) Coordination chemistry of the metal binding site of isopenicillin N synthetase. *Inorg. Chem.* **29**, 1111–1112.

Ming, L.-J., Lynch, J.B., Que, L., Jr. Frolik, C., Kriauciunas, A. and Chen, V.J. (1991) NMR and NOE studies of mononuclear nonheme Fe(II) metalloenzymes. *J. Inorg. Biochem.* **43**, 555.

Moore, G.R. and Pettigrew, G.W. (1990) *Cytochrome c*, Vol. 2, Springer-Verlag, Berlin-Heidelberg.

Moore, G.R. and Rogers, N.K. (1985) *J. Inorg. Biochem.*, **23**, 219–226.

Moore, G.R., Williams, R.J.P., Peterson, J., Thomson, A.J. and Mathews, F.S. (1985) *Biochim. Biophys. Acta*, **829**, 83–96.

Moore, G.R., Mann, S. and Bannister, J.V. (1986) Isolation and properties of the complex non-haem-iron containing cytochrome b_{557} (bacterioferritin) from *Pseudomonas aeruginosa*. *J. Inorg. Biochem.* **28**, 329–336.

Moore, G.R., Cheesman, M.R., Kadir, F.H.A., Thomson, A.J., Yewdall, S.J. and Harrison, P.M. (1992a) Spectroscopic identification of the haem axial ligands of haemoferritin and location of possible haem binding sites in ferritin by molecular modelling. *Biochem. J.*

Moore, G.R., Kadir, F.H.A. and Al-Massad, F. (1992b) Haem binding to ferritin and possible mechanisms of physiological iron uptake and release by ferritin. *J. Inorg. Biochem.*, **47**, 175–181.

Morgan, J. and Wikström, M. (1991) *Biochemistry*, **30**, 948–958.

Müllner, E.W., Neupert, B. and Kühn, L.C. (1989) A specific mRNA binding factor regulates the iron-dependent stability of cytoplasmic-transferrin receptor mRNA. *Cell*, **58**, 373–382.

Nakamoto, M., Tanaka, K. and Tanaka, T. (1988) *J. Chem. Soc., Chem. Commun.*, 1422–1423.

Nakamura, A. and Ueyama, N. (1989) *Adv. Inorg. Chem.*, **33**, 39–67.

Navaratnam, S., Feiters, M.C., Al-Hakim, M., Allen, J.C., Veldink, G.A. and Vliegenthart, J.F.G. (1988) Iron environment in soybean lipoxygenase-1. *Biochim. Biophys. Acta*, **956**, 70–76.

Neilands, J.B. (1990) Molecular aspects of regulation of high affinity iron absorption in micro-organisms, in *Metal-Ion Induced Regulation of Gene Expression*, **8**, (eds G.L. Eichhorn and L.G. Marzilli), Elsevier, New York, Amsterdam, pp. 63–90.

Neilands, J.B., Konopka, K., Schwyn, B., Coy, M., Francis, R.T., Paw, B.H. and Bagg, A. (1987) Comparative biochemistry of microbial iron assimilation, in *Iron Transport in Microbes, Plants*

and Animals, (eds G., Winkelmann, D. van der Helm and J.B. Neilands), VCH, Weinheim, pp. 3–33.
Nelson, L.G., Lo, F.Y.-K., Rae, A.D. and Dahl, L.F. (1982) *J. Organometal. Chem.*, **225**, 309–329.
Noda, I., Snyder, B.S. and Holm, R.H. (1986) *Inorg. Chem.*, **25**, 3851–3853.
Noodleman, L. (1991) *Inorg. Chem.*, **30**, 246–256.
Noodleman, L., Case, D.A. and Aizman, A. (1988) *J. Am. Chem. Soc.*, **110**, 1001–1005.
Nordlund, P., Sjöberg, B.-M. and Eklund, H. (1990) Three dimensional structure of the free radical protein of ribonucleotide reductase. *Nature*, **345**, 593–598.
Nordlund, P., Uhlin, U., Hajdu, J. and Eklund, H. (1991) The structure of the iron centre of ribonucleotide reductase and its implications for the formation of a stable free tyrosine radical. *J. Inorg. Biochem.*, **43**, 534.
Ohlendorf, D.H., Lipscomb, J.D. and Weber, P.C. (1988) Structure and assembly of protocatechuate 3,4-dioxygenase. *Nature*, **336**, 403–405.
Ohno, R., Ueyama, N. and Nakamura, A. (1990) *Inorg. Chim. Acta*, **169**, 253–255.
Olson, J.S., Mathews, A.J., Rohlfs, R.J., Springer, B.A., Egeberg, K.D., Sligar, J.T., Renaud, J-P. and Naigai, K. (1988) *Nature*, **336**, 265–266.
O'Sullivan, T. and Millar, M.M. (1985) *J. Am. Chem. Soc.*, **107**, 4096–4097.
Palermo, P.E., Power, P.P. and Holm, R.H. (1982) *Inorg. Chem.*, **21**, 173–181.
Palermo, R.E., Singh, R., Bashkin, J.K. and Holm, R.H. (1984) *J. Am. Chem. Soc.*, **106**, 2600–2612.
Perutz, M.F. (1990a) *Ann. Rev. Physiol.*, **52**, 1–25.
Perutz, M.F. (1990b) *Mechanisms of Cooperativity and Allosteric Regulation in Proteins*, Cambridge University Press, Cambridge.
Perutz, M.F., Fermi, G., Luisi, B., Shaaman, B. and Liddington, R.C. (1987) *Accounts of Chemical Research*, 309–320.
Petrouleas, V., Koulougliotis, D., Sanakis, Y., Deligiannakis, Y., Chisholm, D.A., Diner, B.A. and Martinez Lorente, M.A. The iron of the ferroquinone complex of photosystem II. (1991) *J. Inorg. Biochem.*, **43**, 561.
Pettigrew, G.W. and Moore, G.R. (1987) *Cytochrome c*, Vol. 1, Springer-Verlag, Berlin-Heidelberg.
Pickett, C.J. (1985) *J. Chem. Soc., Chem. Commun.*, 323–326.
Pierce, J.R. And Earhart, C.F. (1986) *Escherichia coli* K12 envelope proteins specifically required for ferrienterobactin uptake. *J. Bacteriol.*, **166**, 930–936.
Pohl, S. and Bierback, U. (1991) *Z. Naturforsch*, **46b**, 68–74.
Poulos, T.L., Freer, S.T., Alder, R.A., Edwards, S.L., Yonetani, T. and Kraut, J. (1980) *J. Biol. Chem.*, **255**, 575.
Poulos, T.L., Finzel, B.C., Gunsalus, I.C., Wagner, G.C. and Kraut, J. (1988) *J. Biol. Chem.*, **260**, 16122–16130.
Pressler, U., Staudenmaier, H., Zimmerman, L. and Braun, V. (1988) Genetics of the iron dicitrate transport system of *Escherichia coli*. *J. Bacteriol.*, **170**, 2716–2724.
Priggemeyer, S., Rompel, A., Eggers-Borkenstein, P., Krebs, B., Henkel, G., Nolting, H.-F., Hermes, C., Körner, M. and Witzel, H. (1991) The active site of purple acid phosphatase from red kidney beans as determined by X-ray absorption spectroscopy. *J. Inorg. Biochem.*, **43**, 543.
Que, L., Anglin, J.R., Bobrik, M.A., Davison, A. and Holm, R.H. (1974) *J. Am. Chem. Soc.*, **96**, 6042–6048.
Que, L., Bobrik, M.A., Ibers, J.A. and Holm, R.H. (1974) *J. Am. Chem. Soc.*, **96**, 4168–4178.
Que, L. Jr., Lipscomb, J.D., Münck, E. and Wood, J.M. (1977) Protocatechuate 3,4-dioxygenase inhibitor studies and mechanistic implications. *Biochim. Biophys. Acta*, **485**, 60–74.
Que, L. Jr., and True, A.E. (1990) *Prog. Inorg. Chem.* **38**, 97.
Ralston, D.M. and O'Halloran, T.V. (1990) Metalloregulatory proteins and molecular mechanisms of heavy metal signal transduction, in *Metal-Ion Induced Regulation of Gene Expression*, **8**, (eds, G.L. Eichhorn and L.G. Marzilli), Elsevier, New York, Amsterdam, pp. 1–31.
Ravi, N., Moura, I., Tavares, P., LeGall, J., Huynh, B.H. and Moura, J.J.G. (1991). *J. Inorg. Biochem.*, **43**, 252.
Reddy, M.B., Chidambaram, M.V. and Bates, G.W. (1987) Iron Bioavailability, in *Iron Transport in Microbes, Plants and Animals* (eds. G. Winkelmann, D. van der Helm and J.B. Neilands), VCH, Weinham, pp. 429–444.
Reichard, P. (1988) Interactions between deoxyribonucleotide and DNA synthesis. *Ann. Rev. Biochem.* **57**, 349–374.
Reynolds, M.S. and Holm, R.H. (1988) *Inorg. Chem.*, **27**, 4494–4499.

Ristau, O., Rein, H., Jänig, G.G. and Ruckpaul, K. (1978) *Biochim. Biophys. Acta*, **536**, 226–234.
Robbins, A.H. and Stout, C.D. (1985) *J. Biol. Chem.*, **260**, 2328–2333.
Robbins, A.H. and Stout, C.D. (1989) *Proc. Natl. Acad. Sci., U.S.A.*, **86**, 3639–3643.
Rohrer, J.S., Islam, Q.T., Watt, G.D., Sayers, D.E. and Theil, E.C. (1990) Iron environment in ferritin with large amounts of phosphate, from *Azotobacter vinelandii* and horse spleen, analyzed using extended X-ray absorption fine structure (EXAFS). *Biochemistry*, **29**, 259–264.
Rouault, T.A., Stout, C.D., Kaptain, S., Harford, J.B. and Klausner, R.D. (1991) Structural relationship between an iron-regulated RNA-binding protein (IRE-BP) and aconitase: functional implications. *Cell*, **64**, 881–883.
Ruckpaul, K., Rein, H. and Blanck, J. (1989) in *Basis and Mechanisms of Regulation of Cytochrome* P_{450}, (Eds K. Ruckpaul and H. Rien), Taylor and Francis, London.
Saak, W., Henkel, G. and Pohl, S. (1984) *Angew. Chem. Intl. Ed. Engl.*, **23**, 150–151
Saak, W. and Pohl, S. (1988) *Z. Naturforsch*, **43b**, 813–817.
Schäffer, S., Hantke, K. and Braun, V. (1985) Nucleotide sequence of the iron regulatory gene *fur*. *Mol. Gen. Genet.*, **200**, 110–113.
Schejter A. and Plotkin, B. (1988) *Biochem. J.*, **255**, 353–356.
Shendan, R.P., Allen, L.C. and Carter, C.W. (1981) *J. Biol. Chem.*, **256**, 5052–5057.
Sheriff, S., Hendrickson, W.A. and Smith, J.L. (1987) Structure of myohemerythrin in the azidomet state at 1.7–1.3 Å resolution. *J. Mol. Biol.*, **197**, 272–293.
Shimizu, T., Ito, O., Hatano, M. and Fujii-Furiyama, Y. (1991) *Biochemistry*, **30**, 4659–4662.
Slappendel, S., Malmström, B.G., Petersson, L., Ehrenberg, A., Veldink, G.A. and Vleigenthart, J.F.G. (1982) On the spin and valence state of iron in native soybean lipoxygenase. *Biochem. Biophys. Res. Commun.*, **108**, 673–677.
Smith, J.M.A., Ford, G.C., Harrison, P.M., Yariv, J. and Kalb, A.J. (1989) Molecular size and symmetry of bacterioferritin of *Escherichia coli*. *J. Mol. Biol.* **205**, 465.
Snasdeit, H., Krebs, B. and Henkel, G. (1984) *Inorg. Chem.*, **23**, 1816–1825.
Snyder, B.S., Reynolds, M.S., Noda, I. and Holm, R.H. (1988) *Inorg. Chem.*, **27**, 595–597.
Snyder, B.S. and Holm, R.H. (1988) *Inorg. Chem.*, **27**, 2339–2347.
Snyder, B.S., Reynolds, M.S., Holm, R.H., Papaefthymiou, G.C. and Frankel, R.B. (1991) *Polyhedron*, **10**, 203–213.
Snyder, B.S. and Holm, R.H. (1990) *Inorg. Chem.*, **29**, 274–279.
Snyder, S.H. and Bredt, D.S. (1992) *Sci. Am.*, **266**, 28–35.
Sola, M., Cowan, J.A. and Gray, H.B. (1991) *Biochem*, **28**, 5261–5268.
St. Pierre, T.G., Webb, J. and Mann, S. (1988) Ferritin and hemosiderin: structural and magnetic studies of the iron core, in *Biomineralisation* (eds S. Mann, J. Webb and R.J.P. Williams), VCH, Weinham, pp. 295–344.
St. Pierre, T.G., Bell, S.H., Dickson, D.P.E., Mann, S., Webb, J., Moore, G.R. and Williams, R.J.P. (1986) Mössbauer spectroscopic studies of the cores of human, limpet and bacterial ferritins. *Biochim. Biophys. Acta.*, **870**, 127–134.
Stephens, P.J., Mckenna, M.C., Ensign, S.A., Bonami, D. and Ludden, P.W. (1989) *J. Biol. Chem.*, **264**, 16347–16350.
Stevens, W.C. and Kurtz, D.M. (1985) *Inorg. Chem.*, **24**, 3444–3449.
Stiefel, E.I. and Watt, G.D. (1979) *Azotobacter* cytochrome $b_{557.5}$ is a bacterioferritin. *Nature*, **279**, 81–83.
Stoddard, B.L., Howell, P.L., Tinge, D. and Petsko, G.A. (1990) The 2.1.-Å resolution structure of iron superoxide dismutase from *Pseudomonas ovalis*. *Biochemistry*, **29**, 8885–8893.
Stout, C.D. (1988) *J. Biol. Chem.*, **263**, 9256–9260.
Stout, C.D. (1989) *J. Mol. Biol.*, **205**, 545–555.
Stout, G.H., Turley, S., Sieker, R.C. and Jensen, L.H. (1988) *Proc. Natl. Acad. Sci., U.S.A.*, **85**, 1020–1022.
Strittmatter, P. (1964) in *On Rapid Mixing and Sampling Techniques in Biochemistry*, (Eds B. Chance, R.H. Eisenhardt, Q.H. Gibson and K.K. Lunberg), Academic Press, New York, pp. 71–84.
Stubbe, J.A. (1989) Protein radical involvement in biological catalysis. *Ann. Rev. Biochem.*, **58**, 257–285.
Theil, E.C. (1983) Ferritin: structure, function, and regulation. *Adv. Inorg. Biochem.*, **5**, 1–38.
Theil, E.C. (1990) Regulation of ferritin and transferrin receptor mRNAs. *J. Biol. Chem.*, **265**, 4771–4772.

Theil, E.C. and Aisen, P. (1987) The storage and transport of iron in animals cells, in *Iron Transport in Microbes, Plants and Animals*, (eds G. Winkelmann, D. van der Helm and J.B. Neilands), VCH, Weinheim, pp. 491–520.

Thomson, A.J. (1985) in *Metalloproteins*, Part 1, (ed P.M. Harrison) Macmillan, London, pp. 79–120.

Thomson, A.J., George, S.J., Armstrong, F.A., Hatchikian, E.C. and Yates, M.G. (1990) *Pure Appl. Chem.*, **62**, 1071–1074.

Traylor, T.G. and Sharma, V.S. (1992) *Biochemistry*, **31**, 2847–2849.

Treffry, A., Sowerby, J.M. and Harrison, P.M. (1978) Variable stoichiometry of Fe(II)-oxidation in ferritin *FEBS Lett.*, **95**, 221–224.

Tremel, T., Hoffmann, R. and Jemmis, E.D. (1989) *Inorg. Chem.*, **28**, 1213–1224.

Tron, T., Crimi, M., Colson, A.M. and Degli Eposti, M. (1991) *Eur. J. Biochem.*, **199**, 753–760.

True, A.E., Scarrow, R.C., Holz, R.C. and Que, L. Jr. (1991) X-ray absorption spectroscopy of uteroferrin and its oxidation derivatives. *J. Inorg. Biochem.*, **43**, 545.

Ueyama, N., Kajiwara, A., Terakawa, T., Ueno, S. and Nakamura, A. (1985) *Inorg. Chem.*, **24**, 4700–4704.

Ueyama, N. and Nakamura, A. (1988) in *Metalloproteins*, Vol. 8, (eds S. Otsuka, T. Yamanaka), Elsevier, Amsterdam, pp. 153–265.

Van Bauren, K.J.H., Van Gelder, B.F., Wilting, J. and Braams, R. (1974) *Biochim. Biophys. Acta*, **333**, 421–426.

Vincent, J.B., Crowder, M.W. and Averill, B.A. (1992) Hydrolysis of phosphate monoesters: a biological problem with multiple chemical solutions. *TIBS*, **17**, 105–110.

Vliegenthart, J.F.G. and Veldink, G.A. (1982) Lipoxygenases, in *Free Radicals in Biology*, **5**, Academic Press, New York, pp. 29–64.

Watt, G.D., Frankel, R.B. and Papaefthymiou, G.C. (1985) Reduction of mammalian ferritin. *Proc. Natl. Acad. Sci. USA.*, **82**, 3640–3643.

Watt, G.D., Frankel, R.B., Papaefthymiou, G.C., Spartalian, K. and Stiefel, E.I. (1986) Redox properties and Mössbauer spectroscopy of *Azotobacter vinelandii* bacterioferritin. *Biochemistry*, **25**, 4330–4336.

Watt, G.D., Jacobs, D. and Frankel, R.B. (1988) Redox reactivity of bacterial and mammalian feritin: is reductant entry into the ferritin interior a necessary step for iron release? *Proc. Natl. Acad. Sci. USA.*, **85**, 7457–7461.

Weaver, J. and Pollack, S. (1989) Low-M_r iron isolated from guinea pig tericulocytes as AMP-Fe and ATP-Fe complexes. *Biochem. J.*, **261**, 787–792.

Weber, G. (1972) *Ann. Rev. Biophys. Bioeng.*, **1**, 553–570.

Weigel, J.A., Srivastava, K.K.P., Day, E.P., Münck, E. and Holm, R.H. (1990) *J. Am. Chem. Soc.*, **112**, 8015–8023.

Weigel, J.A. and Holm, R.H. (1991) *J. Am. Chem. Soc.*, **113**, 4184–4189.

Weinberg, E.D. (1989) Cellular regulation of iron acquisition *Quart. Rev. Biol.*, **64**, 261–290.

Weir, M.P., Gibson, J.F. and Peters, T.J. (1984) Haemosiderin and tissue damage. *Cell Biochemistry and Function*, **2**, 186–194.

Whittaker, J.W. and Lipscomb, J.D. (1984) ^{17}O-Water and cyanide ligation by the active site iron of protocatechate 3,4-dioxygenase. *J. Biol. Chem.*, **259**, 4487–4495.

Wikström, M. (1989) *Nature*, **338**, 776–778.

Williams, R.J.P. (1982) Free manganese(II) and iron(II) cations can act as intracellular cell controls. *FEBS Lett.*, **140**, 3–10.

Wilson, M.T. and Greenwood, C., in press, in *The Cytochrome c Source Book*, (Eds R. Scott and G. Mauk).

Winkelmann, G., van der Helm, D. and Neilands, J.B. (1987) *Iron Transport in Microbes, Plants and Animals*, VCH, Weinheim.

Wolff, T.E., Power, P.P., Frankel, R.B. and Holm, R.H. (1980) *J. Am. Chem. Soc.*, **102**, 4694–4703.

Wong, G.B., Bobrik, M.A. and Holm, R.H. (1978) *Inorg. Chem.*, **17**, 578–584.

Xu, B. and Chasteen, N.D. (1991) The stoichiometry of iron(II) oxidation in ferritin. *Transactions of the 10th International Conference on Iron and Iron Proteins*, Oxford, p. 018.

You, J.F., Snyder, B.S., Papaefthymiou, G.C. and Holm R.H. (1990) *Inorg. Chem.*, **112**, 1067–1076.

Zähringer, J., Baliga, B.S. and Munro, H.N. (1976) Novel mechanism for translational control in regulation of ferritin synthesis by iron. *Proc. Natl. Acad. Sci. USA.*, **73**, 857–861.

Zhang, Y.-P., Bashkin, J.K. and Holm, R.H. (1987) *Inorg Chem.*, **26**, 694–702.

7 Models for iron biomolecules

A.K. POWELL

7.1 Models for iron biomolecules

Iron is found in a variety of different chemical environments in biology. These iron sites are usually classified according to certain structural characteristics rather than function. For example, haemoproteins all contain the familiar iron-porphyrin group but chemical 'fine-tuning' allows the iron centre to assume roles as varied as oxygen transport and storage, enzymatic reactions, and electron transfer. In addition, iron biomolecules can be divided into those which utilise iron to effect a biological function (O_2 transport, electron shuttling, etc.) and those which transport and store iron.

Chemists have long tried to produce small molecule analogues of the iron sites present in the huge range of iron proteins. These models may be corroborative or speculative depending on whether or not the structural details of the metal site are known. Some attempts have been made to synthesise compounds which reproduce the function of the biomolecule. Good working models often require compromise on structural accuracy as the chemist working *in vitro* endeavours to reproduce the stereochemical flexibility invested in the metal site by the protein structure. The various approaches to synthesising model compounds of biological molecules have been reviewed in several articles (Ibers and Holm, 1980; Holm, 1975, 1977; Lever, 1972). In this article the vast battery of small molecule analogues for iron biosites is drawn upon in order to illustrate some of the successes of chemists in reproducing the stereochemistries of such sites. There are fewer examples of functional models and it seems that the challenge for chemists now is to find ways of using *in vitro* systems to help unravel the remaining mysteries in iron biochemistry, such as the intermediate steps of enzyme cycles; how iron is assimilated into and removed from ferritin, and how nature has learned to deal with 'rust'.

7.2 Haem proteins

The haem group is a common motif in biological systems and consists of an iron atom coordinated in an approximately square planar fashion by the

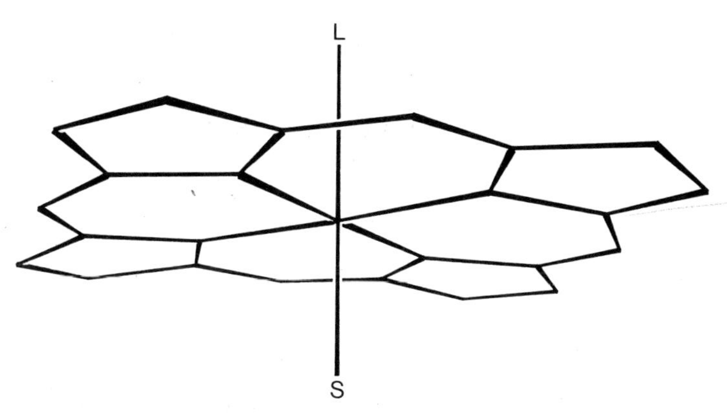

Figure 7.1 The Fe environment in haem proteins. For oxygen carriers and activators L = his or cys and S = substrate. For electron carriers L and S are amino acid residues.

pyrrole nitrogens of a porphyrin ring with one or two further axial ligands as illustrated in Figure 7.1. The nature of the axial ligands and the spin and redox state of the iron affect function so that haem proteins can be divided into three broad groups:

(1) Oxygen carriers such as myoglobin and haemoglobin where the iron is in the +2 oxidation state and cycles between a high and low spin electronic configuration.
(2) Oxygen activators such as peroxidases, oxidases and cytochromes P_{450} where both high and low spin states and oxidation states from +2 to +6 may be important.
(3) Electron transfer proteins such as the cytochromes a, b and c where low spin configurations of iron in the +2 and +3 oxidation states are important.

Although it is relatively easy to prepare porphyrin complexes of both Fe(II) and Fe(III) these simple analogues do not reproduce any of the other characteristics of the proteins such as spin state and coordination geometry of the metal and, indeed, function of the protein itself. This led to the development of a rich area of metal porphyrin chemistry with many elegant solutions found to the challenge of producing complexes which model the protein stereochemistry, electronic properties and function more closely. The models for the haem sites of proteins of types 1 and 2 have been particularly successful in helping to elucidate the structures and functions of these. This has largely been achieved by the development of functionalised porphyrin ligands which can provide stereochemical control at the iron atom so that the coordinative unsaturation of the iron is preserved, the irreversible bimolecular oxidation of the iron to give an oxo bridged di-iron(III) species is avoided, and lastly substrate recognition is realised (e.g. O_2 vs. CO in haemoglobin and myoglobin models). The functionalised porphyrins are classified according to the environment they provide for the iron site, for example the 'picket-fence', 'pocket', 'picket-pocket', 'tailed picket-fence', 'capped', and 'strapped' porphyrins shown in Figure 7.2 which have been used to great effect in modelling both oxygen carriers and activators (Collman, 1977, and references therein; Morgan and Dolphin, 1987, and references therein).

(a)

(b)

(c)

(d)

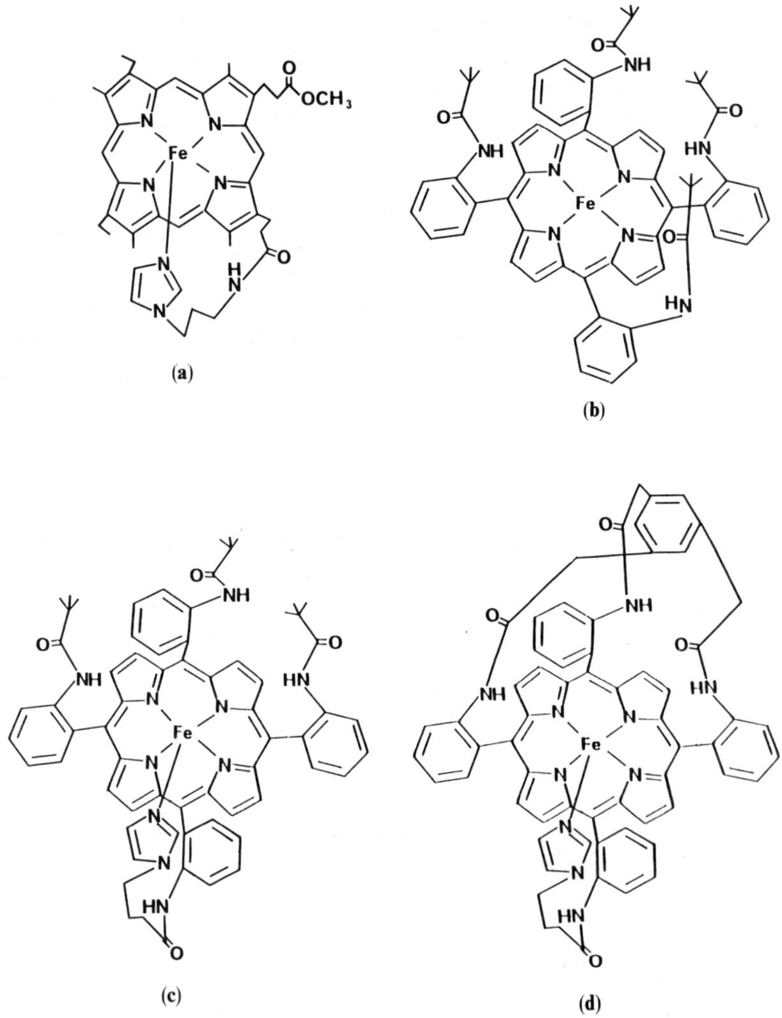

Figure 7.2 Examples of functionalised porphyrins used to model haem sites: (a) chelated haem; (b) picket fence haem; (c) tailed picket fence haem; (d) tailed pocket haem—replacing the tail with a picket yields the picket-pocket porphyrin.

7.2.1 *Models for oxygen carriers*

The iron in this class of haem proteins is in the +2 oxidation state and is tightly bound by the porphyrin ring nitrogens and an axial imidazole ligand from a histidyl residue. The other axial site is used to bind oxygen reversibly; this process is generally regarded as being accompanied by a change in spin state but not in oxidation state of the iron (Newton and Hall, 1984, and references therein). Simple iron(II) porphyrin complexes are irreversibly oxidised by molecular oxygen. One of the challenges for the synthetic chemist trying to model the biological situation is to produce an iron(II) porphyrin complex which is capable of reversible oxygenation (Suslick and Reinert, 1985, and references therein). The preparation of tailed picket-fence porphyrins with protecting groups providing an enclosure for the iron site as well as supplying the axial imidazole ligation on one side of the ring yielded such iron(II) porphyrin complexes as shown in Figure 7.3 (Collman *et al.*, 1973, 1975). These structurally well-defined compounds helped in elucidating the mode of dioxygen binding in the native systems by providing detailed spectroscopic data, such as vibrational spectra, which could be compared directly with the natural systems. In this way it has been established that the dioxygen binds in a bent end-on rather than side-on fashion (Perutz, 1979 and references therein). Another goal has been to produce models which are capable of differentiating between substrates such as O_2 and CO as is the case in the

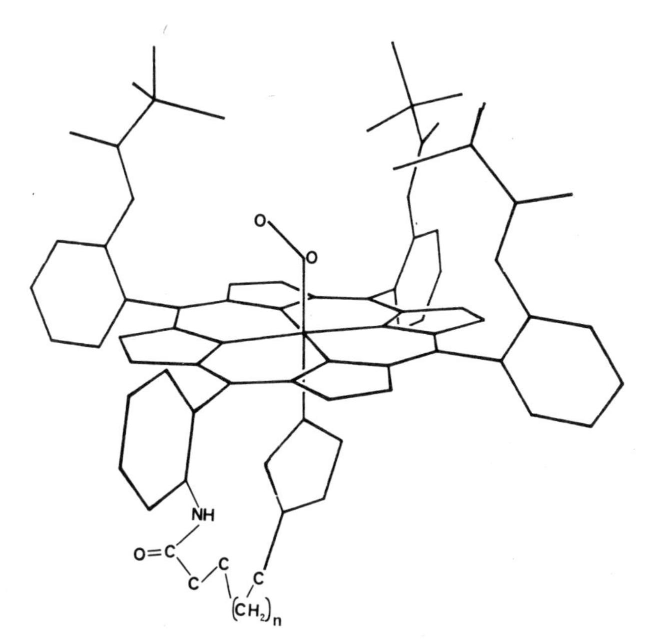

Figure 7.3 Model for oxyHb and oxyMb synthesised by oxygenation of a tailed picket-fence porphyrin Fe(II) complex.

native systems, where CO is discriminated against. The origin of this substrate differentiation is thought to be steric hindrance, hydrogen bonding, or porphyrin ruffling (Johnson *et al.*, 1991, and references therein; Kim and Ibers, 1991, and references therein) so that the CO molecule is forced to bind in an energetically unfavourable fashion with either the Fe–C–O bond bent at the C or else a tilted but linear Fe–C–O bond (Kim *et al.*, 1989; Li and Spiro, 1988). Attempts to model this situation have been most successful using the sterically demanding capped porphyrins, but even so, the large degree of distortion from linearity of the iron to carbonyl oxygen vector which is required to model the natural system properly has not yet been achieved.

7.2.2 *Models for oxygen activators*

These are haem proteins which are five coordinate with the sixth site of the iron available to a substrate molecule such as hydrogen peroxide in peroxidases, oxygen in oxidases, or even carbon from the substrate to give an iron-carbon bond as in the case of cytochromes P_{450}. In all cases the catalytic cycle involves a high oxidation state intermediate where the iron is in the +IV or +V state (Dawson, 1988). It appears that the 'fine-tuning' provided by the fifth axial ligand is responsible for the specificity of the enzymes to catalyse different reactions. In general oxygenases and cytochromes P_{450} catalyse dioxygen activation using two electrons and two protons with the insertion of one oxygen atom from O_2 into a substrate and the formation of water:

$$RH + O_2 + 2e^- + 2H^+ \rightarrow ROH + H_2O$$

Cytochrome P_{450} dependent oxygenases are involved in many steps of biosynthesis and biodegradation of endogenous compounds such as steroids and fatty acids. They are also important in the metabolism of exogenous compounds such as drugs. The use of iron-porphyrin model systems for cytochromes P_{450} has been reviewed in two recent articles (Mansuy, 1987; Mansuy *et al.*, 1989). The models serve three purposes:

(1) Elucidation of the nature of the iron complexes involved as intermediates in the catalytic cycle of dioxygen activation and substrate hydroxylation.
(2) Elucidation of the mechanisms of cytochromes P_{450} via the preparation and characterisation of models for the cytochrome P_{450}-iron-metabolite complexes formed during the metabolism of some drugs or exogenous compounds. This in turn helps in identifying the chemical basis for the formation of inhibitory complexes which is of relevance to pharmacology and toxicology.
(3) Production of catalytically active systems capable of reproducing the main reactions catalysed by cytochromes P_{450}.

Considerable success has been achieved in all three areas. Model iron-porphyrin complexes for all but one of the suggested intermediates of the catalytic cycle of dioxygen activation and substrate hydroxylation have been synthesised and characterised by X-ray crystallography (Schappacher *et al.*, 1985, and references therein). Their spectroscopic properties have also been measured and found to be very similar to those of the enzymatic complexes. In this way, the model compound approach has contributed considerably to the elucidation of the details of the catalytic cycle. An X-ray crystal structure determination of a bacterial cytochrome P_{450} from *Pseudomonas putida* grown on camphor (P_{450}cam) has shown that the axial ligand is a cysteinate residue (Poulos *et al.*, 1987) and the evidence from spectroscopic data is that the cysteinate-iron-porphyrin part of the active site is present in all cytochromes P_{450}. In addition, the dioxygen activation cycles and the classes and mechanisms of reactions catalysed (e.g. C–H bond hydroxylation, alkene epoxidation, aromatic ring hydroxylation) are very similar, leading to the conclusion that the chemoselectivity of the oxidations catalysed is dependent on the cys–iron–porphyrin part and the intrinsic reactivity of the active oxygen complex formed at the iron. In contrast the substrate specificity and regioselectivity of substrate oxidations depend on the nature of the amino acid residues from the protein which are in the active site and which can be different for different cytochromes P_{450}. In this way, it has been the modelling of the formation of the active oxygen species at the iron and the development of catalytic systems displaying the gross chemoselectivity of cytochromes P_{450} which have been most successful to date. The use of iron tetrameso-aryl porphyrins which are very resistant towards oxidative degradation has resulted in catalytic turnovers from 10 to 300 per second for alkene epoxidation being measured with the catalysts lasting for at least 100 000 turnovers. This is comparable to the natural systems and offers a real possibility of using the model systems in organic synthesis (Mansuy, 1987).

The question of how the axial ligands can tune the properties of the oxygen activating haem enzymes has been discussed recently with reference to catalase models (Robert *et al.*, 1991, and references therein). The use of a variety of iron-porphyrin model compounds to catalyse the dismutation of hydrogen peroxide and a competition study on the oxygenase activity of the models revealed that in the model systems, an N-base as a proximal ligand enhances the dismutation, although in the natural systems this ligand is a tyrosinate residue, and that oxygenase activity could only be observed for Mn based models and was also greatest for the N-base proximal ligand systems. It was concluded that in the natural systems it might be the lack of the potentially destructive oxygenase activity in the tyrosinate coordinated catalase sites which favours this coordination geometry. It should also be noted that catalase activity is easily reproduced with a low turnover *in vitro* using an iron(III)/trien based system in which it is likely that the nitrogen donors of the trien ligand provide an approximately square planar equatorial geometry for the iron site (Wang, 1955, 1970; Ochiai and Williams, 1979).

7.2.3 Models for electron transfer proteins

Cytochromes a, b and c are electron transfer proteins. They all possess distinctive three banded electronic absorption spectra in which the Soret bands are the most characteristic feature for each protein. The iron is coordinated by two axial ligands in addition to the porphyrin ring and generally cycles between low spin Fe(II) and low spin Fe(III). The effect of the axial ligands on the redox potential of the iron in such systems was investigated using a tailed porphyrin to provide one axial imidazole ligand. The redox potential for the complex formed with pentamethylene sulphide as the 6th ligand was found to be much more positive than for the complex formed with another imidazole (Marchon *et al.*, 1982). In some cases, intermediate iron(III) spin states may be important and the recent work of Scheidt and coworkers has investigated the role that the orientation of the ligands plays in regulating spin state (Hatano *et al.*, 1991, and references therein).

7.3 Models for iron-sulphur proteins

These proteins contain iron sites where the iron atoms are either bound to sulphur derived from the thiol groups of cysteine residues of the polypeptide chain, known as rubredoxins (type 1 in Figure 7.4), or bound both to cysteine residues and sulphide to give iron-sulphur clusters (types 2–4 in Figure 7.4).

Iron-sulphur proteins are widespread in organisms and can be classified into simple proteins containing only one type of iron-sulphur cluster, and

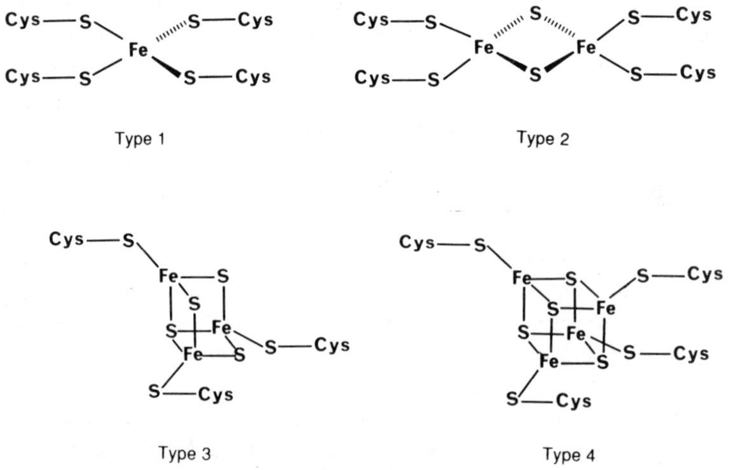

Figure 7.4 Coordination of iron in iron–sulphur proteins. Type 1, $Fe(SR)_4$; type 2, $Fe_2S_2(SR)_4$; type 3, $Fe_3S_4(SR)_3$; type 4, $Fe_4S_4(SR)_4$.

complex proteins which contain other moieties such as flavins, molybdenum, and haem. These proteins fulfil a variety of functions ranging from the electron transfer and transport roles demonstrated by rubredoxins and ferredoxins, to enzymes with both redox and non-redox roles, such as succinate dehydrogenase, nitrogenase and aconitase (Berg and Holm, 1982; Armstrong, 1988). The irons exist in the +2 or +3 or both formal oxidation states. The first iron-sulphur proteins were identified in the 1960s from their characteristic EPR signals and this stimulated efforts to prepare discrete non-biological iron-sulphur species as models. Undoubtedly, the work of Holm and his coworkers in producing new iron-sulphur species *in vitro* was an important stimulus for other synthetic groups working in the area as well as being invaluable in furthering the understanding of the active sites and functions of the proteins themselves. Holm has reviewed the progress of this work in two important accounts (Holm, 1977; Holm *et al.*, 1990). Perhaps the single most striking fact to emerge from this work is that until the existence of iron-sulphur proteins was first recognised in the 1960s, iron-sulphur species such as those shown in Figure 7.4 were unknown to man. The confirmation of the $[Fe_4S_4(Cys)_4]$ 'cubane' type core (type 4 in Figure 7.4) from protein crystallography in 1972 (Carter *et al.*, 1972, 1974; Adman *et al.*, 1973) prompted the efforts which led to the discovery of simple synthetic routes to inorganic analogues of these sites by a number of workers (Holm, 1975, 1977; Schrauzer, 1974; Christou and Garner, 1980). Recently, more attention has been paid to synthesising models which reproduce other properties of these active sites, such as redox potential.

7.3.1 *Models for rubredoxins*

The structural features of the active sites of *Clostridium pasteurianum* and *Desulfovibro vulgaris* rubredoxins (type 1 in Figure 7.4) have been elucidated in X-ray diffraction studies (Watenpaugh *et al.*, 1979). The tetrahedral FeS_4 site in each consists of an Fe(II/III) ion surrounded by two Cys-X-Y-Cys sequences with the resulting $[Fe(S\text{-}Cys)_4]$ core having an approximately D_{2d} symmetry. The complex formed with the chelating thiolato ligand $(S_2\text{-o-xyl})^{2-}$ and iron(III), $[Fe(S_2\text{-o-xyl})_2]^-$, which adopts C_2 symmetry, provided an early simple model for the Fe(III) site (Holm, 1977). The Fe(II) site analogue was subsequently prepared by Holm and coworkers (Holm *et al.*, 1977) using the same bidentate thiolato ligand as before. Although the geometry of the FeS_4 core of the oxidised rubredoxin is almost identical to that of the oxidised form of the model, a large discrepancy between their redox potentials has been observed (Lane *et al.*, 1977). Other simple thiolate Fe(III) complexes with nonchelating ligands are often thermally unstable unless very bulky organic groups are present on the thiolates, as in the case of tetrakis (2,3,5,6-tetramethylbenzenethiolato)ferrate(III) (Millar *et al.*, 1982). This highlights the problems facing the synthetic chemist in trying to model the details of the metal site environment in metalloproteins; although the stereochemistry can often be modelled fairly well, the physical properties such as redox potential of the metal, electron configuration, and reactivity are not so easily reproduced. Although these differences between the native proteins and model complexes are often attributed to the protein environment of the metal sites, it has been suggested recently by Nakamura and Ueyama that it is the exact nature of the peptide environment which is most important in influencing reactivity and redox potential (Nakamura and Ueyama, 1989). They synthesised a range of oligopeptide model complexes of Fe(II) and found that their stabilities and redox potentials varied depending on the amino acid sequence, -X-Y-. They concluded that the electrostatic interactions between the FeS_4 core and the peptide surround may be enhanced by hydrophobic peptide environments. Also the chelation of the Cys-X-Y-Cys peptides can force the peptide chain to form NH–S hydrogen bonds in a suitable conformation to produce the required stability and redox properties by influencing Fe–S bond lengths and torsion angles.

The optical spectroscopic and magnetic properties of structurally characterised $[Fe(SR)_4]^{2-}$ for $R = (2,3,5,6\text{-}(CH_3)_4C_6H)$ and $(2\text{-}(Ph)C_6H_4)$ complexes and their Fe(III) counterparts have been investigated in order to provide a bonding description for the iron site and to gauge the effect of the R groups on the electronic structure (Gebhard *et al.*, 1991). It was concluded that the orientation of the α-C is important in providing a mechanism by which the protein can control the energy of the orbital which donates or accepts the electron in the redox processes involved in electron transfer.

7.3.2 *Models for ferredoxins*

Ferredoxins comprise iron-sulphur proteins that contain an equal number of iron and labile sulphur (sulphide) atoms and display exclusively electron-carrier activity but not classical enzyme function (Armstrong, 1988). They generally possess [2Fe-2S] or [4Fe-4S] cores (types 2 and 4 in Figure 7.4) which are structural types invariant in a wide range of such iron-sulphur proteins. The redox potentials of these proteins can vary enormously, and again this is usually attributed to the exact nature of the protein environment of the iron-sulphur cores (Nakamura and Ueyama, 1989 and references therein).

7.3.2.1 Models for [2Fe-2S] plant-type ferredoxins. The structure of the oxidised form of *Spirulina platensis* ferredoxin has been determined in an X-ray study (Tsukihara *et al.*, 1978) and it is found that three Cys thiolates in the sequence Cys-A-B-C-D-Cys-X-Y-Cys plus one further cysteinate residue bind the $Fe_2S_2^{2+}$ core. Holm and coworkers produced a reasonable structural model for this site in 1981 (Mascharak *et al.*, 1981) using their $[S_2\text{–o–xyl}]^{2-}$ ligand.

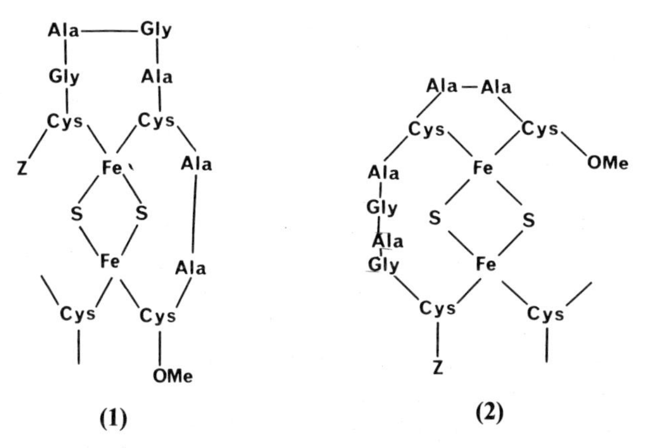

Figure 7.5 Proposed isomers of $[Fe_2S_2$(z-Cys-Gly-Ala-Gly-Ala-Cys-Ala-Ala-Cys-OMe)-(z-Val-Cys-Val-OMe)$]^{2-}$. The ferredoxin from *Spirulina platensis* has structure (**1**).

As in the case of the rubredoxins, the redox potentials of the model systems are found to be much more negative than in the native proteins and Nakamura and Ueyama have tested the effect of different amino acids in model compounds of the general formula $[Fe_2S_2(\text{Z-Cys-X-Y-Cys-OMe})_2]^{2-}$ (Nakamura and Ueyama, 1989; Ueno *et al.*, 1986; Ueyama *et al.*, 1987). Again they concluded that redox potential can be affected by a combination of NH–S hydrogen bond formation and peptide conformation. Their results with an $[Fe_2S_2$(Z-Cys-Gly-Ala-Gly-Ala-Cys-Ala-Ala-Cys-OMe) (Z-Val- Cys-Val-O Me)$]^{2-}$ complex (Ueno *et al.*, 1985) indicated that two isomers were formed. These probably differ in the way in which the Cys-A-B-C-D-Cys-X-Y-Cys chain coordinates to the $[Fe_2S_2]$ core as shown in Figure 7.5. However, as before, the structures of these compounds have only been inferred from spectroscopic measurements and not substantiated by X-ray crystal structure analyses. The synthesis of good structural models for the $[Fe(II)_2S_2(SR)_4]^{4-}$ and $[Fe(II)Fe(III)S_2(SR)_4]^{3-}$ cores has been hampered by the redox instability of these iron sulphur cores.

7.3.2.2 Models for [4Fe-4S] ferredoxins. The [4Fe-4S] 'cubane' type core (type 4, Figure 7.4) can be synthesised remarkably easily *in vitro*, a fact which was not appreciated by chemists until they attempted to prepare synthetic analogues of the iron-sulphur cores found in $HiPIP_{red}$ protein from *Chromatium* and the Fd_{ox} from *Peptococcus aerogenes*. The cluster core of the proteins can exist in one of three physiologically active oxidation levels corresponding to $3Fe^{3+}/Fe^{2+}$ ($[Fe_4S_4]^{3+}$), $2Fe^{3+}/2Fe^{2+}$ ($[Fe_4S_4]^{2+}$), and $Fe^{3+}/3Fe^{2+}$ ($[Fe_4S_4]^{+}$). The HiPIP proteins found in photosynthetic bacteria have very positive redox potentials and contain a single [4Fe-4S] core with the oxidised form corresponding to the $3Fe^{3+}/Fe^{2+}$ oxidation level and the reduced form to $2Fe^{3+}/2Fe^{2+}$, while for ferredoxins, which have negative redox potentials and contain one or two [4Fe-4S] units, the $2Fe^{3+}/2Fe^{2+}$ level corresponds to the oxidised form and the $Fe^{3+}/3Fe^{2+}$ level to the reduced form. The first synthetic analogues for [4Fe-4S] clusters prepared in so-called 'self-assembly' reactions, corresponded to the $2Fe^{3+}/2Fe^{2+}$ oxidation level with the general formula $[Fe_4S_4(SR)_4]^{2-}$, where $(SR)^-$ is an alkyl or aryl thiolate ligand. More recently, the $[Fe_4S_4(SR)_4]^-$ and $[Fe_4S_4(SR)_4]^{3-}$ models have been characterised (Holm *et al.*, 1990). Three principal methods are used in their synthesis (Garner, 1980):

(1) Reaction of a methonolic solution of $FeCl_3$ with NaSR followed by addition of NaHS in methanol (Averill *et al.*, 1973).
(2) Reaction of $FeCl_3$ in methanol with LiSR followed by the addition of Li_2S (Schauzer *et al.*, 1974).
(3) Reaction of $FeCl_3$ or $FeCl_2.4H_2O$ in methanol with NaSR and elemental sulphur (Christou and Garner, 1980).

7.3.3 *Models for other simple iron-sulphur proteins*

The enzyme aconitase is a rare example of an iron-sulphur protein fulfilling a catalytic rather than electron transfer role. It transforms citrate to isocitrate in the Krebs cycle. The inactive form is obtained by aerobic purification and contains an $[Fe_3S_4]^0$ centre (type 3, Figure 7.4) as proved by X-ray crystallography (Robbins and Stout, 1989). The aconitase can be reactivated with Fe^{2+} under reducing conditions to give an $[Fe_4S_4]^{2+}$ core (Kent *et al.*, 1982). Recent studies indicate that the fourth iron atom of the cluster is directly involved in coordinating one carboxyl group of the substrate (Kennedy *et al.*, 1987). $[Fe_3S_4]^0$ cores have also been identified by protein crystallography in *Azotobacter vinelandii* (Stout, 1989 and references therein) and *Desulfovibrio gigas* (Kissinger *et al.*, 1988, 1989). These 'voided cubane' structures are thought to be important in what Holm describes as the 'subsite specific' reactions of certain iron-sulphur clusters such as those which catalyse non redox transformations (Holm *et al.*, 1990 and references therein). Iron(III) species can be bound into the void leading to a differentiation in the iron subsites. In this way, transformations can be effected by the protein bound cluster due to the different binding properties of the iron subsites. In order to model this situation, 4Fe-4S and other cubane type clusters have been prepared from the trithiol $L(SH)_3$ (Figure 7.6) (Stack and Holm, 1987, 1988; Stack *et al.*, 1990; Ciurli and Holm, 1989). This results in a 3:1 iron subsite differentiation as illustrated in Figure 7.6b and the ligand binding properties, structural and electronic features specific to the unique subsite have been investigated (Holm *et al.*, 1990 and references therein; Weigel and Holm, 1991). The reactions of the cluster with various reagents were monitored by ^1H NMR and it was found that biologically relevant monodentate ligands such as RS^-

(a)

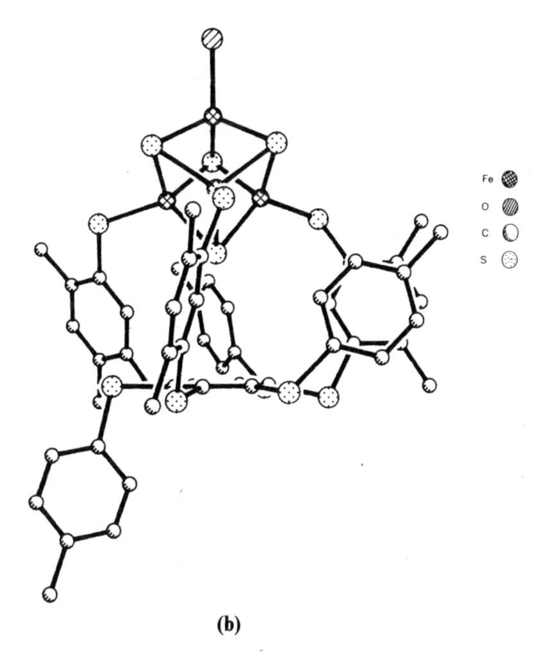

(b)

Figure 7.6 (a) $L(SH)_3$ ligand used to produce the subsite differentiated 4Fe–4S cluster (b).

and RO^- yield 4-coordinate subsites whereas chelating ligands could give 5 and 6 coordinate subsites. Small but measureable changes in $[Fe_4S_4]^{2+/1+}$ redox potentials were also found for the different ligands. A hydroxide cluster with Fe–OH at the unique subsite in equilibrium with a μ_2-oxo bridged double cluster was postulated on the basis of electrochemical evidence (Weigel and Holm, 1991). The hydroxo species would represent the deprotonated form of an aquo species such as is found in the resting state of aconitase (Robbins and Stout, 1989).

7.3.4 *Models for nitrogenase*

There are two metal containing components to the nitrogen fixation enzyme nitrogenase. Much effort has been expended on trying to elucidate the composition and structure of the Mo-Fe (or V-Fe) containing component. Model compounds have played an important role in furthering the understanding of certain aspects of the nitrogen fixation cycle, but progress has been hampered by lack of definitive structural data on the native Fe-Mo (or Fe-V) site. Coucouvanis has recently reviewed progress in the area of small molecule models for this site (Coucouvanis, 1991).

7.4 Proteins containing monomeric iron sites

Tightly bound monomeric iron centres are found at the catalytic sites of a number of non-haem non-iron-sulphur enzymes such as certain oxygenases, catecholases and lipoxygenases (Cox *et al.*, 1988, and references therein). Dioxygenases catalyse the cleavage of molecular dioxygen with subsequent incorporation of both oxygen atoms into organic substrates as part of the biodegradation of aromatic compounds (White *et al.*, 1984). The way in which these enzymes are able to activate oxygen is poorly understood mostly because of the lack of information about the nature of their active sites. Enzymes containing high spin ferric sites, such as catechol 1,2-dioxygenase and protocatechuate 3,4-dioxygenase, catalyse the intradiol cleavage of catechols to *cis,cis*-muconic acids via a mechanism which appears to involve a high spin iron(III) centre exclusively, with no change in spin or oxidation state of the iron observed in a variety of spectroscopic studies (Que *et al.*, 1987, and references therein). Other, such as protocatechuate 4,5-dioxygenase and phenylalanine hydroxylase, which appear to involve an extradiol cleavage mechansim, are less accessible spectroscopically because the active site involves a ferrous centre (Mabrouk *et al.*, 1991, and references therein). These proteins belong to a general class of metalloproteins where phenolate coordination is important, other iron examples being transferrin and the phosphatases (Que, 1983).

7.4.1 *Models for Fe(III) centres*

The X-ray structure of the protocatechuate 3,4-dioxygenase of *Pseudomonas aeruginosa* has been determined to 0.28 nm resolution (Ohlendorf *el al.*, 1988). The iron(III) site is coordinated by two Tyr and two His from the protein and the approximately trigonal bipyramidal pentacoordination in the resting state is completed by one bound water molecule. It is likely that this site is common to many of the other Fe(III) oxygenases. Systems containing copper (Demmin *et al.*, 1981, and references therein), ruthenium (Matsumoto and Kuroda, 1982), vanadium (Tatsumo *et al.*, 1982, 1983), iridium (Barbaro *et al.*, 1991) and iron (Funabiki *et al.*, 1986; Nishida *et al.*, 1984; Russo *et al.*, 1986; Weller and Weser, 1982, 1985) which produce the desired catechol cleavage reaction have been reported. The system originally developed by Weller and Weser in 1982 based on an Fe(III)/nta complex which is capable of catalysing the intradiol reaction thought to be important in the mechanism of the cleavage of catechols in catechol 1,2-dioxygenase and protocatechuate 3,4-dioxygenase was further investigated by Que and coworkers who isolated and characterised by X-ray crystallography the catecholate complex of {Fe(nta)} and also demonstrated its ability to react with dioxygen to afford the desired intradiol cleavage product (White *et al.*, 1984). These workers extended their study to Fe(III)/catecholate complexes with a range of other tetradentate ligands derived from nta in order to investigate the role played by the iron coordination environment and to help elucidate the mechanism involved in the catalytic cycle (Que *et al.*, 1987).

The modelling of other monomeric non-haem iron protein sites such as in lipoxygenases has been hampered by the lack of detailed structural information. It is thought that this site may be similar to that of the iron centre of a bacterial photosynthetic system (*Rhodobacter sphaeroides*) for which a crystal structure analysis has shown that the iron is hexacoordinated with four histidyl residues and a bidentate glutamate (Feiters *et al.*, 1990; Allen *et al.*, 1988) and consequently, a tetradentate polyimidazole ligand has been synthesised in order to produce models to test this hypothesis using comparative spectroscopy (Mulliez, 1989). The ligand forms monomeric iron(II) complexes of general formula $[Fe(tim)X]^+$, where X can be halide or carboxylate and the X-ray structure for $X = Cl^-$ reveals a distorted trigonal bipyramidal geometry at the iron(II). Oxidation of these produced the iron(III) complexes (Mulliez *et al.*, 1991).

The synthesis of model compounds with imidazole-based ligands will help in the investigation of iron/histidyl interactions, which are known to be important in structurally characterised proteins such as iron-based superoxide dismutase (Carlioz *et al.*, 1988) as well as the many less well-defined proteins with monomeric iron sites.

7.5 Oxo-bridged di-iron centres

The Fe–O–Fe motif is found in a variety of chemical and biochemical situations. The $[Fe^{III}-O-Fe^{III}]^{4+}$ unit is important in the hydrolysis of iron(III) and can be trapped by supplying ligands to block the other five hydrolysis sites on the iron centres. A substantial number of complexes containing this unsupported oxide bridge unit has been reported and crystallographically characterised (Murray, 1974; Kurtz, 1990). However, none of these reproduces the spectroscopic properties of the oxo-bridged di-iron centres known to exist in proteins. These appear to require support to the oxide bridge in the form of further bridging ligands such as carboxylate or phosphate (Qué and True, 1990 and references therein). This realisation stimulated efforts to produce multiply-bridged di-iron centres. The best documented and most successful endeavours have been in producing models for the haemerythrin (Hr) active site where protein crystallography and extensive magnetic and spectroscopic studies have also been performed (Vincent *et al.*, 1990 and references therein; Stenkamp *et al.*, 1984, 1985). More recently the crystal structure of ribonucleotide reductase (RR) has been determined revealing important differences in the coordination of the iron centres (Nordland *et al.*, 1990). Two other proteins, purple/pink acid phosphatase (PAP) and methane monooxygenase (MMO) are thought to contain di-iron centres on the basis of spectroscopic and magnetic evidence. Lack of definitive structural data for these has encouraged the exploration of the chemistry and properties of di-iron models bridged by a range of bidentate ligands (Turowski *et al.*, 1990; Sanders-Loehr *et al.*, 1989; Drueke *et al.*, 1989).

7.5.1 *Haemerythrins*

Haemerythrins transport oxygen in marine invertebrates. Spectroscopic and crystallographic studies have shown that the deoxy-form consists of two Fe(II) centres bridged by one hydroxide and two carboxylate functions (from aspartate and glutamate residues) with histidyl residues completing the coordination spheres of the two irons such that one is six and the other five coordinate with one site vacant. The oxy-form exhibits similar coordination, but with an oxide bridge, a hydroperoxide ion occupying the vacant coordination site, and both irons in the +3 oxidation state. The met-form of this protein is produced by oxidation of the deoxy-form so that both irons are in the +3 oxidation state, and the met X^- form results on addition of anions such as azide (N_3^-) to this. These are illustrated in Figure 7.7. It was this latter form which first yielded crystals suitable for the structure determination appearing in 1984 (Stenkamp *et al.*, 1984) which revealed that the models for this reported in 1983 which were synthesised virtually simultaneously in Germany (Wieghardt *et al.*, 1983) and the USA (Armstrong and Lippard, 1983), $[Fe_2^{III}O(O_2CMe)_2(TACN)_2]^{2+}$ and $[Fe_2^{III}O(O_2CMe)_2(HB(pz)_3)_2]$ respectively, reproduce all the important structural, spectroscopic and magnetic properties of the protein site. Subsequently other models containing the $[Fe_2^{III}O(O_2CR)_2]^{2+}$ unit have been characterised (Lippard, 1988 and references therein; Kurtz, 1990 and references therein) including $[Fe_2O(O_2CPh)_2(bipy)_2(N_3)_2]$ which is a model for metazido Hr (Vincent *et al.*, 1988). A model for the biologically more relevant deoxy-form of Hr has also been

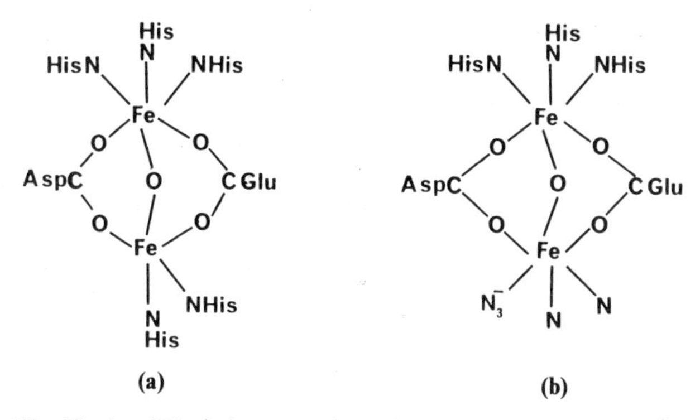

Figure 7.7 The iron(III) environments in methaemerythrin and azidohaemerythrin.

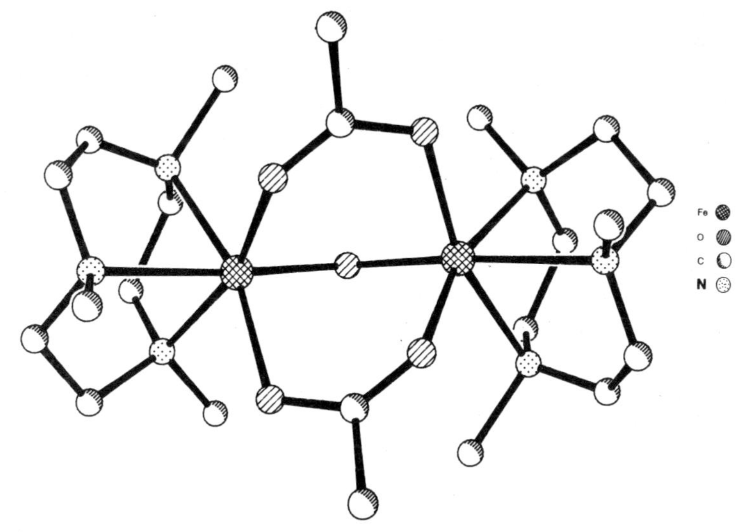

Figure 7.8 The structure of the $[Fe_2^{II}(OH)(O_2CMe)_2(Me_3TACN)]^+$ cation which models the active site of deoxyhaemerythrin.

prepared: $[Fe_2^{II}(OH)(O_2CMe)_2(Me_3TACN)]^+$ (Chaudhuri *et al.*, 1985) Figure 7.8.

Two compounds containing diferrous units are reported to interact irreversibly with dioxygen (Ménage *et al.*, 1990; Tolman *et al.*, 1991) but this irreversible reaction together their compositions makes them better models for the iron centres of RR and MMO (see below).

7.5.2 *Ribonucleotide reductases*

Although the B2 subunit in reductases contains two oxo-bridged iron atoms, the 2.2 Å resolution crystal structure of the *E. coli* RR B2 protein reported in

1990 shows that the coordination of the metals is rather different from met Hr (Nordland *et al.*, 1990). Firstly there is only one carboxylate bridge, provided by a glutamate, supporting the oxide bridge, and secondly the coordination of one iron is completed by a bidentate aspartate residue, a histidyl residue and a water molecule while the other has two monodentate glutamate residues, a histidyl and a coordinated water molecule. Variable temperature magnetic studies suggest that the irons are in the +3 oxidation state being high spin and strongly antiferromagnetically coupled (Petersson *et al.*, 1980). This is supported by the fact that on reduction with dithionite the Mössbauer spectrum has parameters typical of iron(II) ($\delta = 1.26$ mm/s, $\Delta E_Q = 3.13$ m/s at 4.2 K) (Atkin *et al.*, 1973).

Prior to the solution of the X-ray structure of the protein, the $[Fe_2^{III}O(O_2CR)]^{3+}$ unit was found to form when the tetradentate N-donor ligand TPA was used (Yan *et al.*, 1989; Norman *et al.*, 1990). A better model in terms of iron coordination results when the mixed N/O donor ligands BPG or PDA are employed. The crystallographically characterised BPG complex has each iron coordinated by three nitrogens and one carboxylate oxygen from the ligand in addition to the bridging oxide and carboxylate (Figure 7.9) (Ménage and Que, 1991). This is still not a very accurate model for the RRB2 iron site. In particular, the carboxylate binding demonstrates several features which will be hard to reproduce in simple models, for while one iron is coordinated by a bidentate carboxylate (a rare situation in iron complex chemistry) the other has two monodentate carboxylates, one of which appears to interact with the neighbouring iron to give a 'semi-bridging' carboxylate arrangement.

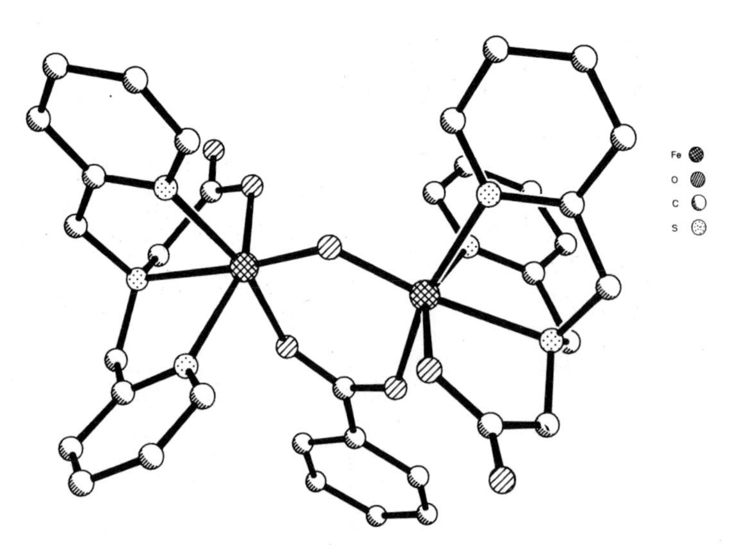

Figure 7.9 The structure of the $[Fe_2^{III}O(BPG)_2(OBz)]^+$ cation which models the active site of RRB2.

7.5.3 *Purple acid phosphatases*

Purple acid phosphatases are characterised by their intense pink or violet colour and their ability to catalyse the hydrolysis of phosphoric acid esters under acidic conditions. The best characterised of these, uteroferrin and beef spleen PAP, contain two iron atoms per molecule, but the details of the metal sites can only be inferred from spectroscopic measurements since attempts to obtain crystals for diffraction studies have been unsuccessful so far. The catalytically active pink form contains iron in the +2 and +3 oxidation states, whereas the oxidised purple forms are diferric and there is some spectroscopic evidence for a relatively close Fe–Fe interaction. The structure proposed for the iron site of the diferric form involves a binuclear diiron core (probably oxo-bridged) and a terminally coordinated phosphato ligand, which is bound to the non reduceable Fe(III) ion (Vincent *et al.*, 1990, and references therein). Diiron(II,III) and diiron(III,III) model compounds containing bridging phosphate and phosphate ester ligands have been synthesised (Bremer *et al.*, 1990; Drueke *et al.*, 1989; Armstrong and Lippard, 1985; Turowski *et al.*, 1990; Yan *et al.*, 1989; Norman *et al.*, 1990). The first complex to contain a terminally coordinated phosphato ligand (Figure 7.10) was prepared by reaction of iron(III) perchlorate with a dinucleating benzimidazole based ligand (Htbpo)

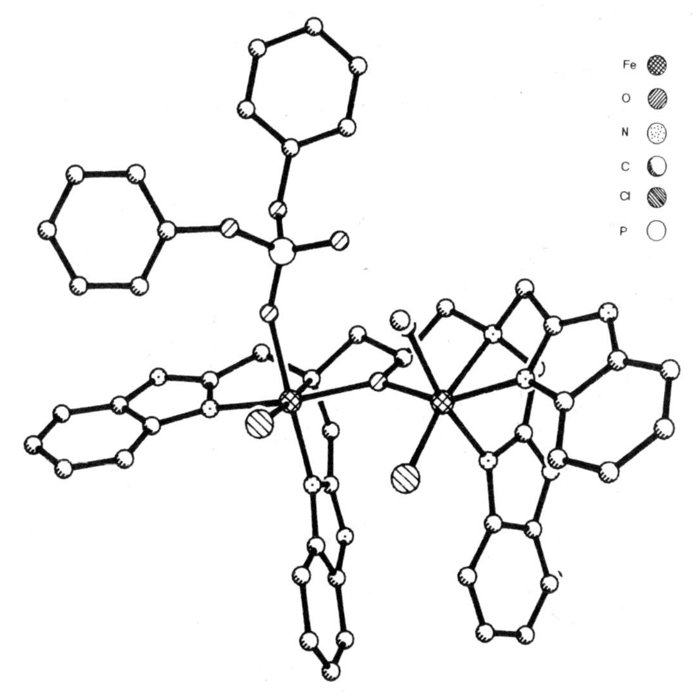

Figure 7.10 The structure of the $[Fe_2^{III}Cl_2\{O_2P(OPh)_2\}(tbpo)(MeOH)]^{2+}$ cation which models the terminal coordination of a phosphate ligand in PAP.

and diphenylphosphate (Bremer *et al.*, 1991). There is strong evidence to suggest PAPs have phenolate interactions at their metal sites (Que, 1983 and references therein) presenting further challenges to biomimetic chemists.

7.5.4 *Methane monooxygenases*

Iron-containing methane monooxygenases are soluble non-haem proteins which contain four irons per protein and two per subunit. It is thought that these are present as oxo-bridged dinuclear units, but the details of the coordination of the irons are unclear and have been inferred only from spectroscopic studies (Vincent *et al.*, 1990 and references therein). The mechanistic aspects of alkane functionalisation in the presence of hydrogen peroxide with various iron-containing model compounds were investigated in order to shed light on the actions of proteins containing diiron oxo-bridged centres (Fish *et al.*, 1991). Esr spectroscopy suggests that mixed valence Fe(II)/Fe(III) and oxidised Fe(III)/Fe(III) species are important in the active form (Woodland *et al.*, 1986 and references therein) which parallels the situation found in the other diiron oxo-bridged proteins discussed here. Therefore models which test the effect of the ligands on the magnetic and spectroscopic properties of the diiron cores should help in assigning the likely coordination of the iron centres. However, two recent studies on diiron model compounds have shown that the nature of the carboxylate bridges, the angle at the μ-oxide and hydrogen bonding interactions with the μ-oxide can all attenuate the magnetic exchange interactions between the iron centres significantly (Beer *et al.*, 1991; Oberhausen *et al.*, 1992). Such studies should also help in gaining an understanding of how proteins can tune their diiron centres to perform their different functions: e.g. reversible dioxygen binding in Hr and oxygen activation in RR and MMO.

7.6 Models for the uptake and transport of iron

Iron uptake (reviewed by Weinberg, 1989) for biological systems can pose problems in the oxidising environment of the earth today, where most iron exists in the form of insoluble iron(III) oxides and oxyhydroxides. Plants and bacteria have developed chelating agents, known as siderophores, which are able to solubilise and hence sequester the iron from the environment. This is achieved either by the formation of a tightly bound Fe(III) monomeric chelate structure or by the simultaneous reduction and chelation of the Fe(III) to give an Fe(II) chelate monomer. These molecules are also used to transport the iron within the plant or bacterium. In higher animals iron is taken up from the gut. There are several mechanisms proposed for iron assimilation and in the case of man the various factors contributing to enhanced or inhibited iron uptake have been thoroughly researched due to the vital importance of this metal in human well-being. However, the chemical and biological species which are important in iron uptake from the gut are still not characterised. Iron is transported in most vertebrates by transferrins, which are found in many physiological fluids, such as tears, milk, and serum. Although the basic structural features of these are known from X-ray crystallography, the exact nature of the interaction of the iron and other metals with the binding site are still subject to much research effort.

7.6.1 *Models for siderophores*

Siderophores are high-affinity iron(III)-binding compounds and consequently they tend to contain oxygen or mixed oxygen/nitrogen ligands arranged in such a way to give a hexadentate chelate structure on coordination with the iron(III) ion. More than one hundred naturally occurring siderophores have been isolated and characterised to date (Matzanke *et al.*, 1989). Simple models for the coordination of the iron(III) site are relatively easy to prepare, such as the classic triscatacholato complex $[Fe(cat)_3]^{3-}$ (Raymond *et al.*, 1976), which is a model for enterobactin as shown in Figure 7.11. New ligands have been synthesised which mimic the coordination properties of the naturally

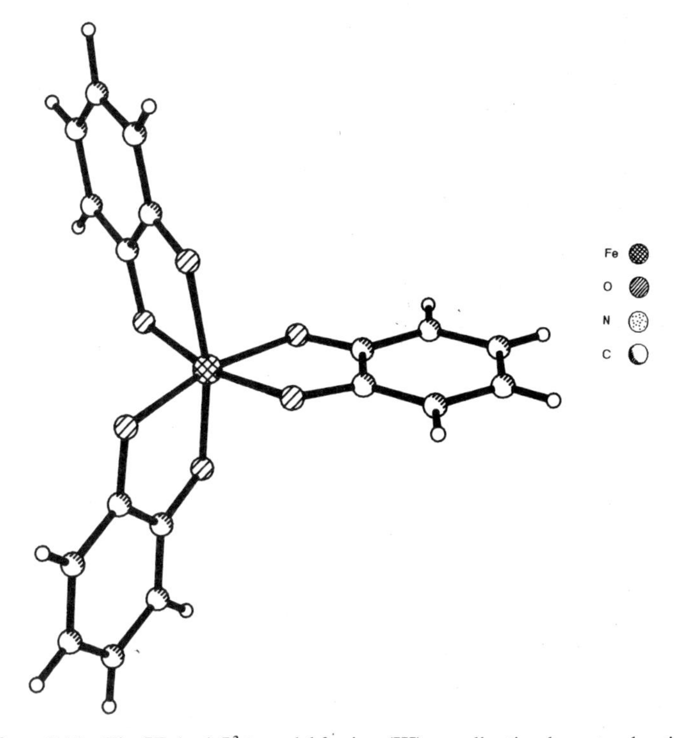

Figure 7.11 The $[Fe(cat)_3]^{3-}$ model for iron(III) coordination by enterobactin.

occurring siderophores, such as stability constants and denticity, more closely. This has been achieved by synthesising macrocyclic ligands which are predisposed to metal complexation because of their steric constraints. Examples are the bicapped MECAM and TRENCAM ligands and the ethane trimer macrocycles based on the 2,3-dihydroxyterephthalamide moiety rather than the 2,3-dihydroxybenzamide subunit of enterobactin shown in Figure 7.12 (Wolfgang and Vögtle, 1984; Stutte *et al.*, 1987; Sun *et al.*, 1986; Rodgers *et al.*, 1985; McMurry *et al.*, 1987). More recent work has involved trying to understand more about the iron uptake and release mechanisms of these siderophores and the research of Raymond, Martell, and Vögtle and their coworkers has been central in this areas, as described in Raymond's recent review (Raymond, 1990) and in the most recent papers from these groups (Scarrow *et al.*, 1991; Garrett *et al.*, 1991; Abu-Dari and Raymond, 1991; Motekaitis *et al.*, 1991). In the case of enterobactin it is likely that hydrolysis of the ligand allows the metal to be released and taken up by the organism. Raymond and coworkers investigated the thermodynamic properties of Fe(III) complexes of enterobactin and its hydrolysis products (Scarrow *et al.*, 1991). They compared the enthalpies and entropies associated with the various

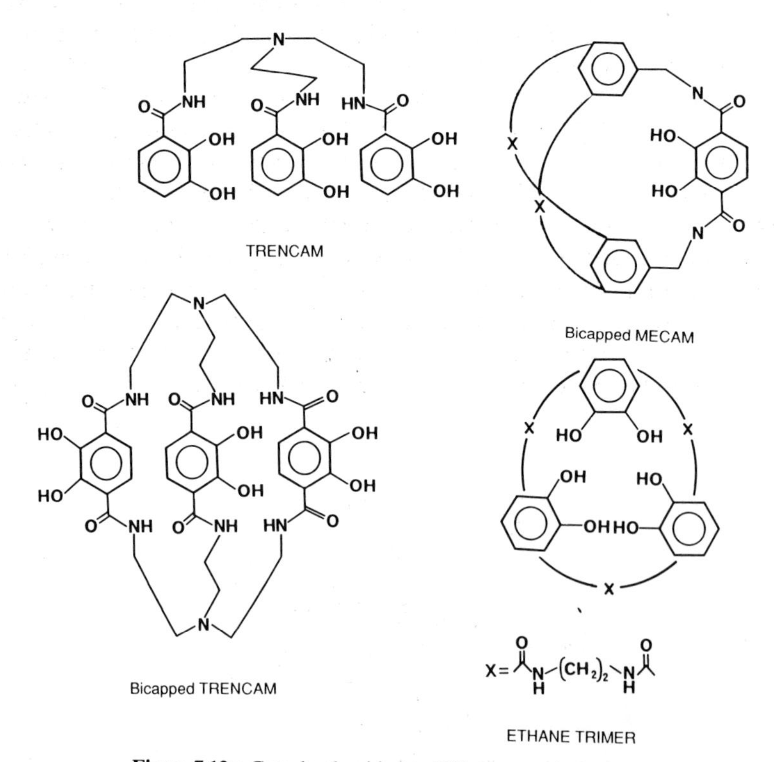

Figure 7.12 Catechoylamide iron(III) sequestering agents.

ligation reactions and compared the results with those for triscatecholate model complexes. They found that the enterobactin hydrolysis products were effective iron uptake mediators and also concluded that the major contribution to the enhanced stability of enterobactin versus the model complexes arises from an entropic effect. This view is supported by thermodynamic studies on the sterically constrained macrocycles.

7.6.2 *Models for transferrins*

Transferrins contain two iron binding sites. X-ray structures of the diferric forms of human lactotransferrin (Anderson *et al.*, 1987, 1989) and of rabbit serotransferrin (Bailey *et al.*, 1988) have revealed the three dimensional structure of the proteins and their iron binding sites to relatively high resolution (0.28 nm and 0.33 nm) and the structures of other transferrins are currently being investigated. The iron binding sites appear identical even in the higher resolution study on human lactoferrin and it is thought that this may well be the case for all transferrins. Each iron atom is coordinated to four protein ligands, 2 Tyr, 1 Asp, 1 His and the coordination is probably completed by a bidentate carbonate (or bicarbonate) ligand (Anderson *et al.*, 1989) as illustrated in Figure 7.13. There is still some question as to the exact identity of the non-protein ligand which binds to the iron, but the best model for the residual electron density found at the non-protein ligand site in difference electron density maps is a carbonate anion, and this view is supported by the observation that the binding of iron requires carbonate or bicarbonate to be present as a synergistic anion (Schlabach and Bates, 1975). Several small molecule models have been synthesised and characterised. Prior to 1987 these were useful in helping to establish the likely nature of the iron(III) coordination site and involved the use of mixed oxygen/nitrogen Schiff base chelating ligands. Spectroscopic studies on native proteins and model compounds strongly suggested that phenolate coordination was important (Chasteen, 1977). Consequently iron(III) complexes formed with tetradentate mixed N and phenolate ligands such as 2-(5-methylpyrazol-3-yl) phenolate)

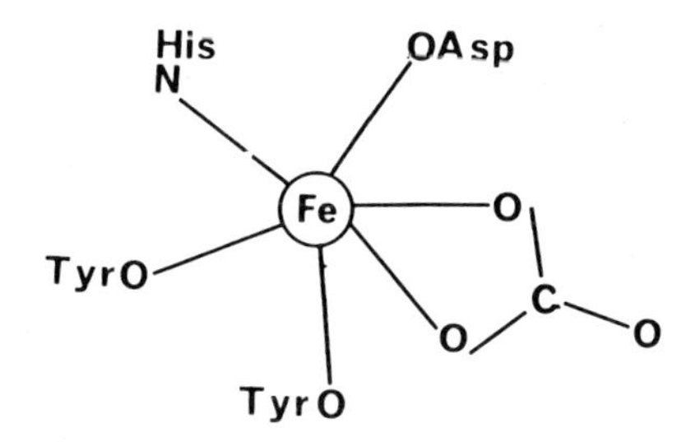

Figure 7.13 The iron binding site in lactoferrin.

(Ainscough *et al.*, 1980); *N*,*N'*-ethylenebis-((*o*-hydroxyphenyl)glycine) (Bailey *et al.*, 1981) and *N*-[2-((*o*-hydroxyphenyl)glycino)ethyl]-salicylidenimine (Carrano *et al.*, 1985) were synthesised and characterised by X-ray crystallography. Our present knowledge of the coordination environment of the iron has led to more sophisticated models being developed. The ligands used for the earlier models favoured an equatorial coordination by the ligand, leaving two *trans* sites available for coordination by smaller molecules, but the transferrin site requires two *cis* sites in order to accommodate a bidentate anion. In addition, the ligation by the protein side chains corresponds to two phenolate, one carboxylate and one imidazole ligand. Ternary iron(III) complexes were synthesised using a variety of bidentate salicylate and phenolate-benzimidazole based ligands by first forming the iron(III) salicylate species and then adding a bidentate phenolate-benzimidazole ligand. The octahedral coordination sphere of the iron is thought to be completed by solvent molecules but crystalline solids could not be isolated from reaction mixtures. Instead, *N*-methylimidazole was added to displace the coordinated water molecules which afforded single crystals of the compound illustrated in Figure 7.14 whose structure was determined by X-ray crystallography (McDevitt *et al.*, 1990). Although the gross iron coordination geometry seems to be well reproduced in these model compounds it turns out that all the spectral features of the electronic absorption spectrum of lactotransferrin cannot be reproduced by any single model system. This may be a result of the distortion imposed by the protein on the iron(III) environment.

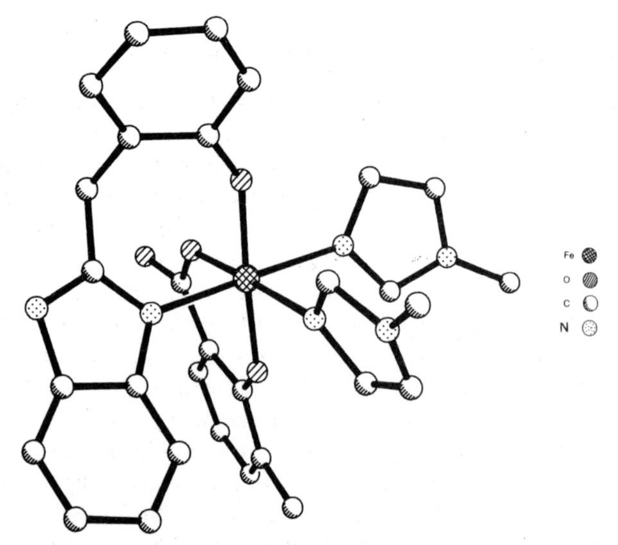

Figure 7.14 Molecular structure of the *N*-methylimidazole adduct of the ternary salicylate, benzimidazole Fe(III) model for transferrin.

7.7 Models for biomineralisation processes

The process by which organisms form inorganic minerals is termed biomineralisation (Lowenstamm and Weiner, 1989 and references therein). Iron oxide and sulphide based biominerals are common in nature and fulfil a variety of functions (Frankel and Blakemore, 1991 and references therein). There has been a growing interest in elucidating the way in which biomineralisation processes occur fuelled by a desire to learn how nature is able to dictate structural type and therefore function of the biomineral. An understanding of these processes offers the possibility of tailoring new materials for industry which can be synthesised to have specific properties.

The biomineralisation processes which produce minerals such as the oxyhydroxide and oxide minerals utilised by various organisms for functions as varied as teeth (e.g. in limpets, goethite, α Fe.O.OH) and navigation (e.g. in magnetotactic bacteria, magnetite, Fe_3O_4) can be modelled in the laboratory by supplying small amounts of anionic species or organic species to hydrolysed iron solutions which modify the crystal morphology and sometimes phase of the mineral formed. An important feature of the biological systems is the way in which organic molecules such as proteins and polysaccharides interact with the inorganic ions on the surfaces of biominerals. The organic molecules control the rate of nucleation, the first stage of crystal growth, which in turn affects the rate of deposition. They also dictate the way in which the inorganic ions interact to build up the crystal structure. Hence the nucleating surface of the organic template may influence the crystal structure of the biomineral (Mann, 1989 and references therein).

7.7.1 Models for ferritin and haemosiderin iron mineral cores

The structures of most iron biominerals have been determined by diffraction methods. However, the iron cores of the iron storage proteins ferritin and haemosiderin are notable exceptions. This is in part due to the fact that the cavity inside ferritin available to the iron core can be approximated by a sphere of diameter 8 nm, which at most can accommodate about 4000 irons in a close-packed iron oxyhydroxide mineral structure. This means that unlike other iron biominerals the iron core of ferritin cannot be regarded as an infinite crystal structure, but must be thought of either as a finite crystallite or else as a trapped portion of an infinite structure. Model compounds have been sought to help in elucidating the structures adopted by the iron mineral cores of iron storage proteins and in some cases the processes by which such biominerals form. It appears that various types of ferritin from different organisms possess different mineral core structures, which can vary in crystallinity, as judged by diffraction pattern and Mössbauer spectroscopy (St Pierre *et al.*, 1989 and references therein) and also composition, as judged by elemental analysis. For example, bacterioferritin contains high levels of inorganic phosphate and relatively low overall levels of iron, whereas horse spleen ferritin (HSF) contains little or only adventitious phosphate and much higher levels of iron. The most research has been carried out on the more crystalline ferritins such as HSF which has been subject to extensive spectroscopic measurements as well as EXAFS analysis (Mackle *et al.*, 1991 and references therein; Webb and Gray, 1974). It was suggested by Towe and Bradley that the mineral ferrihydrite provides a good model for this core on the basis of X-ray powder and electron diffraction studies (Towe and Bradley, 1967). However, this mineral type encompasses a range of phases which form as amorphous ferric hydroxide, $Fe(OH)_3$, and then age and transform into the well characterised minerals goethite (α Fe.O.OH) and haematite (α Fe_2O_3). This has led to a considerable debate not only over the validity of the model, but also about the structure of the mineral model itself (Eggleton and Fitzpatrick, 1990 and references therein). In common with other naturally occurring iron minerals which have been suggested as structural models for the ferritin core, the ferrihydrite model also suffers from being an infinite rather than finite structure. The so-called Saltman Spiro balls which appear to form in

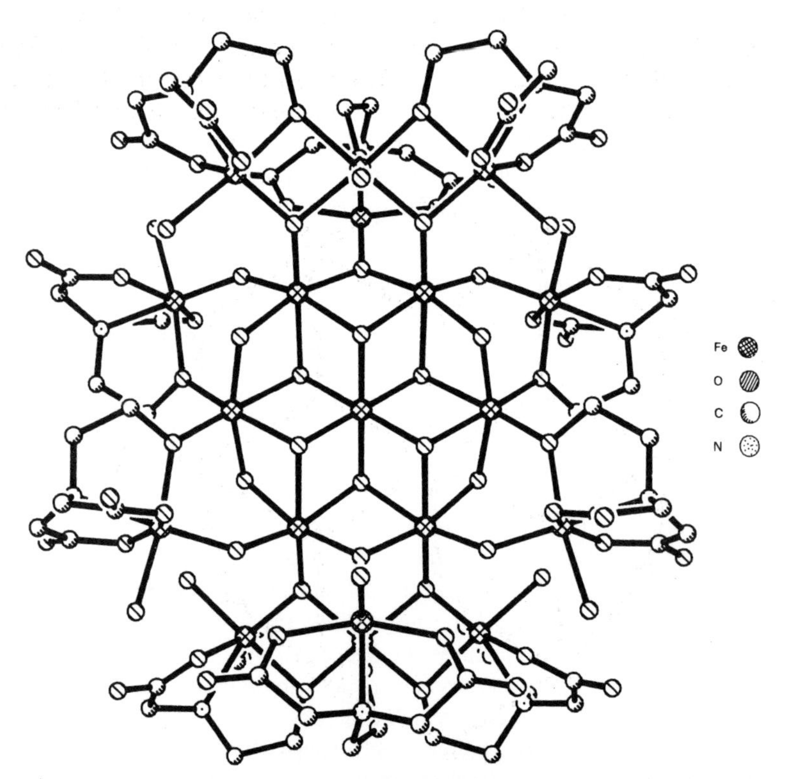

Figure 7.15 The molecular structure of $[Fe_{19}(\mu_3\text{-}O)_6(\mu_3\text{-}OH)_6(\mu_2\text{-}OH)_8(\text{heidi})_{10}(H_2O)_{12}]^+$ (H atoms omitted for clarity).

hydrolysed solutions of Fe(III) in the presence of small quantities of ligands such as citrate might provide excellent models for ferritin cores as they are thought to be composed of an iron oxyhydroxide sphere containing between 2000 and 4000 irons with a surface of ligands holding them together (Spiro *et al.*, 1967). Unfortunately, a lack of structural data and reproducibility has meant that this particular avenue has been abandoned by most chemists. At the other end of the scale, relatively small clusters of oxide and hydroxide bridged iron(III) ions have been isolated and characterised by X-ray crystallography. Although a variety of interesting polyiron complexes was reported prior to 1992 (e.g. Wieghardt *et al.*, 1984; Gorun and Lippard, 1986; Gorun *et al.*, 1987; Lippard, 1988 and references therein; Taft and Lippard, 1990; McCusker *et al.*, 1991 and references therein) and in some cases spectroscopic comparisons with native systems (Islam *et al.*, 1989), most are not directly relevant to ferritin cores since they do not display close-packed iron oxyhydroxide cores and are produced by non-aqueous routes. The 19 and 17 iron species reported recently which were isolated from hydrolysed iron(III) solutions containing the ligand heidi do however model a number of features of the ferritin cores as revealed by single crystal X-ray diffraction (Heath and Powell, 1992). The structure of the 19 iron compound is illustrated in Figure 7.15. The close-packed iron-hydroxide core of this shown in Figure 7.16 is bonded via oxide and hydroxide bridges to further iron(III) centres (nucleation sites) on the inner surface of the organic ligand coat. The core

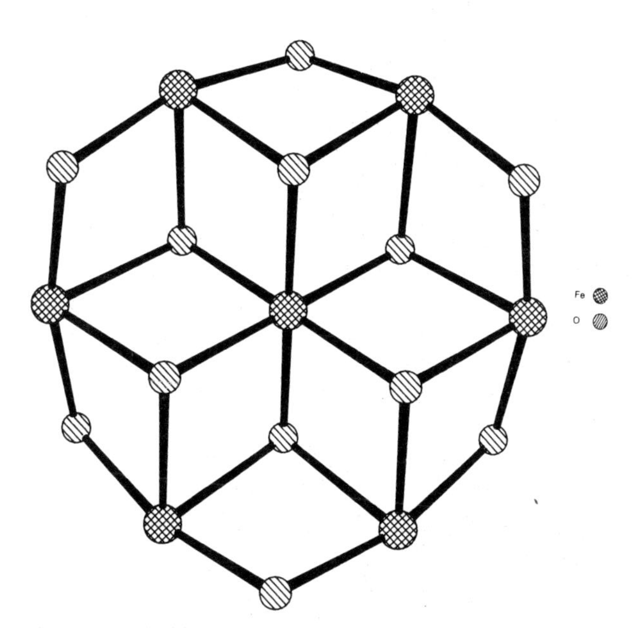

Figure 7.16 The close-packed iron hydroxide core of $[Fe_{19}(\mu_3\text{-}O)_6(\mu_3\text{-}OH)_6(\mu_2\text{-}OH)_8(heidi)_{10}(H_2O)_{12}]^+$ (H atoms omitted for clarity).

derives from a portion of an infinite 2-D $\{Fe(OH)_2{}^+\}_n$ framework, which means that the iron/ligand shell traps the iron in a hydroxide mineral structure not normally observed for the metal in this oxidation state. This raises the question of whether biomineral cores such as those found in ferritin and haemosiderin might possess different structural types from the normal iron(III) minerals in response to the stereochemical demands imposed by the nucleation sites on the protein shell. The comparison of the structural and spectroscopic characteristics of such model systems with the native ones will help to answer such questions.

List of abbreviations

trien	triethylenetetramine
H_3nta	nitrilotriacetic acid
tim	bis,bis(imidazol-2-yl)methane
TACN	1,4,7-triazacyclononane
Me_3TACN	1,4,7-trimethyl-1,4,7-triazacyclononane
HB(pz)$_3$	hydrotris(pyrazolyl)borate
bipy	2,2′-bipyridine
TPA	tris(2-pyridylmethyl)amine
BPG	*N,N′*-bis(2-pyridylmethyl)glycine
PDA	*N*-carboxymethyl-*N*-(2-pyridylmethyl)glycine
Htbpo	*N,N,N′,N′*,-tetrakis(2-benzimidazolylmethyl)-2-hydroxy-1,3-diaminopropane
H_2cat	catechol (1,2-dihydroxybenzene)
H_3heidi	hydroxyethyliminodiacetic acid

Acknowledgement

I would like to thank Sarah Heath for her help in preparing the manuscript and figures.

References

Abu-Dari, K. and Raymond, K.N. (1991) *Inorg. Chem.*, **30**, 519.
Adman, E.T., Sieker, L.C. and Jensen, L.H. (1973) *J. Biol. Chem.*, **248**, 3987.
Ainscough, E.W., Brodie, A.M., Plowman, J.E., Brown, K.L., Addison, A.W. and Gainsford, A.R. (1980) *Inorg. Chem.*, **19**, 3655.
Allen, J.P., Feher, G., Yeates, T.O., Komlya, H. and Rees, D.C. (1988) *Proc. Natl. Acad. Sci. USA*, **85**, 8487.
Anderson, B.F., Baker, H.M., Norris, G.E., Rice, D.W. and Baker, E.N. (1989) *J. Mol. Biol.*, **209**, 711.
Anderson, B.F., Baker, H.M., Norris, G.E., Rice, D.W. and Baker, E.N. (1987) *Proc. Natl. Acad. Sci. USA*, **84**, 1769.
Armstrong, W.H. (1988) *ACS Symp. Ser.*, **372**, 1.
Armstrong, W.H. and Lippard, S.J. (1983) *J. Am. Chem. Soc.*, **105**, 4837.
Armstrong, W.H. and Lippard, S.J. (1985) *J. Am. Chem. Soc.*, **107**, 3730.
Atkin, C.L., Thelander, L., Reichard, P. and Long, G.J. (1973) *Biol. Chem.*, **248**, 7464.
Averill, B.A., Herskovitz, T., Holm, R.H. and Ibers, J.A. (1973) *J. Am. Chem. Soc.*, **95**, 3523.

Bailey, S., Evans, R.W., Garratt, R.C., Gorinsky, B., Hasnain, S., Horsburgh, C., Jhoti, H., Lindley, P.F., Mydin, A., Sarra, R. and Watson, J.L. (1988) *Biochemistry*, **27**, 5804.
Bailey, N.A., Cummins, D., McKenzie, E.D. and Worthington, J.M. (1981) *Inorg. Chim. Acta.*, **50**, 111.
Barbaro, P., Bianchini, C., Mealli, C. and Meli, A. (1991) *J. Am. Chem. Soc.*, **113**, 3181.
Beer, R.H., Tolman, W.B., Bott, S.G. and Lippard, S.J. (1991) *Inorg. Chem.*, **30**, 2082.
Berg, J.M. and Holm, R.H. (1982) in *Iron-Sulfur Proteins*, (ed T.G. Spiro), Wiley-Interscience, New York, chapter 1.
Bremer, B., Schepers, K., Fleischhauer, P., Hasse, W., Henkel, G. and Krebs, B. (1991) *J. Chem. Soc. Chem. Commun.*, 510.
Carlioz, A., Ludwig, M.L., Stallings, W.C., Fee, J.A., Steinman, H.M. and Touati, D. (1988) *J. Biol. Chem.*, **263**, 1555.
Carrano, C.J., Spartalian, K., Appa Rao, G.V.N., Pecoraro, V.L. and Sundaralingam, M. (1985) *J. Am. Chem. Soc.*, **107**, 1651.
Carter, C.W. *et al.* (1972) *Proc. Natl. Acad. Sci. USA*, **69**, 3526.
Carter, C.W. *et al.* (1974) *J. Biol. Chem.*, **249**, 6339.
Carter, C.W., Freer, S.T., Xuong, Ng., Alder, R.A. and Kraut, J. (1972) in *Cold Spring Harbor Symp. Quart. Biol.*, **36**, 381.
Chasteen, N.D. (1977) *Coord. Chem. Rev.*, **22**, 1.
Chaudhuri, P., Wieghardt, K., Nuber, B. and Weiss, J. (1985), *Angew. Chem. Int. Ed. Engl.*, **24**, 778.
Christou, G. and Garner, C.D. (1980) *J. Chem. Soc. Dalton Trans.*, 2360.
Ciurli, S. and Holm, R.H. (1989) *Inorg. Chem.*, **28**, 1685.
Collman, J.P. (1977) *Acc. Chem. Res.*, **10**, 265.
Collman, J.P., Gauge, R.R., Reed, C.A., Halbert, T.R., Lang, G. and Robinson, W.T. (1975) *J. Am. Chem. Soc.*, **97**, 1427.
Collman, J.P., Gauge, R.R., Halbert, T.R., Marchon, J-C. and Reed, C.A. (1973) *J. Am. Chem. Soc.*, **95**, 7868.
Coucouvanis, D. (1991) *Acc. Chem. Res.*, **24**, 1
Cox, D.D., Benkovic, S.J., Bloom, L.M., Bradley, F.C., Nelson, M.J., Que, L. and Wallick, D.E. (1988) *J. Am. Chem. Soc.*, **110**, 2026.
Dawson, J.H. (1988) *Science*, **240**, 433.
Demmin, T.R., Swerdloft, M.D. and Rogic, M.M. (1981) *J. Am. Chem. Soc.*, **103**, 5795.
Drueke, S., Wieghardt, K., Nuber, B. and Weiss, J. (1989) *Inorg. Chem.*, **28**, 1414.
Eggleton, R.A. and Fitzpatrick, R.W. (1990) *Clays Clay Miner.*, **38**, 335.
Feiters, M.C., Boelens, S., Veldink, G.A., Vliegenthart, J.F.G., Navaratnam, S., Allen, J.C., Nolting, H.F. and Hermes, C. (1990) *Recl. Trav. Chim. Pays-Bas*, **109**, 133.
Fish, R.H., Konings, M.S., Oberhausen, K.J., Fong, R.H., Yu, W.M., Christou, G., Vincent, J.B., Coggin, D.K. and Buchanan, R.M. (1991) *Inorg. Chem.*, **30**, 3002.
Frankel, R.B. and Blakemore, R.P., (Eds.), (1991) in *Iron Biominerals*, Plenum Press, New York.
Funabiki, T., Mizoguchi, A., Sugimoto, T., Tada, S., Tsuji, M., Sakamoto, H. and Yoshida, S. (1986) *J. Am. Chem. Soc.*, **108**, 2921.
Garner, C.D. (1980) in *Transition Metal Clusters*, (ed. B.F.G. Johnson), Wiley-Interscience, New York, chapter 4.
Garrett, T.M., McMurry, T.J., Hosseini, N.W., Reyes, Z.E., Hahn, F.E. and Raymond, K.N. (1991) *J. Am. Chem. Soc.*, **113**, 2965.
Gebhard, M.S., Koch, S.A., Millar, M., Devlin, F.J., Stephens, P.J. and Solomon, E.I. (1991) *J. Am. Chem. Soc.*, **113**, 1640.
Gorun, S.M., Papaefthymiou, G.C., Frankel, R.B. and Lippard, S.J. (1987) *J. Am. Chem. Soc.*, **109**, 3337.
Gorun, S.M. and Lippard, S.J. (1986) *Nature*, **319**, 666.
Hatano, K., Safo, M.K., Walker, F.A. and Scheidt, W.R. (1991) *Inorg. Chem.*, **30**, 1643.
Heath, S.L. and Powell, A.K. (1992) *Angew. Chem. Int. Ed. Engl.*, **31**, 191.
Holm, R.H. (1975) *Endeavour*, **34**, 38.
Holm, R.H. (1977) *Acc. Chem. Res.*, **10**, 427.
Holm, R.H., Ciurli, S. and Weigel, J.A. (1990) in *Progress in Inorganic Chemistry*, (Ed. S.J. Lippard), Wiley-Interscience, New York, **38**, 1.
Ibers, J.A. and Holm, R.H. (1980) *Science*, **209**, 223.
Islam, Q.T., Sayers, D.E., Gorun, S.M. and Theil, E.C. (1989) *J. Inorg. Biochem.*, **36**, 51.

Johnson, M.R., Seok, W.K. and Ibers, J.A. (1991), *J. Am. Chem. Soc.*, **113**, 3998.
Kennedy, M.C., Werst, M., Telser, J., Emptage, M.H., Beinert, H. and Hoffman, B.M. (1987) *Proc. Natl. Acad. Sci. USA*, **84**, 8854.
Kent, T.A., Dreyer, J-L., Kennedy, M.C., Huynh, B.H., Emptage, M.H., Beinert, H. and Munck, E. (1982) *Proc. Natl. Acad. Sci. USA*, **79**, 1096.
Kim, K. and Ibers, J.A. (1991) *J. Am. Chem. Soc.*, **113**, 6077.
Kim, K., Fettinger, J., Sessler, J.L., Cyr, M., Hugdah, J., Collman, J.P. and Ibers, J.A. (1989) *J. Am. Chem. Soc.*, **111**, 403.
Kissinger, C.R., Adman, E.T., Sieker, L.C. and Jensen, L.H. (1988) *J. Am. Chem. Soc.*, **110**, 8721.
Kissinger, C.R., Adman, E.T., Sieker, L.C., Jensen, L.H. and LeGall, J. (1989) *FEBS. Lett.*, **244**, 447.
Kurtz, D.M. (1990), *Chem. Rev.*, **90**, 585.
Lane, R.W., Ibers, J.A., Frankel, R.B., Papaefthymiou, G.C. and Holm, R.H. (1977) *J. Am. Chem. Soc.*, **99**, 84.
Lever, A.B.P. (1972) *J. Chem. Ed.*, **49**, 656.
Li, X.Y. and Spiro, T.G. (1988) *J. Am. Chem. Soc.*, **110**, 6024.
Lippard, S.J. (1988) *Angew. Chem. Int. Ed. Engl.*, **27**, 344.
Lowenstamm, H.A. and Weiner, S. (1989) *On Biomineralization*, OUP, Oxford.
Mabrouk, P.A., Orville, A.M., Lipscomb, J.D. and Solomon, E.I. (1991) *J. Am. Chem. Soc.*, **113**, 4053.
Mackle, P., Garner, C.D., Ward, R.J. and Peters, T.J. (1991) *Biochim. Biophys. Acta*, **1115**, 145.
Mann, S. (1989) in *Biomineralization*, (eds S. Mann, J. Webb, R.J.P. Williams), VCH, Weinheim, pp. 35–62.
Mansuy, D. (1987) *Pure. Appl. Chem.*, **59**, 759.
Mansuy, D., Battioni, P. and Battoni, J-P. (1989) *Eur. J. Biochem.*, **184**, 267.
Marchon, J.C., Mashiko, T. and Reed, C.A. (1982) in *Electron Transport and Oxygen Utilisation*, (Ed. C. Ho), Elsevier, North Holland, P. 67.
Mascharak, P.K., Papaefthymiou, G.C., Frankel, R.B. and Holm, R.H. (1981) *J. Am. Chem. Soc.*, **103**, 6110.
Matsumoto, M. and Kuroda, K. (1982) *J. Am. Chem. Soc.*, **104**, 1433.
Matzanke, B.F., Müller-Matzanke, G. and Raymond, K.N. (1989) in *Iron Carriers and Iron Proteins*, (ed. T.M. Loehr), VCH, New York, p. 1.
McCusker, J.K., Vincent, J.B., Schmitt, E.A., Mino, M.L., Shin, K., Coggin, D.K., Hagen, P.M., Huffman, J.C., Christou, G. and Hendrickson, D.N. (1991) *J. Am. Chem. Soc.*, **113**, 3012.
McDevitt, M.R., Addison, A.W., Sinn, E. and Thompson, L.K. (1990) *Inorg. Chem.*, **29**, 3425.
McMurry, T.J., Rodgers, S.J. and Raymond, K.N. (1987) *J. Am. Chem. Soc.*, **109**, 3451.
McMurry, T.J., Hosseini, M.W., Garrett, T.M., Hahn, F.E., Reyes, Z.E. and Raymond, K.N. (1987) *J. Am. Chem. Soc.*, **109**, 7196.
Ménage, S., Brennan, B.A., Juarez-Garcia, C., Münck, E. and Que, L. (1990) *J. Am. Chem. Soc.*, **112**, 6423.
Ménage, S. and Que, L. (1991) *New J. Chem.*, **15**, 431.
Millar, M., Lee, J.F., Koch, S.A. and Fikar, R. (1982) *Inorg. Chem.*, **21**, 4106.
Morgan, B. and Dolphin, D. (1987) *Struct. Bonding (Berlin)*, **64**, 115.
Motekaitis, R.J., Sun, Y. and Martell, A.E. (1991) *Inorg. Chem.*, **30**, 1554.
Motekaitis, R.J., Sun, Y., Martell, A.E. and Welch, M.J. (1991) *Inorg. Chem.*, **30**, 2737.
Mulliez, E. (1989) *Tetrahedron Lett.*, 30, 6169.
Mulliez, E., Guillot, G., Leduc, P., Bois, C. and Chottard, J.C. (1991) SAMBAS, Poster 51.
Murray K.S. (1974) *Coord. Chem. Rev.*, **12**, 1.
Nakamura, A. and Ueyama, N. (1989) *Adv. Inorg. Chem.*, **33**, 39.
Newton, J.E. and Hall, M.B. (1984) *Inorg. Chem.*, **23**, 4627.
Nishida, Y., Shimo, H. and Kida, S. (1984) *J. Chem. Soc. Chem. Commun*, 1611.
Nordland, P., Sjöberg, B-M. and Eklund, H. (1990) *Nature*, **345**, 593.
Norman, R.E., Yan, S., Que, L., Backes, G., Ling, J., Sanders-Loehr, J., Zhang, J.H. and O'Connor, C.J. (1990) *J. Am. Chem. Soc.*, **112**, 1554.
Oberhausen, K.J., Richardson, J.E., O'Brien, R.J., Buchanan, R.M., McCusker, J.K., Webb, R.J. and Hendrickson, D.N. (1992) *Inorg. Chem.*, **31**, 1123.
Ochiai, E. and Williams, D.R. (1979) *Laboratory Introduction to Bioinorganic Chemistry*, Macmillan Press.
Ohlendorf, D.H., Lipscomb, J.D. and Weber, P.C. (1988) *Nature*, **336**, 403.

Patch, M.G., Simolo, K.P. and Carrano C.J. (1983) *Inorg. Chem.*, **22**, 2630.
Perutz, M.F. (1979) *Ann. Rev. Biochem.*, **48**, 327.
Petersson, L., Graslund, A., Ehrenberg, A., Sjöberg, B-M. and Reichard, P. (1980) *J. Biol. Chem.*, **225**, 6706.
Poulos, T.L., Finzel, B.C. and Howard, A.J. (1987) *J. Mol. Biol.*, **195**, 687.
Que, L. (1983) *Coord. Chem. Rev.*, **50**, 73.
Que, L., Kolanczyk, R.C. and White, L.S. (1987) *J. Am. Chem. Soc.*, **109**, 5373.
Que, L. and True, A.E. (1990) in *Progress in Inorganic Chemistry*, (ed. S. Lippard), Wiley-Interscience, New York, **38**, 97.
Raymond, K.N., Isied, S.S., Brown, L.D., Fronczek, F.R. and Nibert, J.H. (1976) *J. Am. Chem. Soc.*, **98**, 1766.
Raymond, K.N. (1990) *Coord. Chem. Rev.*, **105**, 135.
Robbins, A.H. and Stout, C.D. (1989) *Proc. Natl. Acad. Sci. USA*, **86**, 3639.
Robert, A., Loock, B., Momenteau, M. and Meunier, B. (1991) *Inorg. Chem.*, **30**, 706.
Rodgers, S.J., Ng. C.Y. and Raymond, K.N. (1985) *J. Am. Chem. Soc.*, **107**, 4094.
Russo, U., Vidali, M., Zarli, B., Purrello, R. and Maccarrone, G. (1986) *Inorg. Chim. Acta.*, **120**, L11.
Sanders-Loehr, J., Wheeler, W.D., Shiemke, A.K. and Lippard, S.J. (1989) *J. Am. Chem. Soc.*, **111**, 8084.
Scarrow, R.C., Ecker, D.J., Ng, C.Y., Liu, S. and Raymond, K.N. (1991) *Inorg. Chem.*, **30**, 900.
Schappacher, M., Weiss, R., Montiel-Montoga, R., Trautwein, A. and Tabard, A. (1985) *J. Am. Chem. Soc.*, **107**, 3736.
Schlabach, M.R. and Bates, G.W. (1975) *J. Biol. Chem.*, **250**, 2182.
Schrauzer, G.N., Kiefer, G.W., Tano, K. and Doemeny, P.A. (1974) *J. Am. Chem. Soc.*, **96**, 641.
Spiro, T.G., Pape, L. and Saltman, P. (1967) *J. Am. Chem. Soc.*, **109**, 3337.
St. Pierre, T.G., Webb, J. and Mann, S. (1989) in *Biomineralization*, (Eds S. Mann, J. Webb, and R.J.P. Williams), VCH, Weinheim, pp. 295–344.
Stack, T.D.P. and Holm, R.H. (1987) *J. Am. Chem. Soc.*, **109**, 2546.
Stack, T.D.P. and Holm, R.H. (1988) *J. Am. Chem. Soc.*, **110**, 2484.
Stack, T.D.P., Weigel, J.A. and Holm, R.H. (1990) *Inorg. Chem.*, **29**, 3745.
Stenkamp, R.E., Sieker, L.C., Jensen, L.H., McCallum, J.D. and Sanders-Loehr, J. (1985) *Proc. Natl. Acad. Sci. USA*, **82**, 713.
Stenkamp, R.E., Sieker, L.C. and Jensen, L.H. (1984) *J. Am. Chem. Soc.*, **106**, 618.
Stout, C.D. (1989) *J. Mol. Biochem.*, **205**, 545.
Stutte, P., Kiggen, W. and Vögtle, F. (1987) *Tetrahedron*, **43**, 2065.
Sun, Y., Martell, A.E. and Motekaitis, R.J. (1986) *Inorg. Chem.*, **25**, 4780.
Suslick, K.S. and Reinert, T.J. (1985) *J. Chem. Ed.*, **62**, 974.
Taft, K.L. and Lippard, S.J. (1990) *J. Am. Chem. Soc.*, **112**, 9629.
Tatsuno, Y., Tatsuda, M. and Otsuka, S. (1982) *J. Chem. Soc. Chem. Commun.*, 1100.
Tasuno, Y., Tatsuda, M., Otsuka, S. and Tani, K. (1983) *Inorg. Chim. Acta.*, **79**, 104.
Tolman, W.B., Liu, S., Bentsen, J.G. and Lippard, S.J. (1991) *J. Am. Chem. Soc.*, **113**, 152.
Towe, K.M. and Bradley, W.F. (1967) *J. Colloid Interface Sci.*, **24**, 384.
Tsukihara, T., Fukuyama, K., Tahara, H., Katsube, Y., Matsuura, Y., Tanaka, N., Kakudo, M., Wada, K. and Matsubara, H. (1978) *J. Biochem. (Tokyo)*, **84**, 1645.
Turowski, P.N., Armstrong, W.H., Roth, M.E. and Lippard, S.J. (1990) *J. Am. Chem. Soc.*, **112**, 681.
Ueno, S., Ueyama, N., Nakamura, A. and Tsukihara, T. (1986) *Inorg. Chem.*, **25**, 1000.
Ueno, S., Ueyama, N. and Nakamura, A. (1985) *Pept. Chem.*, **1984**, 269.
Ueyama, N., Ueno, S. and Nakamura, A. (1987) *Bull. Chem. Soc. Jpn.*, **60**, 283.
Vincent, J.B., Olivier-Lilley, G.L. and Averill, B.A. (1990) *Chem. Rev.*, **90**, 1447.
Vincent, J.B., Huffman, J.C., Christou, G., Li, Q., Nanny, M.A. Hendrickson. D.N., Fong, R.H. and Fish, R.H. (1988) *J. Am. Chem. Soc.*, **110**, 6898.
Wang, J.H. (1955a) *J. Am. Chem. Soc.*, **77**, 822.
Wang, J.H. (1955b) *J. Am. Chem. Soc.*, **77**, 4715.
Wang, J.H. (1970) *Acc. Chem. Res.*, **3**, 90.
Watenpaugh, K.D., Sieker, L.C. and Jensen, L.H. (1979) *J. Mol. Biol.*, **131**, 509.
Webb, J. and Gray, H. (1974) *Biochim. Biophys. Acta*, **351**, 224.
Weigel, J.A. and Holm, R.H. (1991) *J. Am. Chem. Soc.*, **113**, 4184.
Weinberg, E.D. (1989) *Quart. Rev. Biol.*, **64**, 261.

Weller, M.G. and Weser, U.J. (1982) *J. Am. Chem. Soc.*, **104**, 3725.
Weller, M.G. and Weser, U.J. (1985) *Inorg. Chim. Acta.*, **107**, 243.
White, L.S., Nilsson, P.V., Pignolet, L.H. and Que, L. (1984) *J. Am. Chem. Soc.*, **106**, 8312.
Wieghardt, K., Pohl, J. and Gebert, W. (1983) *Angew. Chem. Int. Ed. Engl.*, **22**, 727.
Wieghardt, K., Pohl, K., Jibril, I. and Huttner, G. (1984) *Angew. Chem. Int. Ed. Engl.*, **23**, 77.
Wolfgang, K. and Vögtle, F. (1984) *Angew Chem. Int. Ed. Engl.*, **23**, 71.
Woodland, M.P., Patil, D.S., Cammack, R. and Dalton, H. (1986) *Biochim. Biophys. Acta*, **873**, 237.
Yan, S., Cox, D.D., Pearce, L.L., Juarez-Garcia, C., Que, L., Zhang, J.H. and O'Connor, C.J. (1989) *Inorg. Chem.*, **28**, 2507

8 Iron chelators of clinical significance

R.C. HIDER and S. SINGH

Until relatively recently chelating agents have been considered to form a small esoteric group of pharmaceuticals with little relevance to general medicine. Penicillamine, for instance, has been used to treat Wilson's disease for over 35 years, and desferrioxamine has found application for the treatment of transfusion-induced iron overload in thalassaemic patients for almost 20 years. This limited interest is surprising as Fe, Cu, Zn, Ca and Mg all have a critical involvement in cell biochemistry and it would seem likely that chelating agents might find roles in rectifying abnormal distribution of these metals and enhancing their absorption from the gastrointestinal tract.

With the postulated involvement of hydroxyl radicals in normal physiological responses as well as a range of disease states (Halliwell and Gutteridge, 1989), clinical interest in the redox active metals Cu and Fe has increased. In principle, selective chelating agents can be designed which scavange non-protein bound forms of these metals, thereby minimising undesirable hydroxyl radical formation. Such compounds are currently being investigated for the treatment of a wide range of disease states and in this chapter we outline some of these new developments.

8.1 Iron transport and distribution

Iron has two important chemical properties which have rendered it a critically important element to virtually all life forms. The redox potential between the two common oxidation states of iron, iron(II) and iron(III), is such that oxidation processes centred on the iron atom can be readily coupled to metabolic reactions. Iron also possesses a high affinity for oxygen atoms. These two properties are utilised widely by iron-containing proteins, for instance as electron transfer proteins in the mitochondria, as hydroxylating enzymes and as oxygen transport proteins such as haemoglobin (Wrigglesworth and Baum, 1980). These important properties also endow iron with the potential of being toxic, this being particularly true should the iron be non-specifically bound to the surface of proteins and membranes. In the presence of molecular oxygen, such weakly coordinated iron can redox cycle between the two oxidation states thereby generating a range of oxygen radicals, including O_2^- and $\cdot OH$ (Halliwell and Gutteridge, 1988). The hydroxyl radical ($\cdot OH$), is highly reactive and is capable of interacting with most types of organic molecule, including sugars, lipids, proteins and nucleic acids. Under *in vivo* conditions, $\cdot OH$ probably possesses a mean free path of less than 10 Å, due to its extreme reactivity. Consequently, the production of hydroxyl radicals is undesirable and there are a number of protective measures adopted by cells to protect against its formation; foremost among these is the tight control of iron storage, transport and distribution within multicellular organisms.

In man the absorption of iron from the gastrointestinal tract is carefully regulated, normal individuals absorbing between 1 and 2 mg each day. An equal amount of iron is lost by exfoliating cells per day. Iron is only absorbed from the duodenum and upper jejunum and is either immediately transferred to the liver or the blood iron transport protein, apotransferrin. Apotransferrin is a globular glycoprotein with a molecular weight of 76 000 and possesses two binding sites each with high but unequal affinity for iron(III) ($K_{eq} \sim 10^{-28}$ M) (Crichton, 1991). The concentration of free iron(III) in the presence of apotransferin in less than 10^{-12} M and at this concentration iron is unable to trigger damage via hydroxyl radical formation. The normal plasma concentration of transferrin is 30 μM. Cells which require iron for maturation or division express high densities of transferrin receptors on the surface of their plasma membrane (Huebers and Finch, 1987). The transferrin–receptor complex is internalised by endocytosis and the iron released by protonation in the relatively acidic endosome (May and Cuatrecasas, 1985). Consequently, iron is specifically directed from the blood to tissue where it is required.

Once iron has been released from transferrin, apotransferrin is secreted back into the extracellular space (Harding *et al.*, 1983) where it is recharged with iron originating from either liver or reticuloendothelial cells. On average, each molecule of transferrin is believed to undertake over 200 such cycles before being degraded. Hepatocytes unlike reticulocytes can also take up transferrin iron without internalisation of the protein-receptor complex. This uptake process requires reduction of ferric to ferrous iron by a membrane bound NADH-dependent reductase (Thorstensen, 1988).

Typically only 30% of the iron-binding sites of transferrin are occupied. It retains sufficient excess binding capacity to ensure that it is never completely saturated, except in chronic iron overload disorders. This is largely to ensure the compartmentalisation and strict regulation of iron mobilisation, which in the free form is potentially toxic. Thus, the level of potentially toxic non-transferrin bound iron (NTBI), in the extracellular space, is vanishingly low in the normal individual. Among other roles, transferrin may also contribute to the defence against infections by depriving microorganisms of iron which is required for their growth and reproduction (Hershko *et al.*, 1988).

In similar fashion, intracellular iron levels are carefully controlled. Iron in excess of functional requirements is stored in two forms, either as soluble ferritin or as the insoluble haemosiderin deposits (Munro and Linder, 1978).

These iron stores can be found in nearly all cells but are present in large amounts in cells with large iron requirements such as the erythroid marrow. Ferritin, which has a molecular weight of approximately 450 000, consists of 24 equivalent subunits. Iron is stored as a polynuclear ferric hydroxide–phosphate complex. Ferritin possesses a hollow centre where up to 4500 iron(III) atoms can be stored in the form of a regular lattice (Ford *et al.*, 1984). In this form the potential to interact with oxygen and to generate oxygen radicals is minimal.

Clearly, in the normal individual, iron levels are under extremely tight control (Figure 8.1) and there is little opportunity for iron-catalysed free radical generating reactions to occur. However, there are situations when the iron status can change, either locally, as in ischaemic tissue or systemically, as with idiopathic haemochromatosis or transfusion-induced iron overload. In such circumstances, abnormal levels of iron can induce toxic symptoms.

8.2 Systemic iron overload

8.2.1 Hyperabsorption of iron leading to iron overload

There are a number of inherited disease states which are associated with the gradual accumulation of iron, for instance, β-thalassaemia and thalassaemia intermedia in the Middle and Far East and idiopathic haemaochromatosis worldwide. In the former disease state, patients are anaemic due to defective haemoglobin synthesis and therefore hyperabsorb iron to provide for the apparent increased requirements of the erythroid marrow to compensate for the defect in erythropoiesis. In the latter disease, patients hyperabsorb iron as they lack the normal control mechanisms for iron absorption resulting in a steady build up of the metal in the body (Figure 8.1). In some regions of the world thalassaemia genes are relatively common. In South East Asia, for instance, approximately 100 000 homozygotes are born each year suffering from different forms of thalassaemia.

The normal body iron content is 4–5 per adult (Figure 8.1) but in some thalassaemic patients, iron levels may accumulate to 50–70 g. This becomes life threatening due to the facile redox chemistry of iron. Iron salts accumulate in highly perfused tissue including the endocrine organs thereby inducing the early onset of diabetes and preventing the onset of puberty (Weatherall and Clegg, 1981). Elevated heart iron levels lead to serious myocardial and aerial damage which in turn cause infarction (Zurlo *et al.*, 1989).

In principle these serious symptoms can be avoided by removal of the excess iron. With haemochromatosis patients, who have normal erythropoietic function, this can be readily achieved by regular plebotomy. Unfortunately, this is not the situation with thalassaemic patients and the application of specific iron chelators is the only effective therapy currently available for the

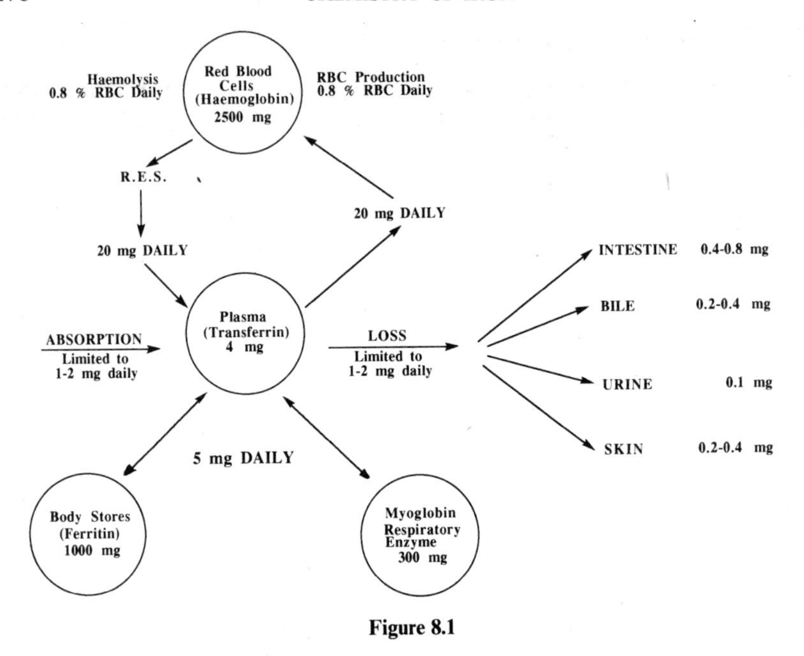

Figure 8.1

treatment of transfusional siderosis and hyperabsorption of iron associated with the disease.

8.2.2 *Blood transfusion leading to iron overload*

Some forms of β-thalassaemia are more critical than those outlined in the previous section. Indeed, for babies who are homozygotes for the defective β-gene and are therefore unable to synthesise any haemoglobin β-subunits, death will ensue soon after the cessation of fetal haemoglobin production. This is typically between 2–6 months after birth (Weatherall and Clegg, 1981). Fortunately, such children can be successfully treated by blood transfusion. Maintenance of a high haemoglobin level slows endogenous erythrocyte production and therefore reduces dietary iron absorption and ensures the survival of the child. One major problem with life-long blood transfusion regimes, apart from expense and inconvenience, is the inevitable accumulation of iron. With every unit of blood, 250 mg of iron is introduced into the body and like the hyperabsorbing patients, regularly transfused children rapidly accumulate large quantities of iron. Consequently they suffer from identical symptoms to those patients discussed in the previous section.

Bone marrow transplant therapy can also be considered for such children, but at the present time the long-term prospects of such patients are unclear and there is always a failure rate, albeit low, with this type of therapy.

8.2.3 Selective iron chelation therapy

Desferrioxamine (**1**) has been used for the treatment of iron overload for over 20 years. Follow-up of patients on lifelong transfusion programmes has shown

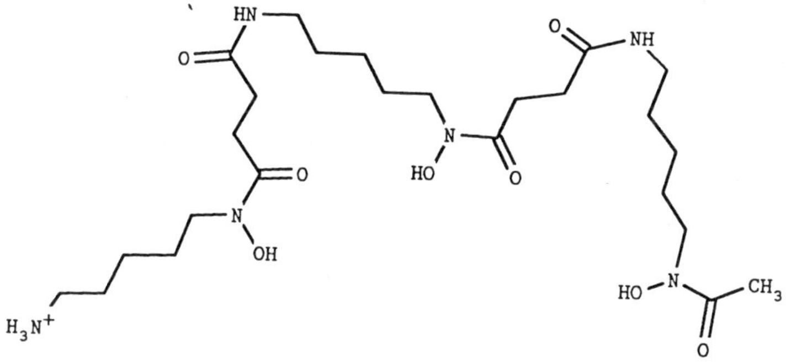

(1)

that the regular use of the drug prevents the development of iron overload and its pathological consequences, while patients not taking desferrioxamine succumb to iron toxicity in their late teens or early twenties. However, desferrioxamine suffers from the disadvantage that it is inactive when administered orally, and only causes sufficient iron excretion to keep pace with the transfusion regimes when given either subcutaneously or intravenously over 8–12 h several times per week. For this reason, many patients find it difficult to comply with the treatment, and some even stop taking the drug altogether, subsequently developing the complications of iron overload. There is, therefore, no doubt that an orally active chelating agent is needed to treat patients on lifelong transfusion programmes.

The ideal properties for a therapeutically effective iron chelator are given in Table 8.1. Desferrioxamine meets most of these requirements and is a highly effective drug with low toxicity when used with care (Porter and Huehns,

Table 8.1 Properties required of an ideal iron chelator

Clinically an iron chelator needs to:
(i) be absorbed when given by mouth;
(ii) have a high affinity for iron(III) with a high chelating efficiency *in vivo*;
(iii) chelate iron specifically without significant affinity for other metals;
(iv) penetrate tissues, cells and subcellular sites with abnormal iron accumulation;
(v) be excreted easily via the kidneys and liver only when combined with iron;
(vi) not redistribute iron to potentially more toxic sites in the body;
(vii) be itself relatively non-toxic;
(viii) not be metabolically degraded to non-chelating metabolites;
(ix) be sufficiently active to achieve negative iron balance in mildly iron overloaded patients;
(x) be unable to encourage bacterial growth

1989). However, it is not orally active and being a natural siderophore (Hider, 1984) delivers iron to some bacteria and fungi, thereby enhancing their pathogenicity (Robins-Browne and Prpic, 1985).

The design of clinically useful iron chelators has been discussed in detail elsewhere (Porter *et al.*, 1989) essentially the choice is currently focused on the amino-carboxylate and the pyridinone families and both bidentate and hexadentate ligands. The stability of an iron complex is influenced by the number of covalently linked arms associated with the chelating agent. Thus, hexadentate ligands form more stable iron complexes than the corresponding bidentate ligands, and therefore have greater scavenging ability at low concentrations ($\leqslant 20\ \mu M$). A further potential disadvantage for bidentate ligands is that they can form partially dissociated complexes (e.g., 2 ligand:1 iron) which in principle are capable of generating hydroxyl radicals. Thus, from chemical considerations alone, hexadentate chelators are likely to have greater therapeutic potential. However, the biodistribution of the larger hexadentate ligands (molecular weight $\geqq 500$) is markedly different to that of the smaller bidentate molecules (molecular weight ~ 200); oral bioavailability, for instance, is critically dependent on molecular size.

Aminocarboxylate ligands have been widely investigated for iron chelation, indeed DTPA (**2**) is used in patients who develop toxic side-effects with desferrioxamine (Jackson *et al.*, 1983) DTPA is not orally active and due to its relative lack of selectivity for iron(III) leads to zinc depletion. In an attempt to enhance the selectively of this class of chelator for iron(III), Martell and co-workers (Martell *et al.*, 1986) have synthesised several analogues which contain both carboxyl and phenolic ligands. A particularly useful compound is HBED (**3**) (Lau *et al.*, 1983). Unfortunately, this molecule is not efficiently absorbed via the oral route and by virtue of the two carboxyl functions, retains a relatively high affinity for zinc. The dimethylester of HBED is orally active and has been highlighted as being of particular interest (Hershko *et al.*, 1984). HBED is a hexadentate ligand, chelating iron with each pair of carboxyl, amino and phenol functions.

Hydroxypyridinone ligands have also been widely investigated for iron(III) chelation. Hydroxypyridin-4-ones and hydroxamates are the only currently available iron(III)-selective ligands that are uncharged under physiological conditions both as the free ligand and as the iron complex (Porter *et al.*, 1989). This property is probably central to their success. Bidentate 3-hydroxypyridin-4-ones (**4**) have a high bioavailability (Kontoghiorghes *et al.*, 1990) and are relatively selective for iron(III) under *in vivo* conditions (Hider and Hall, 1991). The 1,2-dimethyl- (CP20, L1) and 1,2-diethyl derivatives (CP94) have both been extensively studied in iron overloaded animal models (Hider *et al.*, 1990; Porter *et al.*, 1990) and in thalassaemic patients (Klein *et al.*, 1991). They remove iron relatively efficiently at high doses, CP94 being more effective than CP20. In chronic studies, toxic side-effects of both compounds appear between 50 and 100 $mg\,kg^{-1}$. Unfortunately, the oral dose of CP20 required to

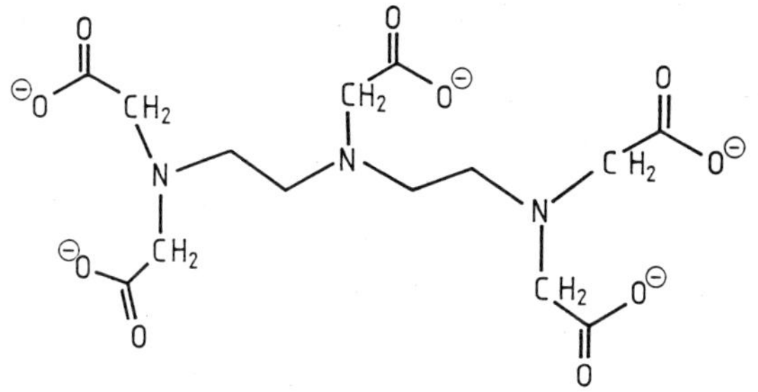

(2)

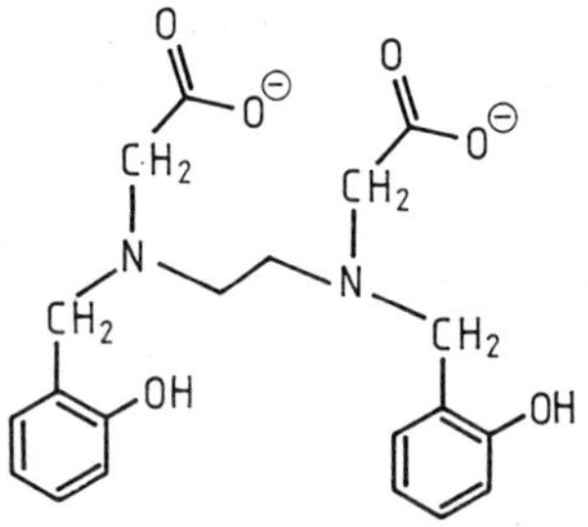

(3)

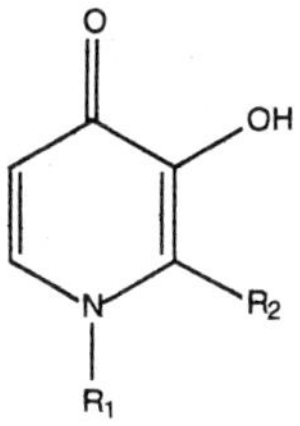

(4)

$R_1 = R_2$ = Me: CP20, L1
$R_1 = R_2$ = Et: CP94

maintain a patient in negative iron balance appears to be in excess of 100 mg kg^{-1} (5 g per 25 h for a 50 kg individual). Not surprisingly therefore side-effects, for instance agranulocytosis, have been observed in patients receiving such high doses ($\geqslant$ 100 mg kg^{-1}) (Bartlett *et al.*, 1990). By virtue of the greater efficacy of CP94 it is hoped that the smaller dose (25 mg kg^{-1}; 1.25 g per 24 h for a 50 kg individual) will maintain iron overloaded patients in negative iron balance.

Should bidentate pyridinone ligands fail to be clinically viable, then hexadentate analogues (**5**) could be investigated. CP130 (**5**) is effective at removing

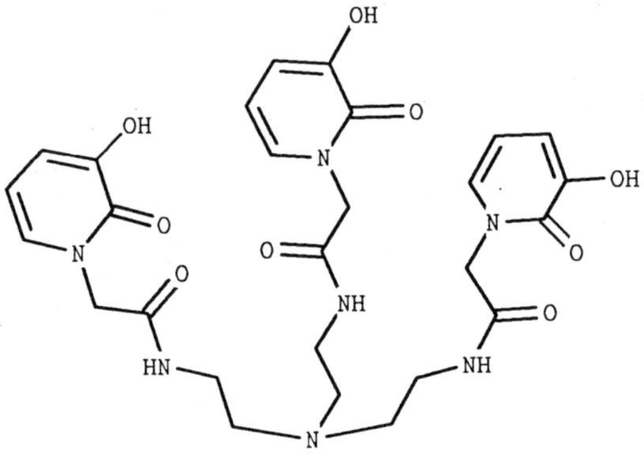

(5)

iron in rodents (Streater *et al.*, 1990), but by virtue of its higher molecular weight, has a lower bioavailability than the smaller bidentate pyridinone ligands when presented via the oral route. Hexadentate pyridinone molecules are likely to be less toxic than their bidentate analogues because of a more restricted biodistribution and in particular lack of ability to cross the blood-brain-barrier due to their relatively high molecular weight (Levin, 1980).

If a sufficiently 'non-toxic' molecule is identified, it will probably find application not only for the treatment of iron-overloaded thalassaemic patients but also for those suffering from sickle cell anaemia and haemochromatosis. The extremely painful crises experienced by many sickle cell patients can be alleviated by blood transfusion and if the associated iron overload could be prevented or even minimised by oral chelation, then such treatment would probably become widespread.

With haemochromatosis, although total body iron levels can be controlled by frequent phlebotomy, there are periods when newly diagnosed patients possess fully saturated transferrin. In such circumstances non-transferrin bound iron (NTBI) is also present in the plasma. This form of iron is considered to be highly damaging and may even be responsible for much of the heart damage associated with this disease. Chelation therapy would remove NTBI (Singh *et al.*, 1990) and hence be expected to protect patients from such damage.

There are an estimated 80 million carriers of thalassaemia genes of one form or another. Thus large numbers of homozygotes are born each year – far too many to be treated by bone marrow transplant or gene therapy. It is clear, therefore, that oral chelation therapy has considerable potential for the treatment of people suffering from this range of haemoglobinopathies.

8.3 Localised and temporary elevation of iron levels

8.3.1 Ischaemic tissue

Over the last decade overwhelming evidence has been gathered which suggests that oxygen-derived free radicals play the major role in producing microvascular and parenchymal damage associated with reperfusion of ischaemic tissue. The above is supported by the ability of superoxide dismutase (Hearse *et al.*, 1986) and other free radical scavangers to reduce tissue injury. Tissue damage is thought to be related to the production of oxygen radicals either within the affected tissues or the associated extracellular fluid via white blood cells and in particular neutrophils. The oxidative damage induced does not occur during the period of ischaemia but subsequently during reperfusion of the tissue. During ischaemia, which can be experimentally induced either by hypoperfusion (as in shock), or by mechanical occlusion of the arteries, oxidative phosphorylation ceases, cellular ATP levels drop and the concentration of both calcium and the low molecular weight iron pool increases (Green *et al.*, 1989). On reperfusion, the rapid increase in oxygen tension results in the production of a large pulse of hydroxyl radicals due to the enhanced labile iron pool (Scheme 8.1). The origin of the redox active iron is likely to be ferritin which is mobilised as a result of an increased cellular reducing potential under the anaerobic conditions associated with ischaemia (Aust and White, 1985). Such pathogenic mechanisms may well hold for ischaemic injury to the small bowel (Hernandez *et al.*, 1987), kidney (Healing *et al.*, 1990), heart (Reddy *et al.*, 1989) and central nervous system (Komara *et al.*, 1986).

There is increasing evidence for the beneficial influence of the presence of iron chelators during reperfusion of ischaemic tissue. Clearly reactions (i), (iii) and (iv) in Scheme 8.1, do not occur in the absence of loosely bound iron and it is such species which are efficiently scavenged by desferrioxamine (**1**) forming the chemically inert iron complex, ferrioxamine. Thus the use of desferrioxmine has been found to reduce brain damage following cardiac arrest (Komara *et al.*, 1986). Indeed when perfused with desferrioxamine (50 mg kg^{-1}, i.v.) the intracellular molecular weight iron pool of dog brain was reduced to 65% of that of the ischaemic level (Table 8.2), but not down to that

Table 8.2 Iron and lipid peroxidation following cardiac arrest

	Low molecular weight iron nmol/mg	Malondialdehyde nmol/100 mg	Conjugated dienes nmol/10 mg
Nonischaemic controls	9.05 ± 0.04	7.32 ± 1.67	0.604 ± 0.121
Standard intensive care	37.04 ± 4.58	12.2 ± 1.90	1.243 ± 0.608
Standard intensive care and desferrioxamine	24.03 ± 3.40	9.4 ± 0.08	0.642 ± 0.521

$$Fe^{2+} + O_2 \rightleftharpoons Fe^{3+} + \dot{O}_2^- \quad \text{(i)}$$

$$2\dot{O}_2^- + 2H^+ \xrightleftharpoons{SOD} H_2O_2 + O_2 \quad \text{(ii)}$$

$$\dot{O}_2^- + H_2O_2 \xrightleftharpoons{\text{Fe catalyst}} O_2 + \dot{O}H + OH^- \quad \text{(iii)}$$

(Haber-Weiss Reaction)

$$Fe^{2+} + H_2O_2 \rightleftharpoons Fe^{3+} + \dot{O}H + OH^- \quad \text{(iv)}$$

(Fenton Reaction)

$$\dot{O}_2^- + NO \rightleftharpoons \dot{O}N\text{-}O\text{-}O^- \quad \text{(v)}$$

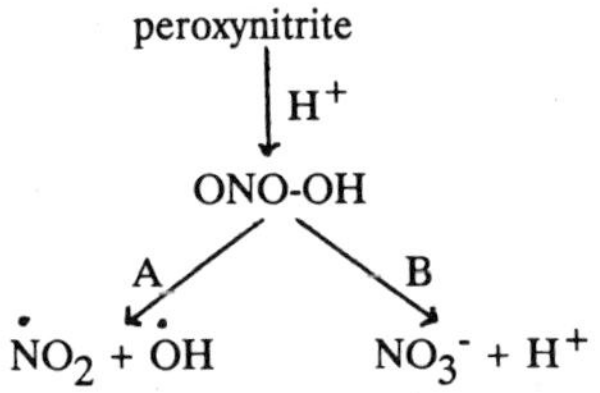

peroxynitrite

$\downarrow H^+$

ONO-OH

A: $\dot{N}O_2 + \dot{O}H$ B: $NO_3^- + H^+$

Scheme 8.1 Reactions (i)–(iv) are likely to be limited to the intracellular space unless the postulated elevated level of ferrous iron (Fe^{2+}) is effluxed from the cell wall via a divalent cation exchange mechanism. Reaction (v) could occur equally well intra- and extracellurarly.

of the non-ischaemic controls. Simultaneously, a marked reduction of both malondialdehyde and conjugated diene levels (monitors of tissue damage) was observed.

In principle, a chelator designed to efficiently permeate the blood–brain barrier would be predicted to be more efficient than desferrioxamine, and the 3-hydroxypyridin-4-ones (**4**) appear to be good candidates. Significantly, 1,2-dimethyl-3-hydroxypyridin-4-one (**4**, CP20) has been reported to prevent post-ischaemic cardiac injury in the rat (van der Kraaij *et al.*, 1989). Iron chelation also has a beneficial effect on myocardial injury following ischaemia and reperfusion. Van der Kraaij and co-workers (1989) have for instance demonstrated a pronounced protective effect with the use of both desferrioxamine and 3-hydroxypyridin-4-ones using perfused rat hearts.

Results of this type suggest that iron chelation could possibly find a role in the first-line treatment of stroke and heart attack, where a single injection of chelator could decrease reperfusion damage to both the brain and heart. Similar problems to those associated with ischaemia also occur during tissue transplantation, and again the presence of desferrioxamine enhances the viability of transplanted tissue (Green *et al.*, 1986).

More recently, the notion that the beneficial effects of iron chelating agents are simply due to chelation of the metal ion has been challenged (Beckman *et al.*, 1990). This is due to the demonstrated ability of the commonly used hydroxamate iron chelator, desferrioxamine, which is capable of acting as a superoxide and hydroxyl radical scavanger (Davies *et al.*, 1987), in the process forming a relatively stable desferrioxamine nitroxide free radical ($T_{\frac{1}{2}} \sim$ 10 min). The above is particularly relevant as recent evidence suggests that nitric oxide (NO), an endothelium derived relaxing factor (EDRF) produced in increased amounts on reperfusion of ischaemic tissue combines with O_2^- to form the peroxynitrite ($ONOO^-$) anion (Scheme 8.1 (v)) ($pK_a = 7.35$) which decays ($T_{\frac{1}{2}} \sim 1$ sec) at physiological pH to form nitrogen dioxide and hydroxyl radicals.

At present it is difficult to distinguish between the different oxygen radical generating systems in relation to ischaemia/reperfusion induced tissue injury. It is likely that the formation of oxygen radicals occurs as a result of metabolic processes and redox active metal catalysed reactions and the beneficial effects of iron chelators are due to both chelation of the metal and their radical scavenging properties.

8.3.2 Inflamed tissue

Disturbance of iron metabolism is a prominent feature of rheumatoid disease (Lawson *et al.*, 1983; Harvey *et al.*, 1983), many rheumatoid arthritis patients having raised levels of synovial fluid ferritin and iron deposits within their synovial tissue (Muirden, 1966; Senator and Muirden, 1968; Blake and Bacon, 1981). Iron deposition in early rheumatoid arthritis is associated with a poor prognosis (Black *et al.*, 1984). Blake and co-workers have recently presented strong evidence to suggest that the persistence of chronic synovitis results from the occurrence of hypoxic-reperfusion injury in the joint (Andrews *et al.*, 1987). An improvement in both acute and chronic inflammation was noted on administration of desferrioxamine (**1**) to animal models (Yoshino *et al.*, 1984). However, when given to rheumatoid patients, desferrioxamine was found to cause a number of undesirable side-effects (Blake *et al.*, 1985).

The hydrophilic members of the 3-hydroxypyridin-4-one family (e.g. **4**, CP20) also possess anti-inflammatory activities in the acute carregeenan-pleurisy model when presented at relatively high doses (Hewitt *et al.*, 1989). Although iron chelators undoubtedly possess anti-inflammatory properties, high concentrations are necessary and selective direction of these molecules to the site of inflammation presents a major problem.

8.3.3 Brain

The regional distribution of iron in the brain has been extensively investigated by histochemical staining and more recently by magnetic resonance imaging.

Table 8.3 The distribution of non-haem iron in different parts of the human brain

	mg iron $100\,g^{-1}$ (fresh weight)
Globus pallidus	21.30 ± 3.49
Red nucleus	19.48 ± 6.86
Substantia nigra	18.46 ± 6.52
Putamen	13.32 ± 3.43
Dentate nucleus	10.35 ± 4.86
Caudate nucleus	9.28 ± 2.14
Thalamus	4.76 ± 1.16
Cerebellar cortex	3.35 ± 0.87
Occipital cortex	4.55 ± 0.67
Sensory cortex	4.32 ± 0.58
Parietal cortex	3.81 ± 0.67
Temporal cortex	3.13 ± 0.57
Prefrontal cortex	2.92 ± 0.41
Frontal white matter	4.24 ± 0.88
Medulla oblongata	1.40 ± 1.16
Meninges	1.02 ± 0.29

Hallgren and Sourander, 1958.
Autopsy cases, 30–100 years of age.

Particularly high levels are located in globus pallidus, red nucleus and substantia nigra with marginally less in the putamen, dentate nucleus and caudate nucleus. Much lower levels are found in the various regions of the cortex and meninges (Table 8.3) (Hallgren and Sourander, 1958).

There is considerable overlap between the iron-rich areas of the brain and those regions where GABA neurones terminate. In these regions, approximately 30% of the iron is bound to ferritin, the remainder being associated with enzymes, respiratory proteins and transferrin. A large proportion of this iron is located within neurones and is probably associated with neurotransmitter metabolism; brain tyrosine hydroxylase for instance, is an iron-containing enzyme (Wrigglesworth and Baum, 1988)

The presence of such high concentrations of iron in certain nuclei of the brain is a potential hazard, particularly as dopamine, an iron chelator, is present in dopamine neurones at a concentration of 50 mM (Anden *et al.*, 1966). Not only is dopamine a chelator of iron but it also enters an internal redox reaction with the metal, thereby generating free radicals (Hider *et al.*, 1981; Mentasti *et al.*, 1976). Thus, dopamine-rich areas of the brain are predicted to be particularly sensitive to imbalance of iron levels. Significantly an increased nigral iron content has been reported for postmortem Parkinsonian brain (Dexter *et al.*, 1987, 1989) (Table 8.4). This elevated level of iron could be at least partially responsible for the observed enhancement of lipid peroxidation that occurs in the substantia nigra of patients dying with Parkinson's disease (Dexter *et al.*, 1989; Sotomatsu *et al.*, 1990; Agid, 1991). Iron-specific chelators could minimise such damage thereby possibly slowing the progression of the disease. Hydroxypyridinones (**4**) are currently being investigated for such properties in animal models.

Table 8.4 Mean iron content in parkinsonian and control brain tissue

	Iron (nmol/g dry weight)	
	Controls	Parkinson's disease
Substantia nigra	10436 (1191) ($n = 9$)	14043 (1186) ($n = 7$)
Cerebellum	4672 (259) ($n = 13$)	4896 (159) ($n = 11$)

Dexter *et al.*, 1987.

8.4 Protection against cellular damage induced by redox cycling chemicals

8.4.1 Paraquat

Paraquat (**6**) is selectively accumulated by lung tissue via the polyamine transport system (Gordonsmith *et al.*, 1983). Once located intracellularly it is

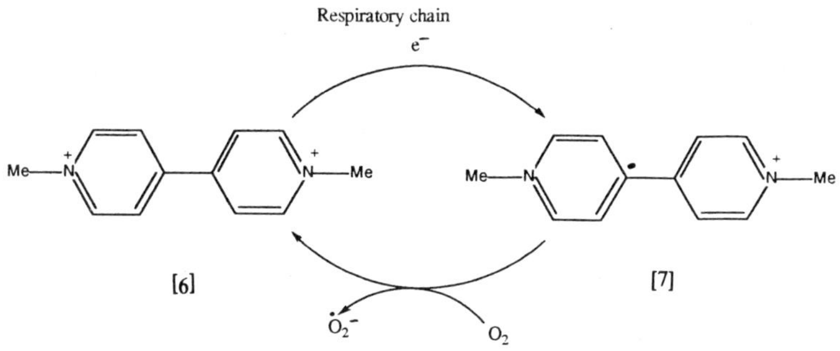

Respiratory chain
e^-
Me—$\overset{+}{N}$ ... $\overset{+}{N}$—Me
[6]
Me—N ... $\overset{+}{N}$—Me
[7]
$\dot{O}_2^-$
O_2

Scheme 8.2

capable of undergoing a one-electron redox cycling mechanism with cytochrome P_{450} reductase (Scheme 8.2). The paraquat radical (**7**) is capable of entering the ferritin core and mobilising iron in the iron(II) state (Aust, 1987). Thus, the intracellular presence of paraquat generates both superoxide anions and elevated levels of iron(II). Consequently, the series of reactions outlined in Scheme 8.1 becomes possible and serious hydroxyl radical-induced damage ensues (Vile *et al.*, 1987). As with ischaemic damage this can in principle be minimised by using selective iron chelators, for instance desferrioxamine (Halliwell and Gutteridge, 1989) and hydroxypyridin-4-ones (van Aspect, personal communication) which occupy all coordination sites on the metal preventing catalysis of free radical generation.

8.4.2 Doxorubicin

Doxorubicin (**8**) and related drugs are capable of interacting with DNA probably via intercalation. In fact the existence of such complexes has been

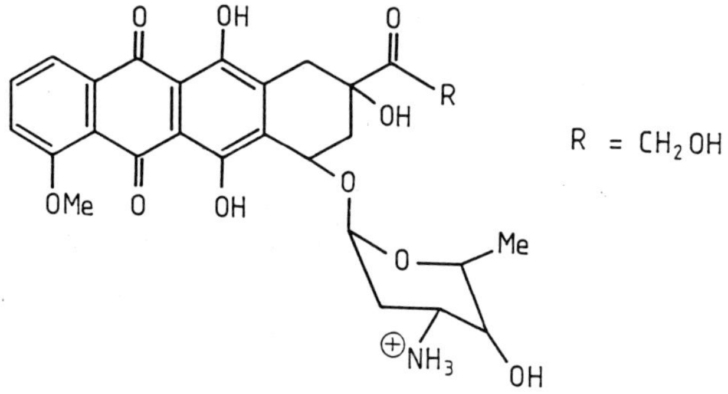

R = CH_2OH

(**8**)

demonstrated by X-ray diffraction methods (Berman and Young, 1981). The precise mode of action of this drug is unclear. It could function like bleomycin leading to DNA nicking, but there is increasing evidence suggesting that it inhibits tropoisomerase. Irrespective of its mode of action, doxorubicin is an extremely successful chemotherapeutic agent, being a first line choice for the treatment of acute leukemias, lymphomas and a variety of solid tumours.

Doxorubicin has an iron chelating site and like paraquat is able to redox cycle (Zweier, 1985; Gianni *et al.*, 1985; Gelvan *et al.*, 1990) and to mobilise iron from ferritin (Thomas and Aust, 1986). This undesirable effect has been associated with cardiac toxicity, doxorubicin being accumulated by heart tissue. Desferrioxamine has been shown to protect tissues against doxorubricin-induced damage and currently, a prodrug of a chelator, ICRF-187 (**9**) (Zinecard) is under clinical investigation for cardiac protection from doxorubricin-induced damage (Myers *et al.*, 1986). ICRF-187 is uncharged and therefore can enter cells, where the cyclic imide links are cleaved by intracellular enzymes to generate a chelator capable of binding, among other cations, iron(III) (Scheme 8.3) (Nolan *et al.*, 1990). The presence of iron(III) may even facilitate the conversion of the prodrug to the hexadentate ligand (**10**).

8.5 Selective inhibition of non-haem containing enzymes

Iron containing metalloenzymes can be divided into haem and non-haem type. In general, chelating agents are not effective inhibitors of haem-containing enzymes for two reasons; the avid interaction between the porphyrin nucleus and iron and bidentate interaction between the ligand and porphyrin-bound iron is not possible. In contrast many non-haem iron-

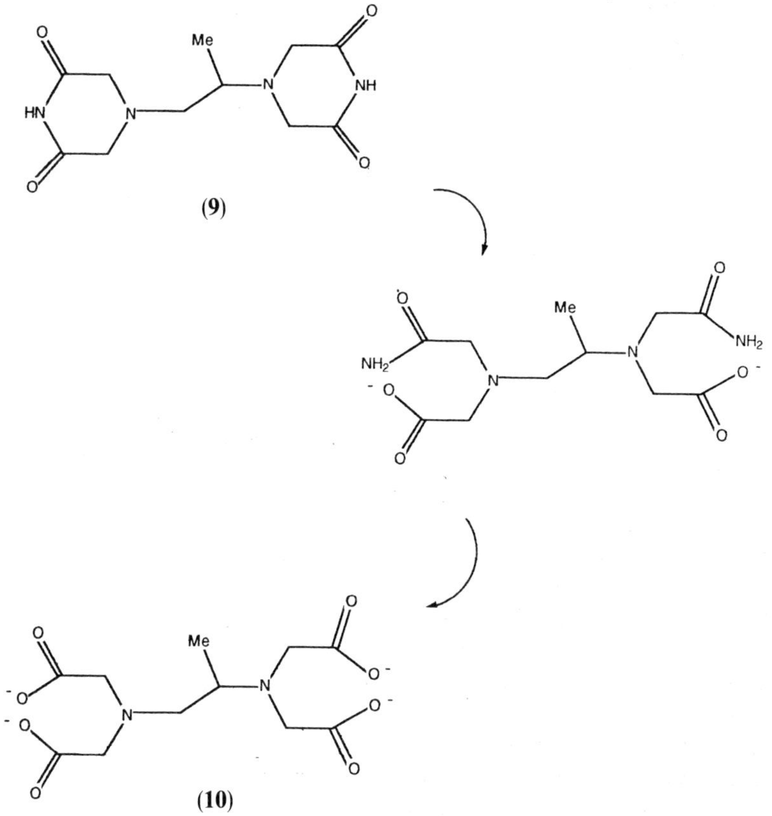

(9)

(10)

Scheme 8.3

containing enzymes are extremely susceptible to inhibition by chelators. Iron centres dominated by oxygen and imidazole ligands are apparently particularly susceptible. Such iron centres are found in ribonucleotide reductase, the lipoxygenase family of enzymes and tyrosine hydroxylase.

8.5.1 Ribonucleotide reductase

Ribonucleotide reductase catalyses the reduction of the four common ribonucleotides to their corresponding deoxyribonucleotides (Scheme 8.4), an essential step in DNA synthesis. All four ribonucleotides are reduced by the same enzyme. The crystallographic structure of the *Escherichia coli* enzyme has been elucidated (Nordlund *et al.*, 1990). The enzyme (molecular weight 250 000) is a complex of two proteins: M_1 which binds the substrate and contains redox active sulphydryl groups and M_2 which contains both a μ-oxo-bridged binuclear iron centre (Figure 8.2) (Nordlund *et al.*, 1990) and

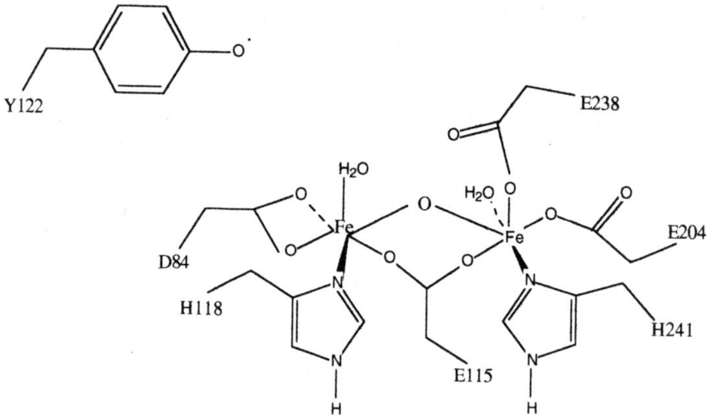

Figure 8.2

a tyrosine moiety side chain which exists as a free radical stabilised by the iron centre (Reichard and Ehrenberg, 1983). This radical, which is only 5.3 Å away from iron centre 1, has access to the substrate binding pocket and is essential for enzyme activity. Electrons for the reduction reaction are supplied from NADPH via thioredoxin, a small redox-active protein (Scheme 8.4).

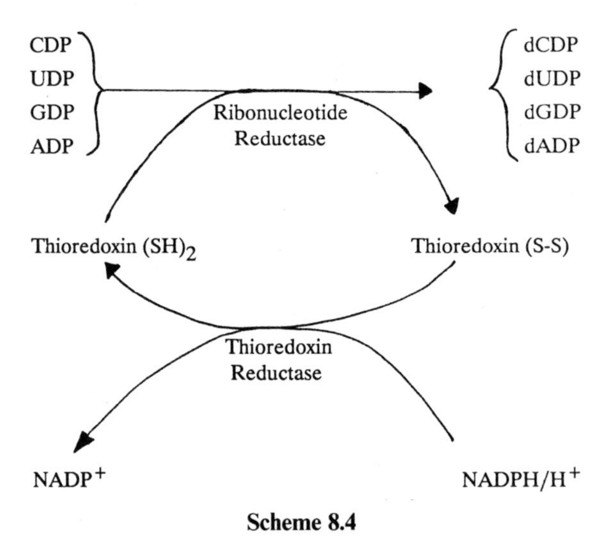

Scheme 8.4

A wide range of iron chelators have been shown to inhibit ribonucleotide reductase (Ganeshaguru *et al.*, 1980; Cory *et al.*, 1981) and this is undoubtedly the reason for cytotoxic properties of such molecules, for instance, tropolone. Some iron chelators may also function as free radicals scavengers, for instance, hydroxyurea which inhibits the enzyme by the latter mode of action. The ability of iron chelators to inhibit ribonucleotide reductase has led to several proposals for therapeutic application.

8.5.1.1 Anti-neoplasmic and antiviral properties. The withholding of iron from cells has been proposed as a strategy for the treatment of neoplastic disease (Weinberg, 1984, 1990). In particular, desferrioxamine has been demonstrated to inhibit the proliferation of a variety of malignant cell lines (Foa *et al.*, 1986; Blatt and Stitely, 1987 and Dezza *et al.*, 1987) and bone marrow neuroblastoma cells (Becton and Bryles, 1988). In addition desferrioxamine has been shown to posses anti-tumour activity in acute neonatal leukemia (Estrov *et al.*, 1987). Iron chelates and in particular oligodentate catecholates have been demonstrated to be potent inhibitors of DNA synthesis in a murine leukemia cell line L1210 (Bergeron *et al.*, 1984), the natural siderophore parabactin **(11d)** being particularly effective (Table 8.5).

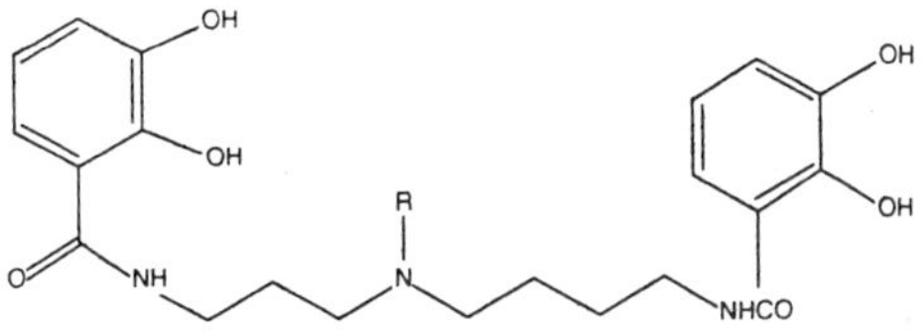

OH OH R N NH O OH OH NHCO

(11a): R = NH_2 CH(ME)OH CO

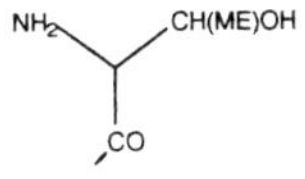

(11b): R = H

(11c): R = OH OH CO $NH(CH_2)_4$

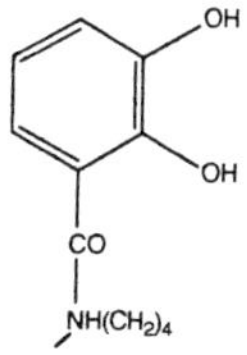

(11d): R = O N CH_3 CO (PARABACTIN)

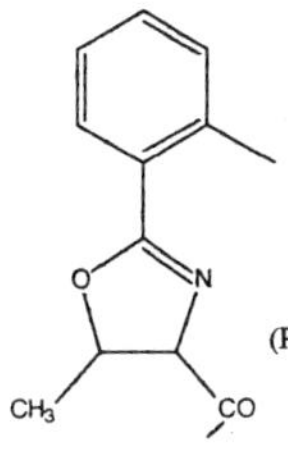

Table 8.5 *In vitro* antileukemic and antiviral activity of spermidine catecholamides and related compounds

Polyamine derivative	Antileukemic activity concentration causing 50% inhibition (IC_{50}) in 48 h (μM)	Antiviral activity IC_{50} (μM) herpes simplex I
2,3-dihydroxybenzoic acid	2800	> 1000
N^1,N^8-bis(2,3-dihydroxybenzoyl)-N^4-threonyl-spermidine (**13a**)	14.0	55.0
N^1,N^8-bis(2,3-dihydroxybenzoyl)-spermidine (**13b**)	7.0	32.0
N^1,N^8-bis(2,3-dihydroxybenzoyl)-N^4-(4-[2,3-dihydroxybenzamido]-butyryl)-spermidine (**13c**)	2.0	18.0
Parabactin (**13d**)	2.0	0.4

Growth inhibition is observed with 5 μM parabactin (**11d**). The same series of compounds are also active against herpes simplex Type 1 virus, the replication of which is also dependent on ribonucleotide reductase for DNA production (Bergeron, 1986).

Unfortunately, such potent iron chelators are toxic to bone marrow and they are only likely to become useful antineoplastic and antiviral agents if they can be specifically directed to these cell types.

8.5.1.2 Anti-malarial activity. Anti-malarial activity of desferrioxamine has been demonstrated in a range of *Plasmodium* species under *in vitro* conditions and more importantly under *in vivo* conditions in rat (Hershko *et al.*, 1991) and man (Gordeuk *et al.*, 1991). Desferrioxamine causes reversible effects during early ring-stage and late schizont stages of the life cycle. In contrast, irreversible cytocidal effects occur when cultures at the late trophozoite/early schizont stage are exposed to the chelator (Raventos-Suarez *et al.*, 1982; Whitehead and Peto, 1990). Primary lesions are associated with the nucleus which is consistent with a perinuclear localisation for ribonucleotide reductase (Gordeuk *et al.*, 1991). The orally active hydroxypyridinones (**4**) have also been demonstrated to possess antimalarial activity in rats (Hershko *et al.*, 1991) and subject to favourable toxicity studies may soon be investigated in man.

Malaria is a worsening problem worldwide. More than 110 million people suffer from infection every year and up to 2 million of these die. A growing number of countries are now affected by drug-resistant strains of the parasites. The introduction of an iron chelator to control such infection is a novel approach, and as chelation is particularly critical at the late trophozoite stage, it may prove possible to limit treatment to relatively short time periods.

8.5.1.3 Psoriasis. Psoriasis involves the hyperproliferation of both keratinocytes and T-lymphocytes and concomitant abnormal keratinocyte differentiation. It can therefore be described as an inflammatory skin disorder. There is no generally satisfactory method of treatment of this disease state at the present time (Menter and Barker, 1991). One approach is the use of an anti-proliferative agent either systemically or topically. In principle, ribonucleotide reductase could provide a suitable target to prevent proliferation and topical use could minimise systemic exposure to potentially toxic molecules. Indeed, an iron chelator prodrug ICRF-159 which is closely related to ICRF-187 (**9**) was found to be remarkably successful for the treatment of psoriasis when presented systemically (Atherton *et al.*, 1980; Horton and Wells, 1983). Unfortunately, continued exposure of patients to ICRF-159 was associated with a relatively high incidence of epitheliomas (Horton *et al.*, 1983) and leukaemia (Lakhani *et al.*, 1984). Consequently it is no longer prescribed.

Other chelators should be investigated for the treatment of psoriasis, indeed the topical presentation of a biodegradable chelator would be predicted to prossess a powerful local effect.

8.5.2 Lipoxygenase enzymes

The lipoxygenase family of enzymes catalyse stereospecific oxygenation reactions of fatty acid substrates. The active site incorporates a non-haem iron centre in as yet an unidentified environment. The active form of the enzyme is in the iron(III) state but is converted to iron(II) as the fatty acid is oxidised (Funk *et al.*, 1990). Catechol-containing compounds inhibit these enzymes by forming a ternary complex with this iron centre and depending on the substituents on the catechol moiety may even facilitate an internal redox reaction leading to the formation of iron(II) and an inactive form of the enzyme (Kemal *et al.*, 1987; Nelson, 1988).

5-Lipoxygenase is the first enzyme in a series involved in the conversion of arachidonic acid to the family of leukotrienes. The inhibition of this enzyme therefore decreases leukotriene production without adversely influencing cyclooxygenase-based prostaglandin production (Cashman, 1985; Taylor and Clarke, 1986).

A wide range of iron chelators have been shown to inhibit this enzyme without inhibiting cyclooxygenase (Cashman, 1985). In particular hydroxamates have been developed as potent, selective 5-lipoxygenase inhibitors (Summers *et al.*, 1987, 1990). Some of these derivatives (**12**, **13**) are reported

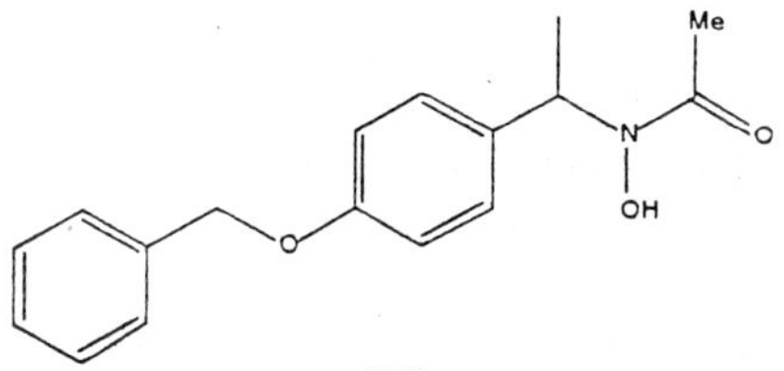

(**12**)

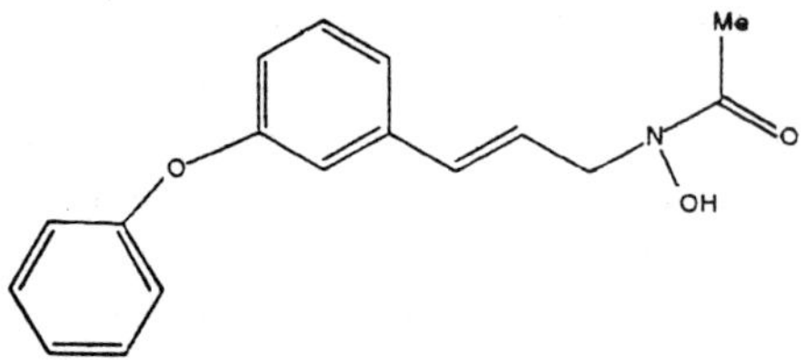

(13)

to possess oral activity (Summers *et al.*, 1988; Jackson *et al.*, 1988). Indeed (**13**) possesses anti-inflammatory properties.

Leukotrienes have been suggested to be important mediators in a wide range of diseases, including asthma, arthritis and psoriasis and consequently chelators with a highly selective action may prove to possess therapeutic potential.

8.6 Treatment of anaemia with iron complexes

The comparison of iron(II) and iron(III) uptake by mammalian intestine has recently been the focus of intense research effort. In a study of 14 different iron preparations in man, Dietzfelbinger showed that the iron(II) preparations, without exception, had a lower bioavailability than iron(II) sulphate and were therefore of dubious therapeutic efficacy (Dietzfelbinger, 1987). Similar conclusions have been reached by others (Kaltwasser *et al.*, 1987). Unfortunately, orally administered iron(II) sulphate generates hydroxyl radicals in the gastrointestinal tract of mammals (Slivka *et al.*, 1986). This property, together with the associated acidity of iron(II) sulphate, may cause irritation and damage to the mucosa. A wide range of side-effects have been reported for iron(II) sulphate (Hallberg *et al.*, 1966). Thus, should an efficiently absorbed iron(III) complex be identified (Geisser and Muller, 1987), it would be of therapeutic benefit.

The non-toxic hydroxypyrone maltol (**14**) binds iron(III) forming a water-soluble complex (Hider and Hall, 1991). In the pH range 4–7, hydroxypyrones possess a lower affinity for iron(III) than EDTA, and by

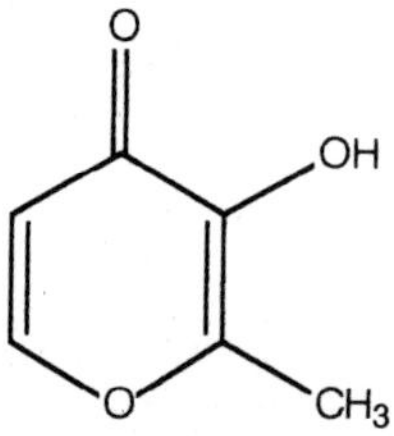

(14)

virtue of the kinetic lability of such complexes, they are able to donate iron to high-affinity binding sites, while minimising nonselective binding of iron to foodstuffs. Thus, iron presented as a maltol complex is relatively well absorbed (Barrand *et al.*, 1987; Levey *et al.*, 1988) and the dissociated maltol is rapidly metabolised (Barrand *et al.*, 1991). Ferric maltol is the only simple iron(III) preparation which compares favourably with iron(II) sulphate (Kelsey *et al.*, 1989, 1991), and in contrast to iron(II) sulphate, there are few, if any, side-effects associated with its oral administration. Furthermore there may be less complications resulting from the co-administration of metal chelating drugs, for instance, quinolones, levodopa and methyldopa (Greene *et al.*, 1990). Consequently, patient compliance is likely to be superior with this iron preparation.

References

Agid, Y. (1991) Parkinson's disease: pathophysiology. *The Lancet*, I, 1321–1327.

Anden, N.E., Fuxe, K., Hamberger, B., Hokfelt, T. (1966) A quantitative study of the nigro-neostriatal dopamine neuron system in the rat. *Acta Physiologica Scandinavica*, **67**, 306–312.

Andrews, F., Morris, C.J., Kondratowicz, G. and Blake, D.R. (1987) The effect of iron chelation on inflammatory joint disease. *Annals of Rheumatic Disease*, **46**, 327–333.

Atherton, D.J., Wells, R.S., Laurent, M.R. and Williams, Y.E. (1980) Razoxane (IRCF 159) in the treatment of psoriasis, *British Journal of Dermatology*, **102**, 307–317.

Aust, S.D. and White, B.C. (1985) Iron chelation prevents tissue injury following ischaemia. *Advances in Free Radical Biology and Medicine*, **1**, 1–17.

Aust, S.D. (1988) Sources of iron for lipid peroxidation in biological systems, in *Oxygen Radicals and Tissue Injury*, (Ed. Halliwell, B.), Federation of American Societies for Experimental Biology, Maryland, pp. 27–33.

Barrand, M.A., Callingham, B.A. and Hider, R.C. (1987) *J. Pharm. Pharmacol.*, **39**, 203–211.

Barrand, M.A., Callingham, B.A., Dobbin, P. and Hider, R.C. (1991) Dissociation of a ferric maltol complex and its subsequent metabolism during absorption across the small intestine of the rat. *Br. J. Pharmacol.*, **102**, 723–729.

Bartlett, A.N., Hoffbrand, A.V. and Kontoghiorghes, G.J. (1990) Long-term trial with the oral iron chelator 1,2-dimethyl-3-hydroxypyrid-4-one (L1). *British Journal of Haematology*, **76**, 301–304.

Beckman, J.S., Beckman, T.W., Chen, J., Marshall, P.A. and Freeman, B.A. (1990) Apparent hydroxyl radical production by peroxynitrite: implications for endothelial injury from nitric oxide and superoxide. *Proceedings of the National Academy of Sciences, USA*, **87**, 1620–1624.

Becton, D.L., and Bryles, P. (1988) Deferoxamine inhibition of human neuroblastoma viability and proliferation. *Cancer Research*, **48**, 7189–7192.

Bergeron, R.J. (1986) Iron: a controlling nutrient in proliferative processes. *Trends in Biochemical Sciences*, **11**, 133–136.

Bergeron, R.J., Gavanaugh, P.F., Kline, S.J., Hughes, R.G., Elliot, G.T. and Porter, C.W. (1984) Antineoplastic and antiherpetic activity of spermidine catecholamide iron chelators. *Biochemical and Biophysical Research Communications*, **121**, 848–854.

Berman, H.M. and Young, P.R. (1981) The interaction of intercalating drugs with nucleic acids. *Annual Reviews in Biophysics and Bioengineering*, **10**, 87–114.

Blake, D.R. and Bacon, P.A. (1981) Synovial fluid ferritin in rheumatoid arthritis: An index of cause of inflammation. *British Medical Journal*, **282**, 189.

Blake, D.R., Gallagher, P.J., Potter, A.R., Bell, M.J. and Bacon, P.A., (1984) The effect of synovial iron on the progression of rheumatoid disease. A histological assessment of patients with early rheumatoid disease. *Arthritis and Rheumatism*, **27**, 495–501.

Blake, D.R., Winyard, P., Lunec, J., Williams, A., Good, P.A., Crewes, S.J., Gutteridge, J.M.C.,

Rowley, D., Halliwell, B., Cornish, A. And Hider, R.C. (1985) Cerebral and ocular toxicity induced by deferioxamine. *Quarterly Journal of Medicine*, **56**, 345–355.

Blatt, J. and Stitely, S. (1987) Anti-neuroblastoma activity of deferioxamine in human cell lines. *Cancer Research*, **47**, 1749–1750.

Cashman, J.R. (1985) Leukotriene biosynthesis inhibitors. *Pharmaceutical Research*, 253–261.

Cory, J.G., Lasater, L., and Sato, A. (1981) Effect of iron-chelating agents on inhibitors of ribonucleotide reductase. *Biochemical Pharmacology*, **30**, 979–984.

Crichton, R.R. (1991) *Inorganic Biochemistry of Iron Metabolism*, Ellis Horwood, London.

Davies, M.J., Donker, R., Dunster, C.A., Gee, C.A., Jones, S. and Willson, R.L. (1987) Deferoxamine (desferal) and superoxide free radicals: Formation of an enzyme-damaging nitroxide. *Biochemical Journal*, **246**, 725–729.

Dexter, D.T., Wells, F.R., Agid, F., Agid, Y., Lees, A.J., Jenner, P. and Marsden, C.D. (1987) Increased nigral iron content in postmortem Parkinsonian brain. *The Lancet*, **II**, 1219–1220.

Dexter, D.T., Wells, F.R., Lees, A.J., Agid, F., Agid, Y., Jenner, P. and Marsden, C.D. (1989a) Increased nigral iron content and alterations in other metal ions occurring in brain in Parkinson's disease. *Journal of Neurochemistry*, **52**, 1830–1836.

Dexter, D.T., Carter, C.J., Wells, F.R., Javoy-Agid, Fe., Agid, Y., Lees, A., Jenner, P. and Marsden, C.D. (1989b) Basal lipid peroxidation in substantia nigra is increased in Parkinson's disease. *Journal of Neurochemistry*, **52**, 381–389.

Dezza, L., Cazzola, M., Danova, M., Carlo-Stella, C., Bergamaschi, G., Brugnatelli, S., Invernizzi, R., Mazzini, G., Riccardi, A. and Ascari, E. (1987) Effects of deferioxamine inhibition of human neuroblastoma viability and proliferation. *Cancer Research*, **48**, 7189–7192.

Dietzfelbinger, H. (1987) *Arzneim-Forsch.* **37**, 105–107.

Dolence, E.K., Minnick, A.A., Lin, C.E. and Miller, M.J. (1991) Synthesis and siderophore and antibacterial activity of N^5-acetyl-N^5-hydroxy-L-ornithine-derived siderophore-β-lactam conjugates: Iron-transport-mediated drug delivery. *Journal of Medicinal Chemistry*, **34**, 968–978.

Foa, P., Maiolo, A.T., Lombardi, L., Villa, L. and Polli, E.E. (1986) Inhibition of proliferation of human leukaemic cell lines by deferoxamine. *Scandinavian Journal of Haematology*, **36**, 107–110.

Ford, G. C., Harrison, P.M., Rice, D.W., Smith, J.M.A., Treffry, A., White, J.L. and Yariv, J. 1984. *Philosophical Transactions of the Royal Society, London* **B304**, 551–565.

Funk, M.O., Carroll, R.T., Thompson, J.F., Sands, R.H. and Dunham, W.R. (1990) Role of iron in lipoxygenase catalysis. *Journal of American Chemical Society*, **112**, 5375–5376.

Ganeshaguru, K., Hoffbrand, A.V., Grady, R.W., Cerami, A. (1980) Effect of various iron chelating agents on DNA synthesis in human cells. *Biochemical Pharmacology* **29**, 1275–1279.

Geisser, P. and Müller, A. (1987) *Arzneim. Forsch.*, **37**, 110–114.

Gelvan, D., Berg, E., Saltman, P. and Samuni, A. (1990) Time-dependent modifiers of ferric-adriamycin. *Biochemical Pharmacology*, **39**, 1289–1295.

Gianni, L., Zweier, J.L., Levy, A. and Myers, C.E. (1985) Characterisation of the cycle of iron-mediated electron transfer from adriamycin to molecular oxygen. *Journal of Biological Chemistry*, **260**, 6820–6829.

Gordeuk, V.R., Thuma, P.E., Brittenham, G.M., Zulu, S., Simwanza, G., Mhangu, A., Flesch, G. and Parry D. (1991) Iron chelation with deferoxamine B in adults with asymptomatic *Plasmodium falciparum* Parasitemia. *Blood*, in press.

Gordonsmith, R.H., Brooke-Taylor, S., Smith, L.L. and Cohen, G.M. (1983) Structural requirements of compounds to inhibit pulmonary diamine accumulation. *Biochemical Pharmacology*, **32**, 3701–3709.

Green, C.J., Gower, J.D., Healing G., Cotterill, L.A., Fuller, B.J. and Simpkin, S. (1989) The importance of iron, calcium and free radicals in reperfusion injury: An overview of studies in ischaemic rabbit kidneys. *Free Radical Research Communications*, **7**, 255–264.

Green, C.J., Healing, G., Simpkin, S., Lunec, J. and Fuller, B.J. (1986) Deferoxamine reduces susceptibility to lipid peroxidation in rabbit kidneys subjected to warm ischaemia and reperfusion. *Comparative Biochemistry and Physiology*, **85B**, 113–117.

Greene, R.J., Hall, A.D. and Hider, R.C. (1990) The interaction of ordally administered iron with levodopa and methyldopa therapy. *J. Pharm. Pharmacol.*, **42**, 502–504.

Hallberg, L., Ryttinger, L. and Solvell, L. (1966) *Acta Med. Scand.*, **181**, 3–10.

Hallgren, B. and Sourander, P. (1985) *Journal of Neurochemistry*, **3**, 41–54.

Halliwell, B., Cornish, A. and Hider, R.C. (1985) Cerebral and ocular toxicity induced by deferoxamine. *Quarterly Journal of Medicine*, **56**, 345–355.

Halliwell, B. and Gutteridge, J.M.C. (1989) *Free Radicals in Biology and Medicine*, 2nd edition, Clarendon Press. Oxford.

Harding, C., Heuser, J. and Stahl, P. (1983) Receptor-mediated endocytosis of transferrin and recycling of the transferrin receptor in rat reticulocytes. *Journal of Cell Biology*, **97**, 329–339.

Hartmann, A., Fielder, H.P. and Braun, V. (1979) Uptake and conversion of the antibiotic albomycin by *Escherichia coli* K-12. *European Journal of Biochemistry*, **99**, 517–524.

Harvey, A.R., Clarke, B.J., Chui, D.H.K., Kean, W.F. and Buchanan, M.W. (1983) Anaemia associated with rheumatoid disease. Inverse correlation between erythropoiesis and both IgM and rheumatoid factor levels. *Arthritis and Rheumatism*, **26**, 28–34.

Healing, G., Gower, J., Fuller, B. and Green, C. (1990) Intracellular iron distribution. An important determinant of reperfusion damage of rabbit kidneys. *Biochemical Pharmacology*, **39**, 1239–1245.

Hearse, D.J., Manning, A.S., Downey, J.M. and Yellon, D.M. (1986) Xanthine oxidase: a critical mediator of myocardial injury during ischaemia and reperfusion? *Acta Physiologica Scandinavica*, **548**, 67–78.

Hernandez, L.A. Grisham, M.B. and Granger, D.N. (1987) A role for iron in oxidant mediated ischaemic injury to intestinal microvasculature. *American Journal of Physiology*, **253**, G49–G53.

Hershko, C., Grady, R.W. and Link, G. (1984) Phenolic ethylenediamine derivatives: A study of orally effective iron chelators. *Journal of Laboratory and Clinical Medicine*, **103**, 337–346.

Hershko, C.H., Peto, T.E.A and Weatherall, D.J. (1988) Iron and infection. *British Journal of Haematology*, **296**, 660–664.

Hershko, C., Theanacho, E.N., Spira, D.T., Peter, H.H., Dobbin, P. and Hider, R.C. (1991) The effect of N-alkyl modification on the antimalarial activity of 3-hydroxypyridin-4-one oral iron chelators. *Blood*, **77**, 637–643.

Hewitt, S.D., Hider, R.C., Sarpong, P., Morris, C.J. and Blake, D. (1989) Investigation of the anti-inflammatory properties of hydroxypyridinones. *Annals of Rheumatic Diseases*, **48**, 382–388.

Hider, R.C. (1984) Siderophore mediated absorption of iron. *Structure and Bonding*, **39**, 25–87.

Hider, R.C. and Hall, A.D. (1991) Clinically useful chelators of tripositive elements. *Progress in Medicinal Chemistry*, **28**, 41–173.

Hider, R.C., Mohd-Nor, A.R., Silver, J., Morrison, I.E.G. and Rees, L.V.C. (1981) Model compounds for microbial iron-transport compounds. *Journal of the Chemical Society, Dalton Transactions*, 609–622.

Hider, R.C., Singh, S., Porter, J.B. and Huehns, E.R. (1990) The Development of hydroxypyridin-4-ones as orally active iron chelators. *Annals New York Academy of Sciences*, **612**, 327–338.

Horton, J.J., MacDonald, D.M. and Wells, R.S. (1983) Epitheliomas in patients receiving razoxane therapy for psoriasis. *British Journal of Dermatology*, **109**, 675–678.

Horton, J.J. and Wells, R.S. (1983) Razoxane: A review of 6 years' therapy in psoriasis. *British Journal of Dermatology*, **109**, 669–673.

Huebers, H. and Finch, C.A. (1987) The physiology of transferrin and transferrin receptors. *Physiological Reviews*, **67**, 520–582.

Jackson, M.J., Brenton, D.P. and Modell, B. (1983) DTPA in the management of iron overload in thalassaemia. *Journal of Inherited Metabolic Diseases*, **6**, Suppl. 2, 97–98.

Jackson, W.P., Islip , P.J., Kneen, G., Pugh, A. and Wates, P.J. (1988) Acetohydroxamic acids as potent, selective orally active 5-lipoxygenase inhibitors. *Journal of Medicinal Chemistry*, **31**, 499–500.

Kaltwasser, J.P., Werner, E. and Niechzial, M. (1987) *Arneim.-Forsch* **37**, 122–129.

Kelsey, S.M., Blake, D.R., Hider, R.C., Gutteridge, C.N. and Newland, A.C. (1989) *Clin. Lab. Haematol.*, **11**, 287–290.

Kelsey, S.M., Hider, R.C., Bloor, J.R., Blake, D.R., Gutteridge, C.N. and Newland, A. C. (1991) Absorption of low and therapeutic doses of ferric maltol, a novel ferric iron compound, in iron deficient subject using a single dose iron absorption test. *Journal of Clinical Pharmacy and Therapeutics*, **16**, 117–212.

Kemal, C., Louis-Flamberg, P., Krupinski-Olsen, R. and Shorter, A.L. (1987) Reduction inactivation of soybean lipoxygenase 1 by catechols: A possible mechanism for regulation of lipoxygenase activity. *Biochemistry*, **26**, 1064–7072.

Klein, J., Damani, L.A., Chung, D., Epemolu, O., Olivieri, N. and Koren, G. (1991) A high-performance liquid chromatogrphic method for the measurement of the iron chelator 1,2-dimethyl-3-hydroxypyridin-4-one in human plasma. *Therapeutic Drug Monitoring*, **13**, 51–54.

Komara, J.S., Nayini, N.R., Bialick, H.A., Indrien, R.J., Evans, A.T., Garritano, A.M., Hoehner, T.J., Jacobs, W.A., Huang, R.R., Krause, G.S., White, B.C. and Aust, S.D. (1986) Brain iron delocalisation and lipid peroxidation following cardiac arrest. *Annals of Emergency Medicine*, **15**, 384–389.

Kontoghiorghes, G.J., Bartlett, A.N., Hoffbrand, A.V., Goddard, J.G., Sheppard, L., Barr, J. and Nortey, P. (1990) Long-term trial with the oral iron chelator 1,2-dimethyl-1-3-hydroxypyrid-4-one (L1). *British Journal of Haematology*, **76**, 295–300.

Lakhani, S., Davidson, R.N., Hiwaizi, F. and Marsden, R.A. (1984) Razoxane and leukaemia. *The Lancet*, **II** 288–289.

Lau, E.H., Cerny, E.A., Wright, B.J. and Rahman, Y.E. (1983) Improvement of iron removal from the reticuloendothelial system by liposome encapsulation of N,N′-bis[2-hydroxybenzyl]-ethylenediamine-N,N'-diacetic acid (HBED). *Journal of Laboratory and Clinical Medicine*, **101**, 806–816.

Lawson, A.A.H., Owen, E.T. and Mowat, A.G. (1983) Nature of anaemia in rheumatoid disease. Storage of iron in rheumatoid disease. *Annals of the Rheumatic Diseases*, **26**, 552–559.

Levin, V.A. (1980) *Journal of Medicinal Chemistry*, **23**, 682–684.

Levey, J.A., Barrand, M.A., Callingham, A. and Hider, R.C. (1988) *Biochem. Pharmacol.*, **37**, 2051–2057.

Martell, A.E., Matekaitis, R.J. and Clarke, E.T. (1986) Synthesis of N,N′-di(2-hydroxybenzyl)-ethylenediamine-N,N′-diacetic acid (HBED) derivatives. *Canadian Journal of Chemistry*, **64**, 449–456.

May, W.S. and Cuatrecasas, P. (1985) Transferrin receptor: Its biological significance. *Journal of Membrane Biology*, **88**, 205–215.

Mentasti, E., Pelizzetti, E. and Saini, G. (1976) Interactions of Fe(III) with adrenaline, L-dopa and other catechol derivatives. *Journal of Inorganic and Nuclear Chemistry*, **38**, 785–788.

Menter, A. and Barker, J.N.W.N. (1991) Psoriasis in practice. *The Lancet* **II**, 231–234.

Muirden, K.D. (1966) Ferritin in synovial cells in patients with rheumatoid arthritis. *Annals of the Rheumatic Diseases*, **25**, 387–401.

Munro, H. and Linder, M.C. (1978) Ferritin: biosynthesis and role in iron metabolism. *Physiology Reviews*, **58**, 317–396.

Myers, C., Gianni, L., Zweier, J., Muindi, J., Sinha, B.K. and Elior, H. (1986) Role of iron in adriamycin biochemistry. *Federation Proceedings*, **45**, 2792–2797.

Nelson, M.J. (1988) Catecholate complexes of ferric soybean lipoxygenase 1. *Biochemistry*, **27**, 4273–4278.

Nolan, K.B., Murphy, T., Hermanns, R.D., Rahoo, H., Creighton, A. (1990) Chelating agents as anti-tumour pro-drug razoxane. *Inoganica Chemica Acta*, **168**, 283–288.

Nordlund, P., Sjöberg, B.M. and Eklund, H. (1990) Three-dimensional structure of the free radical structure of the free radical protein of ribonucleotide reductase. *Nature*, **345**, 593–598.

Porter, J.B. and Huehns, E.R. (1989) The toxic effects of deferoxamine. *Baillière's Clinical Haematology*, **2**, 459–474.

Porter, J.B., Huehns, E.R. and Hider, R.C. (1988) The development of iron chelating drugs. *Baillière's Clinical Haematology*, **2**, 257–292.

Porter, J.B., Morgan, J., Hoyes, K.P., Burke, L.C., Huehns, E.R and Hider, R.C. (1990) Relative oral efficacy and acute toxicity of hydroxypyridin-4-one iron chelators in mice. *Blood*, **76**, 2389–2396.

Raventos-Suarez, C., Pollack, S. and Nagel, R.L. (1982) *Plasmodium falciparum:* inhibition of *in vitro* growth by deferoxamine. *American Journal of Tropical Medicine and Hygiene*, **31**, 919–922.

Reddy, B.R., Kloner, R.A. and Przyklenk, K. (1989) Early treatment with deferoxamine limits myocardial ischaemic/reperfusion injury. *Free Radical Biology and Medicine*, **7**, 45–52.

Reichard, P. and Ehrenberg, A. (1983) Ribonucleotide reductase – a radical enzyme. *Science*, **221**, 514–519.

Robins-Browne, R.M. and Prpic, J.K. (1985) Effects of iron and deferoxamine on infections with *Yersinia enterocolitica. Infectious Immunology*, **47**, 774–779.

Senator, G.B. and Muirden, K.D. (1968) Concentration of iron in synovial membrane, synovial fluid and serum in rheumatoid arthritis and other joint diseases. *Annals of Rheumatic Diseases*, **27**, 49–53.

Silley, P., Griffiths, J.W., Monsey, D. and Harris, A.M. (1990) Mode of action of GR69153, a novel catechol-substituted cephalosporin and its interaction with the *ton* B-dependent iron transport system. *Anti-microbial Agents and Chemotherapy*, **34**, 1806–1808.

Singh, S., Hider, R.C. and Porter, J.B. (1990) A direct method for quantification of non-transferrin-bound iron. *Analytical Biochemistry*, **186**, 320–323.

Slivka, A., Kang, J. and Cohen, G. (1986) *Biochem. Pharmacol.*, **35**, 553–556.

Streater, M., Taylor, P.D., Hider, R.C. and Porter, J.B. (1990) Novel 3-hydroxy-2(1H)-pyridinones. Synthesis, iron(III)-chelating properties and biological activity. *Journal of Medicinal Chemistry*, **33**, 1749–1755.

Sutomatsu, A., Nakano, M. and Hirai, S. (1990) Phospholipid peroxidation induced by the catechol-Fe^{3+} (Cu^{2+}) complex: A possible mechanism of nigrostriatal cell damage. *Archives of Biochemistry and Biophysics*, **283**, 334–341.

Summers, J.B., Gunn, B.P., Martin, J.G., Mazdiyasni, H., Stewart, A.O., Young, P.R., Goetze, A.M., Bousko, J.B., Dyer, R.D., Brooks, D.W. and Carter, G.W. (1988) Orally active hydroxamic acid inhibitors of leukotriene biosythesis. *Journal of Medicinal Chemistry*, **31**, 3–5.

Summers, J.B., Kim, K.H., Mazdiyasni, H., Holms, J.H., Ratajczyk, J.D., Stewart, A.O., Dyer, R.D. and Carter, G.W. (1990) *Journal of Medicinal Chemistry*, **33**, 992–998.

Summers, J.B., Mazdiyasni, H., Holms, J.H., Ratajczyk, J.D., Dyer, R.D., Carter, G.W. (1987) Hydroxamic acid inhibitors of 5-lipoxygenase. *Journal of Medicinal Chemistry*, **30**, 574–580.

Taylor, G.W. and Clarke, S.R. (1986) The leukotriene biosynthetic pathway: A target for pharmacological attack. *Trends in Pharmacological Sciences*, 100–103.

Thomas, C.E. and Aust, S.D. (1986) Release of iron from ferritin by cardiotoxic anthracycline antibiotics. *Archives of Biochemistry and Biophysics*, **248**, 684–689.

Thorstensen, K. (1988) Hepatocytes and reticulocytes have different mechanisms for the uptake of iron from transferrin. *Journal of Biological Chemistry*, **263**, 16837–16841.

van der Kraaij, A.M.M., van Eijk, H.G. and Koster, J.F. (1989) Prevention of postischaemic cardiac injury by the orally active iron chelator 1,2-dimethyl-3-hydroxy-4-pyridone (L1) and the antioxidant (+)-cyanidanol-3. *Circulation*, **80**, 158–164.

Vile, G.F., Winterbourn, C.C. and Sutton, H.C. (1987) Radical-driven Fenton reactions: studies with paraquat, adriamycin, and anthraquinone-6-sulfonate and citrate, ATP, ADP, and pyrophosplate iron chelates. *Archives of Biochemistry and Biophysics*, **259**, 616–626.

Washington, J.A., Jones, R.N., Allen, S.D., Gerlack, E.H., Koontz, F.P., Murray, P.R., Pfaller, M.A. and Erwin, M.E. (1991) *In vitro* comparison of GR69153, a novel catechol-substituted cephalosporin, with ceftazidime, and ceftriaxone against 5203 recent clinical isolates. *Antimicrobial Agents and Chemotherapy*, **35**, 1508–1511.

Weatherall, D.J. and Clegg, J.B. (1981) *The Thalassaemia Syndromes*, 3rd edition, Blackwell Scientific Publications, Oxford.

Weinberg, E.D. (1984) Iron withholding: A defense against infection and neoplasia. *Physiological Reviews*, **64**, 65–102.

Weinberg, E.D. (1990) Cellular iron metabolism in health and disease. *Drug Metabolism Reviews*, **22**, 531–579.

Whitehead, S. and Peto, T. (1990) Stage-dependent effect of deferoxamine on growth of *Plasmodium falciparum in vitro*. *Blood*, **76**, 1250–1255.

Wrigglesworth, J.M. and Baum, H. (1980) The biochemical functions of iron, in *Iron in Biochemistry and Medicine II*, (eds. A. Jacobs and M. Worwood), Academic Press, London, pp. 29–86.

Wrigglesworth, J.M. and Baum, H. (1988) Iron-dependent enzymes in the brain; in *Brain Iron: Neurochemical and Behavioural Aspects*, (eds Moussa, B.H. and Youdim, B.H.), Taylor and Francis, London.

Yoshino, S., Blake, D.R. and Bacon, P.A. (1984) The effect of deferoxamine on antigen induced inflammation in the rat air pouch. *Journal of Pharmacy and Pharmacology*, **36**, 543–545.

Zähner, H., Diddeus, H., Keller-Schierlein, W. and Naegli, H.U. (1977) *Japanese Journal of Antibiotics*, **30**, 5201–5205.

Zurlo, M.G., De Stefano, P., Borgna-Pignatti, C., Di Palma, A., Piga, A., Melevendi, C., Di Gregorio, F., Burattini, M.G. and Terzoli, S. (1989) Survival and causes of death in thalassaemia major. *The Lancet*, **II** 27–30.

Zweier, J.L. (1985) Iron-mediated formation of an oxidised adriamycin free radical. *Biochemica et Biophysica Acta*, **839**, 209–213.

Index